Texas Used Oil Management: A Practical Guide for Environmental Compliance

ALSO BY RL GEMEINHARDT

Earth Dwellers Guide to Recycling and Environmental Conservation

Texas Used Oil Management: A Practical Guide for Environmental Compliance

R.L. Gemeinhardt

Requests for authorization should be addressed to
VRM Group, LLC
Attn: Ron Gemeinhardt
190—B2 Gulf Freeway, #126
League City, TX 77573

Cover Design by Debra Gemeinhardt
Interior layout and design and eBook conversion by www.writingnights.org
Book preparation by Chad Robertson
Edited by Dr. Don Gemeinhardt

ISBN-978-1-64467-396-6

Printed in the United States of America on acid free paper.

24 23 22 21 20 19 18 17 8 7 6 5 4 3 2 1

Disclaimer and Terms of Use:

PLEASE READ THESE TERMS AND CONDITIONS CAREFULLY BEFORE ACCESSING OR USING THE INFORMATION IN THIS Texas Used Oil Management: A Practical Guide for Environmental Compliance (Guide). The Author and VRM Group, LLC own all right, title and interest in this Guide. By accessing and using this resource in any form you hereby agree to be legally bound by the terms and conditions set forth herein. If you do not agree to abide by the following terms and conditions, please do not use the Guide. If you have questions about the following terms and conditions, please contact VRM Group, LLC at Rong@VRMUS.com. VRM Group, LLC or the Author reserves the right, at its sole discretion, to change, modify or otherwise alter these terms and conditions at any time.

The Texas Used Oil Management: A Practical Guide for Environmental Compliance is being provided for information purposes only. It is not intended to take in account any specific needs of your business. The VRM Group, LLC or the Author and all other identified organizations, public or private, individuals, state, federal, county, or city agencies or any other governmental entity make no representations or warranties as to the completeness or accuracy thereof. Persons using this information must make their own determination as to its suitability for their purposes in support of their own environmental health and safety programs and requirements. VRM Group, LLC or the Author and all other participating organizations are in no way responsible for damages of any nature resulting from the use of this information. This Guide does not imply or express any legal advice or professional services advice or council in anyway and no implied contract is expressed or implied by the purchase or use of this Guide in any form.

All Materials on this Guide in any form are provided by VRM Group, LLC and the Author as a service and may be used only for informational purposes.

User acknowledges that the Guide contains information protected by copyrights, trademarks, or other proprietary rights, and these rights are protected in all forms and technologies existing now or hereinafter developed. You may use the material for noncommercial and personal use only provided that you maintain all copyright and other notices contained in such material. You further agree that you will not disassemble, de-compile, reverse engineer or otherwise modify the material in any way at any time. Individual images in this Guide are for viewing purposes only and may not be downloaded or copied for redistribution without prior written permission. Any unauthorized or prohibited use may subject the offender to civil liability and criminal prosecution under applicable federal and state laws.

Reference in this Guide to any specific commercial products, process, service, manufacturer, or company does not constitute its endorsement or recommendation by VRM Group, LLC or the Author.

DEDICATION

To my loving wife Marilyn who has always supported me in all my environmental work no matter where it took me in this great country or for how long it often kept me away from home. Marilyn was especially tolerant of all my long hours in my home office as I wrote both the thesis this book is based on and then the final book form that is now published.

CONTENTS

LIST OF FIGURES

Flowcharts

Forms

Graphs

Plots

Tables

PREFACE

Used oil management in Texas provides significant and unique compliance challenges. In an effort to assist those that are involved in or contemplating participation, this guide has been developed to provide an overview of used oil management environmental compliance requirements associated with Used Oil Handlers (UOH) in Texas. The typical facility covered by this guide would store from 50,000 gallons to 1,500,000 gallons, and perform used oil collection services and used oil filter and absorbent collection with limited processing. Generator, Storage, Processor, Re-Refiner, Transfer Facility, Transporter, Marketer, and Do-it-Yourself Collection Center requirements are addressed. This guide does not address all the requirements associated with operation of a used oil refinery.

The content of this guide addresses the state of Texas and federal requirements associated with defining used oil intended for recycle under the Resource Conservation and Recovery Act (RCRA); permitting; associated compliance; storage; reporting; Texas Commission of Environmental Quality (TCEQ) overview; enforcement; TCEQ support; oil spill prevention associated with Spill Prevention, Control, and Countermeasure (SPCC) and Facility Response Plan (FRP) requirements; Tier II reporting, Environmental Management Systems (EMS); and compliance training associated with permits and plans.

This guide provides a limited overview and possible applicability requirements related to The Comprehensive Environmental Response, Compensation and Liability Act (CERCLA) associated with All Appropriate Inquiry (AAI), due diligence associated with sale or facility acquisition, Clean Air Act (CAA), Clean Water Act (CWA), RCRA, and Texas Industrial And Hazardous Wastes management.

This guide is intended to provide awareness of the typical associated environmental compliance with comments provided by the author from decades of corporate, company on-site compliance management, and consultant services provided to petroleum facilities.

This guide primary content was developed and approved as a thesis requirement for the master's program at Green Mountain College by the author.

ACKNOWLEDGEMENTS

With this book being based on my master's thesis I must especially acknowledge my thesis advisor Jon Leibowitz who without his instructive and insightful guidance in so many ways my thesis and this book would have never come to fruition. Also I must note my second thesis reader Desiderion Mora who's editing was so helpful. On a personal and professional level, Ben Cowart was the person that has inspired and supported my efforts to really understand used oil management and what it means to work in this area of recycling and compliance as I tried to add value to not only the protection of the environment but how important clear guidance is needed for this industry at so many levels. My brother Dr. Don Gemeinhardt who has been not only supportive in getting this project done but provided valuable support and collaboration as the final editor. Debbie Gemeinhardt as the cover designer provided that special cover that expressed that special message about the book from a first glance.

PART I: USED OIL COMPLIANCE

INTRODUCTION

This guide addresses the typical state of Texas requirements for the management of a small used oil facility, and provides typical operational areas identified in the provided Facility Example. As Texas has been authorized by the Environmental Protection Agency (EPA) **(75)** to implement the federal rules associated with compliance of the various acts and regulations, our example facility is subject to the guide and will address the requirements found in the Texas Administrative Code (TAC) and identify the Code of Federal Regulations (CFR) requirements as addressed by reference in the TAC. This includes an overview as well as select detail requirements required for specific permits, operations, and management responsibilities and compliance awareness associated with possible activities in Texas.

The guide is not intended to be a complete reference to all detailed requirements; instead, it is offered to provide an awareness of the environmental compliance and management practices that are typically associated with ownership or management of a Used Oil Storage Facility in Texas. This Guide is not intended to provide any implied or expressed consultant opinion or reliance on information within it. Commentary in italics is offered in many sections to clarify citations or agency positions, or to express additional concern as to why a specific requirement is so important.

Areas of requirement discussion associated with the Facility Example within the guide will be limited to specific compliance requirements under the Resource Conservation Recovery Act (RCRA), which contains the Used Oil regulations that Texas has adopted along with other specific Texas requirements. Additional discussion addresses Clean Water Act (CWA) compliance overview, Clean Air Act (CAA) Permit by Rule overview, Emergency Planning & Community Right-to-Know Act (EPCRA) requirements associated with required TIER II reporting, the Compensation Comprehensive

Environmental Response Compensation and Liability Act (CERCLA) as applied to the use of Phase I and Phase II site assessments utilization for sales and purchasers of facilities, general due-diligence requirements the seller and purchaser could be exposed to, and minimum required Environmental Training related to the permits and plans a typical facility may have.

It should be noted that environmental compliance is never considered complete without ensuring that federal and state requirements are met in any jurisdiction. As indicated above, Texas includes the federal requirements (either specifically or by reference) along with some additions of its own in the Texas Administrative Code (TAC). The EPA **(1)** describes the authorization as a rulemaking process that delegates the primary responsibility of implementing the RCRA hazardous waste program to individual states in lieu of the EPA. This process ensures national consistency and minimum standards while providing flexibility to states in implementing rules. Currently, 50 states and territories have been granted authority to implement the base, or initial, program. State RCRA programs and other authorized programs, such as the CAA or the CWA, must be at least as stringent as the federal requirements, but states can choose to adopt more stringent requirements as well. This is a long and detailed process, but as indicated, it gives the state more flexibility in its management and implementation of the applicable regulations. Texas received its initial final authorization for RCRA (the primary used oil applicable management regulations) on December 26, 1984 (49 FR 48300) **(134)** to implement its Base Hazardous Waste Management Program. As additional federal rules are amended, states then submit for authorization for that new regulation as Texas has done as recently as 2015 **(134)**.

The state legislature has given the Texas Commission on Environmental Quality (TCEQ), the primary agency authority that has oversight of the compliance regulations referenced in this guide associated with the recycle of Used Oil.

The TCEQ as the primary agency for used oil management was not the original agency designation as **(135)** "In 1991, Texas Senate Bill 2 created the Texas Natural Resource Conservation Commission (TNRCC) which combined the functions of the former Texas Water Commission and the former Texas Air Control Board. The transfer of functions to the TNRCC from the two agencies became effective on September 1, 1993. House Bill 2912, Article 18 of the 77th Texas Legislature, 2001, changed the name of the TNRCC to the Texas Commission on Environmental Quality (TCEQ) and directed the TNRCC to adopt a timetable for phasing in the change of the agency's name. The TNRCC decided to make the change of the agency's name to the TCEQ effective September 1, 2002. The change of name became effective September 1, 2002, and the legislative history of the name change is documented at (See, Act of June 15, 2001, 77th Leg. R. S., Ch 965, Section 18.01, 2001 Tex. Gen. Laws 1985). The TCEQ may perform any act authorized by law either as the TNRCC or as the TCEQ."

However, a unique distinction in Texas is that the Railroad Commission (RRC) has jurisdiction over the discharge, storage, handling, transportation, reclamation, or disposal of waste materials (both hazardous and non-hazardous) that result from the activities associated with the exploration, development, or production of oil or gas or geothermal resources and other activities regulated by the RRC. A list of activities that generate wastes that are subject to the jurisdiction of the RRC is found at Texas Health and Safety Code Section 401.415 **(154).** Such wastes are termed "oil and gas wastes." One must be aware of the generating process of the used oil received from any generator to ensure this waste is not subject to regulation by the RRC besides the TCEQ. The RRC requirements and permitting process are not covered in this guide.

The TCEQ jurisdiction over solid waste can be found at Chapter 361, Sections 361.001 through 361.754 of the Texas Health and Safety Code.**(7)** The TCEQ's jurisdiction encompasses hazardous and nonhazardous, industrial, and municipal solid waste. The definition of solid waste can be found at Texas Health and Safety Code Section 361.003(**136**). Even though the EPA has designated the TCEQ as the lead agency to coordinate RCRA activities between the two agencies, the EPA remains responsible for the regulation of any hazardous waste for which TCEQ has not been previously authorized.

The Resource Conservation Recovery Act (RCRA) was enacted on November 29, 1985. EPA developed a used oil program that primarily regulated used oil recycled by being burned for energy recovery. These regulations were codified in Part 266, Subpart E to fulfill the statutory requirement of RCRA §3014. In 1992, the EPA finalized standards in the Federal Register (57 FR 41566) **(137),** establishing requirements for generators, transporters, transfer facilities, collection centers, processors and re-refiners, burners, and marketers in 40 CFR Part 279. The rule also included provisions for cleanup in the case of a spill or release of used oil. The primary oil regulations used in Texas are found in Part 279 **(64),** with other additions by Texas identified in the TAC.

The EPA 1992 standards under this program defined used oil as any oil that has been refined from crude oil (or synthetic oil) that has been used, and as a result of such use, is contaminated by physical or chemical impurities. Types of oils included in this definition are spent automotive lubricating oils, hydraulic fluids, compressor oils from refrigeration units, and metal working oils. The former used oil program defined used oil similarly, except that synthetic oils were not explicitly included in the definition. **(138)**

Guide Overview

The guide is intended to provide an overview of the Texas and Federal environmental compliance requirements associated with the operation of a small Used Oil Handler as identified in the Texas Administrative Code (TAC) §324.2, which defines a Used Oil Handler as "a transporter or an owner or operator of a used oil transfer, processing, re-refiner, or off-specification used oil burning facility." **(118)**

As identified above, this guide is not intended to be a compliance reference for all applicable regulations associated with the operation of the facility; instead, it is intended as an overview of the typical environmental requirements associated with permitting, plans, and reporting, as well as a limited discussion on what to consider when involved in a purchase or sale of a Texas Used Oil Recycling facility.

The writing style for this guide is intended to be along the same lines as the citizen guides written by federal and state agencies and environmental groups. What is also critical for the reader to understand is that this guide offers no implied or expressed guidance, support, or legal opinion of any kind; reliance on any information is at the reader's own risk. One must seek professional environmental compliance, agency, or legal assistance to manage environmental compliance. The guide is only an attempt to provide awareness of possible issues and a regulatory background for typical activities conducted at a Texas Used Oil Handler facility. The bottom line for use of this guide is that the writer or anyone connected with its development for the content or application of identified requirements cannot be held responsible for decisions made based on the information it contains. The guide is only intended to provide a summary of issues to offer an understanding of some of the complexity of used oil management in Texas.

In order to make the point clear on how complex the issues are from an environmental compliance and health and safety standpoint, the issues provided below are sections from a Securities Exchange Commission (SEC) submittal that was developed with assistance from compliance specialists and legal counsel supporting the company; anyone reviewing the business as a public company will be made aware of the compliance issues for a Used Oil Handler in Texas. Not all facilities would be subject to all the compliance issues or management practices identified in the example, but it gives the reader a good idea of the complexity of this activity as a business. This example is used with permission from the company **(139)**; it should be noted that the company operates more than one facility.

CERCLA/RCRA

In the course of our operations, we occasionally generate materials that are considered "hazardous substances" and, as a result, may incur CERCLA liability for cleanup costs. In addition, claims may be filed for personal injury and property damage allegedly caused by the release of hazardous substances or other pollutants. We also generate solid wastes that are subject to the requirements of the United States Resource Conservation and Recovery Act, as amended, or "RCRA," and comparable state statutes. Although we use operating and disposal practices that are standard in the industry, hydrocarbons or other wastes may have been released at properties owned or leased by us now or in the past, or at other locations where these hydrocarbons and wastes were taken for treatment or disposal. Under CERCLA, RCRA and analogous state laws, we could be required to clean up contaminated property (including contaminated groundwater), or to perform remedial activities to prevent future contamination.

Air Emissions

The Clean Air Act (CAA), as amended, and similar state laws and regulations restrict the emission of air pollutants and also impose various monitoring and reporting requirements. These laws and regulations may require us to obtain approvals or permits for construction, modification, or operation of certain projects or facilities and may require use of emission controls.

Global Warming and Climate Change

While we do not believe our operations raise climate change issues different from those generally raised by the commercial use of fossil fuels, legislation or regulatory programs that restrict greenhouse gas emissions in areas where we conduct business or that would require reducing emissions from our truck fleet could increase our costs.

Water Discharges

We operate facilities that are subject to requirements of the United States Clean Water Act, as amended, or "CWA," and analogous state laws for regulating discharges of pollutants into the waters of the United States and regulating quality standards for surface waters. Among other things, these laws impose restrictions and controls on the discharge of pollutants, including into navigable waters as well as the protection of drinking water sources. Spill prevention, control, and counter-measure requirements under the CWA require implementation of measures to help prevent the contamination of navigable waters in the event of a hydrocarbon spill. Other requirements for the prevention of spills are established under the United States Oil Pollution Act of 1990, as amended, or "OPA", which amended the CWA and applies to owners and operators of vessels, including barges, offshore platforms, and certain onshore facilities. Under OPA, regulated parties are strictly liable for oil spills and must establish and maintain evidence of financial responsibility sufficient to cover liabilities related to an oil spill for which such parties could be statutorily responsible.

State Environmental Regulations

Our operations involve the storage, handling, transport and disposal of bulk used oil and waste materials some of which contain oil, contaminants and other regulated substances. Various environmental laws and regulations require prevention, and where necessary, cleanup of spills and leaks of such materials and some of our operations must obtain permits that limit the discharge of materials. Failure to comply with such environmental requirements or permits may result in fines and penalties, remediation orders and revocation of permits. In Texas, we are subject to rules and regulations promulgated by the Texas Railroad Commission and the Texas Commission on Environmental Quality, including those designed to protect the environment and monitor compliance with water quality.

Occupational Safety and Health Act

We are subject to the requirements of the United States Occupational Safety and Health Act, as amended, or "OSHA," and comparable state laws that regulate the protection of employee health and safety. OSHA's hazard communication standard requires that information about hazardous materials used or produced in our operations be maintained and provided to employees, state and local government authorities and citizens.

Transportation Regulations

We conduct intrastate motor carrier (trucking) operations and from time to time we may conduct interstate motor carrier (trucking) operations that are subject to federal regulation by the Federal Motor Carrier Safety Administration, or "FMCSA," a unit within the United States Department of Transportation, or "USDOT." The FMCSA publishes and enforces comprehensive trucking safety regulations, including rules on commercial driver licensing, controlled substance testing, medical and other qualifications for drivers, equipment maintenance, and drivers' hours of service, referred to as "HOS." The agency also performs certain functions relating to such matters as motor carrier registration (licensing), insurance, and extension of credit to motor carriers' customers. Another unit within USDOT publishes and enforces regulations regarding the transportation of hazardous materials, or "hazmat."

This guide does not address compliance issues associated with OSHA and Transportation Regulations with limited discussion regarding CERCLA associated with protections provided if you complete certain due-diligence activities as part of a sale or purchase identified as innocent buyers protections provided by Phase I and Phase II Site Assessments

Federal and Texas Milestones

To assist in your understanding of how Texas got to this point of managing its environmental compliance associated with used oil and other related compliance requirements for our Example Facility and how the Texas Commission on Environmental Quality (TCEQ) that is the implementing agency for the majority of used oil regulations and other compliance in Texas was established a history of legislation and regulation a milestone listing is provided below. The listing below of the Federal and State legislation and regulatory development is from the TCEQ Strategic Plan for 2015-2019 **(106)** and is provided to show how Texas got here as an implementing state and legislation was put in place to protect the environment. Significant federal actions and Texas actions since 1976 that impact Texas used oil management and the example facility operations and permitting to be addressed are listed in bold, italicized font.

Federal and Texas Milestones Dates

1953 – Texas Legislature creates the Texas Water Pollution Control Advisory Council in the Department of Health as the first state body charged with dealing with pollution–related issues.

1956 – The U.S. Congress passes the Water Pollution Control Act.

1956 – Texas' first air–quality initiative is established when the state Department of Health begins air sampling.

1961 – The Texas Injection Well Act is passed, authorizing the Texas Board of Water Engineers to regulate waste disposal (other than that from the oil and gas industry) into the subsurface through injection wells.

1961 – The Texas Pollution Control Act establishes the Texas Water Pollution Board, and eliminates the Water Pollution Advisory Council, creating the state's first true pollution–control agency.

1962 – The Texas Board of Water Engineers becomes the Texas Water Commission, with additional responsibilities for water conservation and pollution control.

1962 – The Texas Water Pollution Board adopts its first rules and regulations.

1963 – The U.S. Congress enacts the Clean Air Act.

1965 – The U.S. Congress passes the Solid Waste Disposal Act

1965 – The U.S. Congress passes the Water Resources Planning Act.

1965 – The Texas Water Commission becomes the Texas Water Rights Commission and functions not related to water rights are transferred to the Texas Water Development Board.

1965 – The Texas Clean Air Act establishes the Texas Air Control Board in the Department of Health to monitor and regulate air pollution in the state.

1967 – The Texas Water Quality Act establishes the Texas Water Quality Board (TWQB), assuming all functions of the Water Pollution Control Board. The TWQB adopts its first rules.

1967 – The Texas Air Control Board adopts its first air–quality regulations.

1969 – A U. S. presidential order creates the U.S. Environmental Protection Agency (EPA).

1969 – The Texas Solid Waste Disposal Act authorizes the Texas Water Quality Board to regulate industrial solid waste, and the Texas Department of Health to regulate municipal solid waste.

1969 – Texas takes over most federal air–monitoring responsibilities.

1970 – The federal Clean Air Act is amended, requiring states to develop state implementation plans.

1971 – The EPA adopts National Ambient Air Quality Standards (NAAQS).

1971 – The Texas Air Control Board establishes an air permits program.

1972 – U.S. Congress passes the Clean Water Act.

1972 – The Texas Air Control Board submits its first State Implementation Plan to the EPA. It also deploys the first continuous air–monitoring station.

1973 – The Texas Legislature removes the Texas Air Control Board from the Department of Health, making it an independent state agency.

1974 – U.S. Congress passes the Safe Drinking Water Act.

1976 – U.S. Congress passes the Resource Conservation and Recovery Act to govern the disposal of all types of solid and hazardous wastes; this amended the Solid Waste Disposal Act of 1965.

1977 – The Texas Legislature creates the Texas Department of Water Resources by combining the three existing water agencies. A six–member board is set up as a policy–making body for the new agency. The Texas Water Development Board is retained as the legislative and policymaking body. The Water Rights Commission is renamed the Texas Water Commission and sits as a quasi–judicial body that rules on permits. The Water Quality Board is abolished.

1980 – U.S. Congress passes the Comprehensive Environmental Response, Compensation, And Liability Act (CERCLA), better known as the Superfund bill, to pay for the cleanup of contaminated sites.

1984 – U.S. Congress passes the Hazardous and Solid Waste Amendments (HSWA) to the Resource Conservation and Recovery Act.

1984 – Texas receives final RCRA authorization from the EPA.

1986 – Congress passes the Superfund Amendments and Reauthorization Act (SARA), reauthorizes CERCLA, and creates the Toxic Release Inventory.

1987 – U.S. Congress passes the Water Quality Act.

1990 – U.S. Congress passes the Oil Pollution Act.

1990 – The Texas Water Commission receives initial authority for the federal HSWA.

1991 – The Texas Air Control Board is expanded to implement the 1990 Amendments to the federal Clean Air Act.

1991 – The Legislature, in special session, creates the Texas Natural Resource Conservation Commission, to be effective Sept. 1, 1993. Preparation begins for the consolidation of the Texas Water Commission and the Texas Air Control Board into the TNRCC.

1992 – U.S. EPA finalized on September 10, 1992, standards establishing requirements for generators, transporters, transfer facilities, collection centers, processors and re–refiners, burners, and marketers in 40 CFR Part 279. The rule also included provisions for cleanup in the case of a spill or release of used oil.

1993 – The Texas Natural Resource Conservation Commission begins operation, bringing together for the first time regulatory programs for air, water, and waste.

1998 – The EPA delegates to Texas the National Pollutant Discharge Elimination System (NPDES) program, becoming the Texas Pollutant Discharge Elimination System, administered by the TNRCC.

2001 – The TNRCC is continued for 12 years under HB 2912, which changes the TNRCC's name to the Texas Commission on Environmental Quality by Jan. 1, 2004.

2002 – On Sept. 1, the Texas Natural Resource Conservation Commission formally changes its name and begins doing business as the Texas Commission on Environmental Quality.

2005 – The TCEQ undertakes comprehensive review and overhaul of the state's regulations on municipal solid waste. The commissioners direct TCEQ personnel to begin a comprehensive review, including extensive public involvement, of the agency's enforcement process.

2006 – The TCEQ reviews the extensive public comments on the agency's enforcement process. The commissioners adopt a number of significant revisions to the process and start a pilot field–citation program, and the TCEQ adopts major revisions, streamlining and improving state regulations on municipal solid waste.

2008 – In early 2008, the TCEQ upgrades its electronic permitting system (ePermits) for submissions of applications for the stormwater general permit. After the program upgrade, usage rose from 22 percent to 53 percent.

2011 – The Legislature continues the TCEQ for 12 years, until 2023, under HB 2694.

PART II: USED OIL OVERVIEW IMPORTANCE OF RECYCLING USED OIL AND CHALLENGES TO RECYCLING (133)

Importance of Recycling Used Oil and Challenges to Recycling (133)

The following are facts about used oil in the United States and Texas compiled from several state sources and the EPA (133), starting with what should be a pretty obvious statement to all: Recycling used motor oil helps protect the environment and saves energy.

Negative Effects of Not Recycling Oil

First, the BAD things used oil and used oil filters cause by not being recycled:

- The United States generates a total of 1.3 billion gallons of waste oil each year, of which 800 million gallons are recycled and 500 million are disposed of improperly. From a business perspective, there is a tremendous amount of available oil if we could just get it to the facility. (133)
- Nearly 40% of the pollution in America's waterways is from used motor oil. (157)
- Used motor oil is insoluble, persistent, and can contain toxic chemicals and heavy metals. It's slow to degrade. It sticks to everything from beach sand to bird feathers. Used motor oil is a major source of oil contamination of waterways and can result in pollution of drinking water sources. – *EPA Used Oil Management Program* (158)
- One gallon of used motor oil can contaminate 1 million gallons of fresh water—a year's supply of water for 50 people. (157)

- Oil doesn't dissolve in water. It lasts a long time and sticks to everything from beach sand to bird feathers. Oil and petroleum products are toxic to people, wildlife, and plants. One quart of motor oil can pollute 250,000 gallons of water. Oil that leaks from our cars onto roads and driveways is washed into storm drains, and then usually flows directly into a lake or stream. Used motor oil is the largest single source of oil pollution in lakes, streams, and rivers. Americans spill 180 million gallons of used oil each year into the nation's waters. This is 16 times the amount spilled by the Exxon Valdez in Alaska! **(160)**
- It only takes 1 pint of used motor oil to put an oil sheen on a 1 acre pond. **(157)**
- One gallon of used motor oil that is re-refined will produce 2.5 quarts of lubricating oil, but it takes 42 gallons of crude oil to produce the same 2.5 quarts of lubricating oil. **(157)**
- Used oil poured down your sewer line can damage your community's water treatment system, contaminating your drinking water and costing you money. In most areas, improper disposal of used oil is a crime. – *American Petroleum Institute* **(159)**
- The failure to recycle used motor oil in the United States represents not only an environmental and human health hazard, but a missed opportunity to reduce our nation's dependence on foreign oil. **(157)**
- Each year, the United States generates 425 million used automotive oil filters containing 160,000 tons of iron units and 18 million gallons of oil. **(157)**
- Used oil filters can contain more than 45% used motor oil in weight when removed from the vehicle. **(157)**

Positive Effects of Recycling Oil

Now the good things recycling used oil provides:

- Recycled used motor oil can be re-refined into new oil, processed into fuel oils, and used as raw materials for the petroleum industry. **(157)**
- One gallon of used motor oil that is re-refined will produce 2.5 quarts of lu-

bricating oil, but it takes 42 gallons of crude oil to produce the same 2.5 quarts of lubricating oil. **(157)**

- If we re-refined the 1.3 billion gallons of waste oil generated in the United States each year, we would save 1.3 million barrels of oil per day, or half the daily output of the Alaska pipeline. **(157)**
- Large industrial boilers can efficiently burn the used oil with minimum pollution. As a result, some used oil is sent to power plants or cement kilns to be burned as fuel. On a smaller scale small quantities of used oil are burned in specially designed heaters to provide space heating for small businesses. **(157)**
- Re-refining used motor oil takes only one-third the energy of refining crude oil to lubricant quality. **(157)**
- Recycling all the used oil filters sold annually in the United States would result in the recovery of about 160,000 tons of steel, or enough steel to make 16 new stadiums the size of Atlanta's Turner Stadium. **(157)**

Some Observations on Why People Don't Recycle

The American Petroleum Institute (API) **(159)** identifies that it is relatively common for people to recycle paper, glass, and metals (aluminum and steel). The infrastructure and services for this recycling activity are available and relatively well understood. However, significant to our used oil recycling concerns is that the recycling of petroleum products is less well known and sometimes confusing, indicating that more education of the public is needed. Regulators and recycling agencies have effectively communicated with businesses the need for and benefits of recycling used oil that is very strongly related to a proven cost effectiveness when disposition of used oil as a recycled material and not a waste; this is a tremendous cost and regulatory burden not associated with recycling. The APA also points out that more can be done to communicate with individuals who work on their cars or trucks at home. These individuals generate significant quantities of used petroleum products that need to be recycled, an amount estimated to be at least 500 million gallons per year. As the API and the EPA **(157)** note, many individuals who are unfamiliar with the importance of recycling used oil are unconsciously harming the environment by throwing it away with their normal garbage or emptying their used oil into storm drains. Such actions, especially emptying used oil into storm drains, can cause real harm to the environment in so many ways.

Earth911 (87) is a website that has promoted recycling for more than 20 years. Its vision and mission statement is "Our vision and mission are simple—create a community that helps consumers find their own shade of green, adopt environmentally sound practices and drive impactful environmental changes. We are here to deliver a mix of targeted media content and recycling data so that you can live a happier, healthier, sustainable lifestyle; one that protects this wonderful planet we all call home." Earth 911 recently conducted a survey of the public in various age groups to find out why people don't recycle more; the overall number one reason was "No. 1 reason is convenience – or should I say, inconvenience," an unnamed survey respondent said. "We don't have a [curbside] recycling program where I live. You have to collect all of your items and then drive them to the middle of nowhere to drop them off. It takes too much extra effort." Common issues included storage as a barrier, confusing recycling programs, and an overall perception among all the groups surveyed that pro-recyclers had a superior attitude of many pro-recyclers, which served as a turnoff to all groups. The middle age group also felt that the economics did not make sense to the average individual and the time it took to recycle was not worth the trade-off as most people do it because it feels good—not because of the economic benefit to them that was not worth it to many from their perspective.

What all the perspectives by agencies and interest groups seem to identify is a lack of understanding of the actual facts of the impacts of not recycling has on the environment and your taxes that are used to correct and manage the harm the materials cause to the environment. As agencies, environmentalists, concerned citizens and specific to those in the recycling business we need to support more education and frankly get involved to promote recycling and develop processes. The Do-It-Yourself (DIY) group generating used oil needs some help and not because we can just make an environmental positive impact but it could provide a significant business benefit that is need of a new management approach.

Texas Support to Recycle Education and Assistance

Texas, through its "Take Care of Texas" program (140), specifically promotes the importance of recycling and conservation of water, energy, aluminum cans, and paper or plastic bottles, as well as conserving resources of all types. The TCEQ and other Texas environmental groups go to great lengths to educate the public and children at all grade levels, but many Texas residents still do not truly understand why it is so important to recycle used oil. One additional important TCEQ support section that will be discussed more in this guide is the TCEQ's Small Business and Local Government Assistance Section (141), which offers free, confidential help to small businesses and local governments working to comply with state environmental regulations that include pollution prevention and recycle advice and support. It is available to all small businesses and can really provide significant assistance.

Other Challenges to Recycling

Education and new management approaches to insure more collection are significant but over the last few years a more detrimental impact that significantly impacts the overall industry from a business perspective and that is the cost of crude oil that has plummeted from a high of $105 in 2012 to current prices over the last 3 years from $44 to almost $73 providing much less incentive to all involved in the recycle process (see the NASDAQ chart (Chart 1) below.)

Graph 1 – NASDAQ Crude Oil Chart

Crude Oil

WTI (NYMEX) Price

End of day Commodity Futures Price Quotes for Crude Oil WTI (NYMEX)

Crude Oil Related ETFs: USO, DBO, USL, DNO

NASDAQ Crude Oil Chart retrieved from: (88)

http://www.nasdaq.com/markets/crude-oil.aspx?timeframe=5y

The industry itself is undergoing a significant change in the entire business plan moving from a "pay for oil" approach where businesses were paid for their used oil to charging for the pick-up of the oil the way it was about 10 years ago. Even with the change, recyclers are still challenged and the entire business is undergoing a significant redesign and restructure as pointed out in an article by Mallory Szczepanski (2016)**(88)** where she discusses the challenges. According to Szczepanski, "many waste and recycling companies are feeling the effect of the slump in oil prices and are working to adjust to ensure operations run smoothly. Tactics include implementing fees and sliding scales, staff reductions and redeployments and other measures."

Szcepanski discusses how low oil prices may come as a relief at the pump, but they have had a damaging effect on the waste and recycling industry. Most significantly is that the cost of virgin materials has dropped below that of recycled materials. This is decreasing manufacturer demand for recycled goods, including not just the typical recycled materials like paper and plastic, but used oil, which has lost its value to refiners as a substitute feed. Szcepanski identifies a possible solution based on the approach utilized by leading recycler Clean Harbors Environmental Services Inc., based in Norwell, Mass. This company offers a sliding scale tied to energy prices, which either reduces or increases the surcharge customers pay based on current energy costs as part of the attempt to reinvent the business plan needed to compete in the current market. Szcepanski notes that business must change; managing waste streams is a challenge in itself, but waste and recycling companies are also in competition for business with each other. Therefore, companies need to look at their competition and come up with a creative and thoughtful approach to make their business stand out and generate more business from different sources. Doing so will build the customer base while allowing the company to do everything they can to provide the best service possible and become more responsive to overall customer needs. An increased focus on safety will also attract more industrial customers who view safety as a priority.

The overall state of the used oil recycling is undergoing a consolidation, both in collection and processing to pursue the economics of scale along with providing benefits from generation point to end product cost control to maximize profits and respond to market fluctuations. **(88)**

Used Oil Regulation Development

As indicated above, RCRA §3014 required EPA to decide whether or not to list used oil as regulated hazardous waste under RCRA and to establish standards applicable to used oil recycling. On November 29, 1985, EPA issued a proposed rule to list all used oil as hazardous waste, including petroleum-derived and synthetic oils (50 FR 49164) **(142)**. On the same day, management standards were proposed for recycled used oil and final regulations were issued prohibiting burning off-specification used oil in nonindustrial boilers and furnaces. By November 19, 1986, EPA decided not to list used oil, primarily because of the possibility that characterizing used oil as a hazardous waste might discourage or hinder recycling practices (51 FR 41900) **(143)**.

As is often the situation in regulation implementation, court cases are involved in clarification or establishing certain aspects of regulation application or authority. Used oil criteria is no different regarding technical criteria for listing waste specified in the statute, rather than the effects listing might have on the competing goal of recycling. Consequently, a supplemental proposed rule published in the Federal Register on September 23, 1991, offered three options for regulating used oil recycling (56 FR 48000) **(144)**. The first option was to identify all used oils as listed hazardous waste under RCRA. Alternatively, EPA could have designated only certain used oils (primarily the nonindustrial ones) as hazardous. The third option was to promulgate basic management standards for used oil, and only categorize used oil as RCRA hazardous waste when disposed.

Prior to making a final selection of one of the three approaches to recycled used oil regulation, the final rule of May 20, 1992, stated that used oil destined for disposal would not be listed as a hazardous waste (57 FR 21524) **(145)**. EPA said that most used oil has the potential to be recycled, either through re-refining or burning for energy recovery, and that when destined for disposal, the existing network of regulations provides adequate protection from plausible mismanagement scenarios. Additionally, an exclusion was added under §261.4(b) **(146)** stating that certain used oil filters that have been properly drained may be disposed of as nonhazardous (see Section 3.1 of this module for discussion). The Agency deferred the listing decision on used oil destined for recycling at this time and any decision on management standards for used oil.

On September 10, 1992, EPA promulgated both a final listing decision for recycled used oil and management standards for used oil pursuant to RCRA §3014 **(147)**. The Agency decided not to list recycled used oils because the management standards issued in this rulemaking would adequately protect human health and the environment. These standards affect generators, transporters, burners, marketers, processors, and re-refiners of used oil. The management standards were codified as Part 279. In this rule, the Agency stated that there is a presumption that all used oil will be recycled until the used oil is disposed of or sent for disposal (57 FR 41578) **(155)**. Used oil that is disposed of will need to be characterized like any other solid waste and need to be managed as hazardous if it exhibits a characteristic of hazardous waste or if it is mixed with a listed hazardous waste.

It is extremely important to understand the present program hinges on the presumption that most used oil will be recycled instead of disposed. This means that virtually all used oil will be subject to the same management standards under Part 279 as codified in the TAC. The former program did not make this presumption and therefore imposed variable requirements for used oil management. The current used oil regulations also expanded the requirements that used oil handlers must fulfill.

The recycling presumption as first introduced by the EPA in a 1991 supplemental notice of proposed rulemaking, the Agency first introduced the idea of the recycling presumption (56 FR 48000) **(144)**. According to industry data collected by the Agency, a majority of used oil that enters the recycling system ends up being recycled either by being burned for energy recovery or by being re-refined. These data led to the recycling presumption, which is a regulatory provision that presumes used oil will be recycled. In cases where used oil handlers cannot recycle their used oil, the Agency proposed different criteria for rebutting the recycling presumption. In the September 10, 1992, in the final rule, however, EPA decided that establishing specific criteria was not feasible or necessary. Because the Agency assumes all used oil will be recycled, it is only when used oil is sent for disposal that it is no longer subject to the Part 279 standards. The recycling presumption simplifies the used oil management system by ensuring that handlers need only comply with one set of standards.

Further discussion of what used oil is and how it is defined and managed is discussed later in this Guide. When discussing used oil and waste oil as identified by the regulations, it is important to keep in mind that they are not interchangeable; this is a common misconception. One of the objectives of this guide is to ensure the differences in Texas are understood. The Example Facility referenced in this guide does not manage or accept waste oil, only used oil.

What Is Used Oil as Identified by Texas and the Federal Regulations

Used oil per 40 CFR §279.1.is defined as "any oil that has been refined from crude oil or any synthetic oil that has been used and as a result of such use is contaminated by physical or chemical impurities"**(115)**. TAC Title 30 §324.2 identifies that the "commission incorporates by reference the definitions in 40 Code of Federal Regulations (CFR) §279.1. However, the following words have these additional meanings," none of which change the reference to the definition of used oil.

It is extremely important to understand that the terms *Used Oil* and *Waste Oil* are not interchangeable. **(46)**

In order to identify Used Oil, reference the listing from the TCEQ Used Oil Recyclers Handbook **(109)** to review what the regulators (TCEQ) identify as reasonably clear descriptions. If the material in question does not fit one of these descriptions it may be a hazardous waste; please note this is not an all-inclusive listing. Following the listing is an additional aid in the form of a Flow Chart 1 **(46)** from an EPA guidance manual that can be used to help with the decision based on the material requirement management. When in doubt, always seek assistance, either directly from agency staff or an environmental professional. In addition, remember that errors are not permissible and documentation of the reasons the accepted material meets regulatory requirements is required.

1. TCEQ Used Oil Recyclers: Used oil includes any of the materials listed below.
   a. Any oil, either synthetic or refined from crude oil, that
      i. has been used for its designed and intended purposes; and
      ii. as a result of use, is contaminated by physical or chemical impurities; and
      iii. as a result, has become a spent material (that is, it can no longer be used for its original intended purpose without processing).
   b. Any other material that has physical and chemical properties similar to used oil, is used in normally accepted functions of oil, and has been designated as used oil by the TCEQ.
   c. Used oil that is characteristically hazardous from use (as opposed to oil rendered characteristically hazardous by mixing).
   d. Used oil collected from household DIYs.
   e. Hazardous used oil or used oil mixtures from Conditionally Exempt Small Quantity Generators (CESQGs).
   f. Used oil to be burned for energy recovery.

Flowchart 1 - Used Oil Determination, EPA Guidance Manual (46)

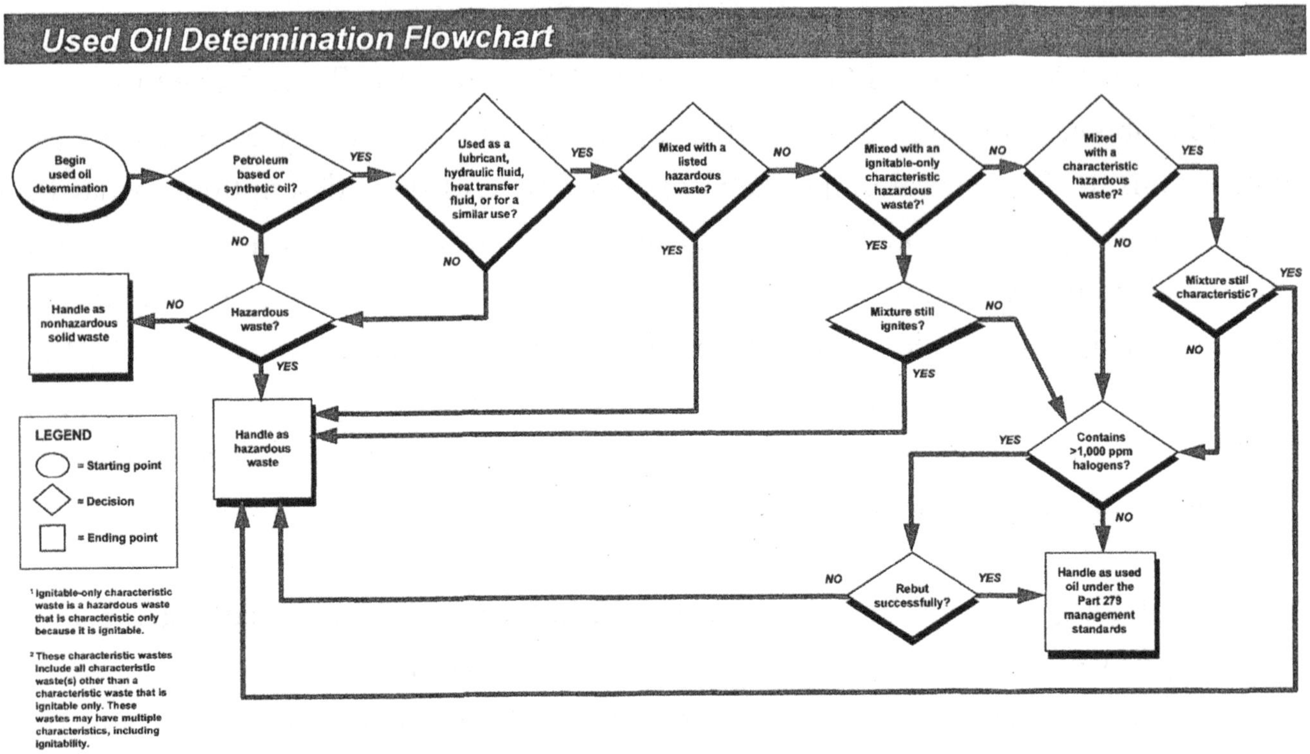

The following are examples of used oil that is destined for recycling, and therefore exempt from the hazardous waste regulations found in the TCEQ used oil recyclers handbook identified above: **(109)**

A key part of the definition of used oil given above is the term, "spent material." The following lists give some examples of oils that have become spent materials through use.

Spent Engine Oil and Vehicle Lubricants.

Used oils include, but are not limited to, the following spent engine lubricating oils and vehicle fluids:

- automotive crankcase oil, including car, truck, marine, and aircraft engine oils not used for engine fuel;
- diesel engine crankcase oil, including car, truck, bus, marine, heavy equipment, and railroad engine oils not used for fuel;
- natural-gas-fired engine oils;
- alternative fuel engine oils;
- transmission fluids;
- brake fluids; and

- power steering fluids.

Spent Industrial Oils.

Used oils also include, but are not limited to, the following spent industrial oils:

- compressor, turbine, and bearing oils;
- hydraulic oils or fluids;
- metalworking oils or oil emulsions, including cutting, grinding, machining, rolling, stamping, quenching, and coating oils;
- electrical insulating oils;
- refrigerator/air conditioning unit oils;
- rubber-making oils;
- cable oils;
- greases; and
- oil-like heat transfer fluids.

> ***Commentary:*** *Please note that both the Spent Engine Oil and Spent Industrial Oils listing statements include the caveat "but are not limited to, the following"* ***(109)****. This indicates the listings are not comprehensive and therefore it is possible that the used oil material in question may not be included in these listings, but is still considered hazardous. It never hurts to ask for clarification, but it can be a very painful lesson if an incorrect assumption is made.*

The following listing from the same TCEQ Handbook above identifies the materials listed that are ***not*** used oil as determined by the TCEQ. However, this does not mean these materials cannot be managed at your facility; only that they are not managed as used oil and are subject to other regulations and possible permitting requirements.

- Used animal or vegetable oils *(they are considered food wastes rather than used oil, because they are not synthetic and not derived from crude oil).*
- Unused contaminated or uncontaminated oils going for reclamation.
- Solid wastes contaminated with used oil *(such as absorbents and scrap metal)* that are not burned for energy recovery and that do not have free-flowing oil.

- Solvents *(such as petroleum spirits, mineral spirits, petroleum ether, acetone, fuel additives, alcohols, paint thinners, brush cleaners, and other cleaners).*
- Substances that cannot readily be recycled in the same processes as used oil.
- Used antifreeze.

Making determinations as to if a material is a used oil is challenging, and when presented with materials that are identified by a generator as a waste oil or oily waste it does not always meet the regulatory definition of used oil to be able to be managed as one. Materials like bottom waste from an oil storage tank or virgin oil spill residue may be contaminated with impurities but still not meet the definition in 40 CFR §279.1 **(65)** because these materials have never been used. If the material in question clearly does not meet one of the above descriptions of a used oil, or it is not a used oil based on the list but is mixed with something that may be a hazardous and solid waste as well as other products, "rebuttable presumption" is used as an objective test. Rebuttable presumption was established by the EPA **(6)** to address mixtures of used oil and hazardous wastes and other materials. As indicated above with the subjective statements "but are not limited to, the following" from a practical standpoint and an enforcement point of view, the agency cannot always determine if used oil has been mixed with a hazardous waste. Therefore, the EPA decided to use the "rebuttable presumption" concept. The process is often difficult but must be done; follow the flow chart based on some minimum knowledge obtained at the start of the management process, and walk through the elements one-by-one as found in a combination of the EPA and the TCEQ guidance documents. Always remember used oil **(70)** with the two generation exceptions from **40 CFR § 279.44 (70)**: 1) The rebuttable presumption does not apply to metalworking oils/fluids containing chlorinated paraffins, if they are processed, through a tolling arrangement as described in § 279.24(c) **(68)**, to reclaim metalworking oils/fluids. Note that the presumption does apply to metalworking oils/fluids if such oils/fluids are recycled in any other manner, or disposed. 2) The rebuttable presumption does not apply to used oils contaminated with chlorofluorocarbons (CFCs) removed from refrigeration units if the CFCs are destined for reclamation. Again, note regarding CFCs that the rebuttable presumption does apply to used oils contaminated with CFCs that have been mixed with used oil from sources other than refrigeration units. With the two generation exceptions, there are three types of generators that are also exempt from the requirement: 1) used oil generated by a household do-it-yourselfer, 2) used oil generated by a CESQG, or 3) used oil generated by a collection center that only accepts household used oil.

All other used oil generators, transporters, processors, transfer facilities, re-refiners, and burners are responsible for ensuring that used oil is not intentionally mixed with a hazardous waste and the proof can be made either by adequate documentation (process knowledge) or by testing (analytical data) the used oil as identified below. As a comment on an EPA question from their EPA hotline regarding the rebuttable presumption **(78),** they discuss that it is a little confusing if you look at 40 CFR § 279.10(b)(ii), § 279.21, § 279.44, § 279.53, § 279.63. These apply to the application of the rebuttable presumption but state that the overall intent of the EPA is to have it apply to all generators with the exceptions noted and other handlers as the TAC identifies.

> ***Commentary:*** *As you review the steps, please note the use of the term "resulting mixture"* ***(78)*** *that is a result of a process generating the material. However, remember you can never intentionally mix used oil with a waste of any kind to avoid regulation; doing so is a serious violation unless you have a hazardous waste treatment permit.*

Rebuttable Presumption Element Review:

You have identified through the use of a field test kit indicating the used oil offered by a generator contains greater than or equal to 1,000 parts per million (ppm) of total halogens is presumed to have been mixed with a regulated halogenated hazardous waste (i.e., spent halogenated solvents), and is therefore subject to applicable hazardous waste regulations.

A person may rebut this presumption by demonstrating through analysis or other documentation that the used oil has not been mixed with halogenated hazardous waste. One way of doing this is to show that the used oil does not contain significant concentrations of halogenated hazardous constituents (50 FR 49176; November 29, 1985 **(46)**. If the presumption is successfully rebutted, meaning the oil is not considered to have been mixed with a regulated hazardous waste, then it will be subject to the used oil management standards instead of the hazardous waste regulations.

Mixtures of Used Oil:

If listed hazardous waste or characteristically hazardous waste does become mixed with used oil, the resulting mixture may have to be managed as hazardous waste. You must perform the following steps for further determination of how to manage the material: (46)

Mixtures of used oil and characteristic waste after testing the resulting mixture is not hazardous, it is managed as a used oil unless:

- Mixtures of used oil and ignitable-only characteristically hazardous waste are managed as used oil, unless the resulting mixture exhibits the characteristic of ignitability. (Less than 60 °C (140 °F) see Table 2 below for analysis method)
- Mixtures of used oil and nonregulated hazardous waste are managed as used oil. (Non-regulated hazardous waste is defined as hazardous waste from households, from farms. (46)
- Mixtures of used oil and listed hazardous waste are managed as listed hazardous waste.
- Mixtures of used oil with hazardous waste from a Conditionally Exempt Small Quantity Generator (CESQG), which are businesses generating 220 pounds or less of hazardous waste per month.
- Mixtures of used oil and fuel products are managed as used oil.
- It also does not apply to a collection center that only accepts household used oil.

Process Knowledge:

Adequate documentation (138) for process knowledge refers to the application of documented descriptions that identifies your "process knowledge" as it refers to a used oil handler's and applicable generator's understanding of the operations and activities that created the used oil at the facility and possible contaminants in the used oil. The word "documented" refers to records of the used oil's origin, use, processes it has undergone, identification of possible contaminants, and other information on used oil activities. If enough documented process knowledge is available to show that mixing did not occur, then the used oil handler may eliminate the cost of analytical testing. It is extremely important that the documentation is complete as possible and you can never have too much documentation, but it needs to be assembled in a clear form to be able to show a regulator or provide to a handler upon request in order to not be required to provide analytical data.

Analytical Data:

When you need to provide analytical data to support the total halogen rebuttable presumption to verify your used oil is exempted as a waste, you must use EPA and TCEQ approved methods that currently reflect the EPA methods. It is important that you use a qualified laboratory who knows what sample preparation is to be used and what method is to be applied. There are many challenges associated with used oil analysis because of the different non-oil materials that can be in them, so the lab's knowledge and experience in running this type of analysis is critical for compliance. *Total halogen* content can be determined by EPA methods (nonspecific determination) listed in Table 1 below. If the total halogen content is equal to or greater than 1,000 ppm, then a more specific test method should be used to determine the level of individual halogenated compounds. If required (see Table 1 EPA Methods below) to determine a *specific halogen* content for example, to determine halogen-containing volatile compounds EPA Method 3585 is used for sample preparation, and Method 8260 or 8021 may be used for sample analysis. For analysis of halogenated pesticides, Method 8081A is recommended. But, again the laboratory may run alternative methods depending on the used oil to be analyzed so long as they are EPA approved.

Table 1 – EPA Methods for Used Oil

Determination	Sample Preparation Method	Sample Analysis Method
Total Halogens	N/A	9075, 9076, 9077
Total Halogens	5050	9056, 9253
Individual Halogens	3585 (liquid organic matrix)	8260 or 8021
Halogenated Pesticides	3580	8081A

Source: EPA Test Methods for Evaluating Solid Waste, Physical/Chemical Methods (SW 846), *3rd ed.*

With your application of the rebuttable presumption and the process knowledge and/or analytical identifies the used oil is at no more than 1000 ppm of total halogens, you have cleared one hurdle. With additional knowledge that it does not contain a listed waste or the mixture does not have the properties of a characteristic waste that includes a characteristic waste that is subject to ignitability property only, so far you have a material that can be managed as a used oil if other requirements are met.

The next step associated with total halogen content involves what additional requirements you must meet depending on how you will use the used oil. You must apply 40 CFR **§ 279.11** Used oil specifications (58) that identifies that used oil burned for energy recovery, and any fuel produced from used oil by processing, blending, or other treatment, is subject to regulation under this part unless it is shown not to exceed any of the allowable levels of the constituents and properties shown in Table 3 found below. The additional analysis requirements to determine the other properties of your used oil are found in Table 2 below.

Table 2 – Additional Possible Analytical Methods for Used Oil Specification (TCEQ UOHH)

Determination	Methods	
	Sample Preparation	Sample Analysis
Multi-element analysis including arsenic, cadmium, chromium, and lead	3051	6010A, 6020
Arsenic	3051	7060A, 7061A, 7062
Cadmium	3051	7130, 7131A
Chromium	3051	7190, 7191
Lead	3051	7420, 7421
Ignitability (Flash Point)	N/A	1010A,1020B

Source: *EPA Test Methods for Evaluating Solid Waste, Physical / Chemical Methods (SW 846),* 3rd ed.

The used oil is no longer subject to this part once used oil that is to be burned for energy recovery has been shown not to exceed any allowable level and the person making that showing complies with § 279.72 (**on-specification used oil fuel) (71),** § 279.73 **(71)** (used oil fuel marketer requirements), and §279.74(b) **(71)** (*On-specification used oil delivery:* A generator, transporter, processor/re-refiner, or burner who first claims that used oil that is to be burned for energy recovery meets the fuel specifications under § 279.11 **(58)** must keep a record of each shipment of used oil to the facility to which it delivers the used oil. Records for each shipment must include identified information.)

The used oil is no longer subject to this part once used oil that is to be burned for energy recovery has been shown not to exceed any allowable level and the person making that showing complies with § 279.72 (**on-specification used oil fuel) (71),** § 279.73 **(71)** (used oil fuel marketer requirements), and §279.74(b) **(71)** (*On-specification used oil delivery:* A generator, transporter, processor/re-refiner, or burner who first claims that used oil that is to be burned for energy recovery meets the fuel specifications under § 279.11 **(58)** must keep a record of each shipment of used oil to the facility to which it delivers the used oil. Records for each shipment must include identified information.)

If the used oil so far that has not been identified as a listed hazardous waste or characteristic waste and meets the allowable levels below you so far either have an on-specification used oil or an off-specification used oil but must still consider two more issues to determine possible final disposition options. The remaining issues are again associated with Total Halogens and the amount equal to or greater than 1000 ppm not to exceed 4000 ppm (A) and the levels of PCBs (polychlorinated biphenyls) (B).

With Total Halogens amount equal to or greater than 1000 ppm not to exceed 4000 ppm you can still apply the rebuttable presumption, ensuring it is not subject to hazardous waste management requirements as long as the used oil is not over 4000 ppm it is still on-specification used oil. If it over 4000 ppm you must identify it as off-specification used oil and a used oil burner may mix off-specification used oil with virgin oil or with on-specification oil. But if the used oil is mixed to produce on-specification used oil fuel, the burner becomes subject to the standards for a used oil processor and possibly to the marketer requirements.

You must know the used oil levels of PCBs (polychlorinated biphenyls) for the next point in your determinations. PCBs are man-made organic chemicals that range in consistency from heavy, oily liquids to waxy solids. They are used for their cooling properties because they only boil at high temperature and do not readily react with other chemicals.

Used oil containing PCBs can be burned for energy recovery, but only in combustion units that thermally degrade the PCBs—for example, rotary kilns, cement kilns, liquid injection incinerators, and high-temperature boilers. Although PCBs have been banned for most applications, they are still occasionally encountered. PCBs are regulated under the federal regulations for the Toxic Substance Control Act (TSCA) at 40 CFR 761 **(148)**. How used oil containing PCBs is regulated depends upon the concentration of PCBs in the used oil. Used oil containing less than the quantifiable level of 2 ppm PCBs is not regulated as a TSCA waste, but rather as used oil under 40 CFR 279. Used oil containing 50 ppm or more PCBs is regulated as a TSCA waste. For used oil containing PCB concentrations of 2 ppm or more, but less than 50 ppm, TSCA requires the used oil to be handled in the same manner as off-specification used oil if you are a Marketer or it is burned for energy recovery. However, if the used oil is to be burned, the PCBs must be completely destroyed. The level of PCBs in the used oil makes a difference on how it is handled. Used oil with PCBs greater than 50 ppm is regulated by the U.S. EPA under the Toxic Substance Control Act (TSCA), codified at 40 CFR 761. Furthermore, if PCB contaminated used oil is mixed with other used oil, the disposal of the entire volume must be based on the concentration of PCBs in the PCB contaminant source, if known. Used oil transporters and processing facilities use retain samples from their customer loads to determine the PCB source and source value. Dilution of the PCB concentration is not allowed under federal law to reduce the levels of PCBs for disposal purposes. If you have material that has 50 ppm or more of PCBs, you do not have an on-specification or an off-specification used oil; it must be managed as a TSCA waste and is not subject to the requirements of Part 279.

Table 3 – Used Oil Constituents

USED OIL NOT EXCEEDING ANY ALLOWABLE LEVEL SHOWN BELOW IS NOT SUBJECT TO THIS PART WHEN BURNED FOR ENERGY RECOVERY 1 **Constituent/property**	**Allowable level**
Arsenic	5 ppm maximum.
Cadmium	2 ppm maximum.
Chromium	10 ppm maximum.
Lead	100 ppm maximum.
Flash point	100 °F minimum.
Total halogens	4,000 ppm maximum. 2
PCBs	<2 ppm or <= 50 ppm
Note: Applicable standards for the burning of used oil containing PCBs are imposed by 40 CFR 761.20(e).	

1. The allowable levels do not apply to mixtures of used oil and hazardous waste that continue to be regulated as hazardous waste (see § 279.10(b)).

2. Used oil containing more than 1,000 ppm total halogens is presumed to be a hazardous waste under the rebuttable presumption provided under § 279.10(b)(1). Such used oil is subject to subpart H of part 266 of this chapter rather than this part when burned for energy recovery unless the presumption of mixing can be successfully rebutted.

Used Oil Mixture Considerations:

With an understanding of what properties qualify your used oil as available for recycle and exempt as a waste as either an on-specification or off-specification used oil when material is collected through typical generation activities not associated with hazardous waste materials, you must still be aware of what is in your mixtures of used oil to ensure you are in compliance. A listing of the different associated considerations is provided below. First, you should be aware of a Texas regulation associated with mixtures of used oil and waste beyond the Federal requirements in 40 CFR Part 279 as it is found in Chapter 7 Section 7.176, Subsections (a) and (c) of the Texas Water Code. **(127)** that impacts used oil management which is identified as follows:

"During the regular session of the 79th Legislature (2005), SB 1299 was passed, relating to an exception to the prohibition against commingling used oil with solid waste if the commingling is incident to the dismantling of scrap, used, or obsolete metals. Chapter 7 Section 7.176, Subsections (a) and (c) of the Texas Water Code were amended to read as follows:

A person commits an offense if the person:

(1) Knowingly mixes or commingles used oil with solid waste that is to be disposed of in landfills or directly disposes of used oil on land or in landfills, unless the mixing or commingling of used oil with solid waste that is to be disposed of in landfills is incident to and the unavoidable result of the dismantling or mechanical shredding of motor vehicles, appliances, or other items of scrap, used, or obsolete metals;

(2) It is an exception to the application of Subsection (a)(2) that the mixing or commingling of used oil with solid waste that is to be disposed of in landfills is incident to and the unavoidable result of the dismantling or mechanical shredding of motor vehicles, appliances, or other items of scrap, used, or obsolete metals." **(127)**

Commentary: *From the above Texas citation please note that it identifies that "a person commits an offense" so this mixture issue is important that it is managed correctly.*

Federal and TCEQ Applications of the Mixture Rules and Management Practices Beyond Mixtures of Hazardous Waste Summarized From Various Guides and Manuals: **(47, 68)**

General Statement on Mixtures:

Any mixture of used oil with solid waste or any material contaminated with or containing used oil is regulated under the Part 279 standards as used oil. Once the used oil has been properly drained or removed from the mixture to the extent possible (i.e., no visible signs of free-flowing oil), the solid waste is no longer subject to regulation as used oil, but rather is classified as a solid waste. Any used oil separated from this type of mixture remains subject to all applicable Part 279 standards. The exception to this is solid waste from which free-flowing used oil has been removed is still subject to regulation as used oil if it is burned for energy recovery.

Mixtures of Used Oil With Products:

The blending of used oil generated in diesel-powered vehicles with diesel fuel is excluded from the processing and re-refining standards when the mixture is used in the generator's own vehicles as a fuel. The mixture itself is also specifically excluded when used in this manner. In addition, as stated earlier, diesel and marine engines are not covered by the standards for used oil burners.

Materials Derived From Used Oil:

Section 279.10(e) dictates when a material is no longer subject to regulation as used oil. Specifically, materials that are reclaimed from used oil and are used beneficially without being burned for energy recovery are not considered used oil and are not subject to the Part 279 standards. These materials are also not considered solid waste and would, therefore, not be subject to regulation under any part of RCRA. Materials produced from used oil that are burned for energy recovery are regulated as used oil and are subject to Part 279. Materials derived from used oil that are disposed of or used in a manner constituting disposal are not considered used oil, but instead are considered solid wastes and are subject to all applicable solid and hazardous waste standards. Please also note that conditional exemptions identified above, such as the one associated with CESQG generators, apply to this general rule.

Blending of Used Oil:

Blending used oil to meet specification is permitted, since the EPA believes blended oil poses no greater hazard than dirty virgin fuel oil. Used oil with over 1,000 ppm total halogens, however, may not be blended to lower the halogen level below 1,000 ppm in order to apply the rebuttable presumption and call it nonhazardous; the rebuttable presumption must be applied to the used oil before it is blended with other materials. A used oil handler blending used oil for purposes of meeting specification would be subject to the processor and re-refiner regulations in Part 279, Subpart F.

Used Oil to the Land:

The use of used oil for road oiling or as a dust suppressant, weed abatement, or other similar uses is prohibited in Texas per TAC §324.4.(2)(D).(115)

Burning Off-Specification Fuel in Certain Units:

Burners may only burn used oil in an industrial boiler or furnace as defined in §260.10, in a hazardous waste incinerator subject to regulation under Part 264/265, Subpart O, or in a used oil-fired space heater. When burning in a space heater unit, the owner or operator must either generate the oil he burns himself or receive it from do-it-yourself oil changers. The maximum capacity of the device may not exceed 0.5 million Btu per hour and combustion gases must be vented to the ambient air (§279.12(c)(2)(iii)).

PART III: USED OIL HANDLERS AND USED OIL CONTAINING MATERIALS MANAGEMENT

Used Oil Transporters and Transfer Facilities: Used oil transporters may consolidate or aggregate used oil for the purposes of transporting and may conduct incidental processing operations that occur during the normal course of transportation (e.g., separating water from used oil that has settled during the aggregation or transportation process). Transporters who otherwise produce used oil-derived products or make these products more amenable for production are considered used oil processors. Used oil transporters must not only notify EPA of their activities, but must also track their shipments and maintain compliance with the rebuttable presumption. The management standards that must be complied with in Part 279, Subpart D are similar to regulations for hazardous waste transporters in many respects, including a provision for used oil transfer facilities. Like generators, used oil transporters who engage in other used oil management practices must follow the applicable standards in Part 279. **(70)** For example, a transporter who processes used oil must also comply with the requirements for used oil processors and re-refiners in Part 279, Subpart F, including transporters who blend used oil with virgin oils to meet specification. Transporters, however, may conduct incidental processing operations and consolidation for the purposes of transportation in compliance with §279.41 **(70)** without being considered a processor or re-refiner.

Storage: Used oil transporters may store used oil at transfer facilities for a limited time during the normal course of transportation. EPA defines transfer facility as any transportation-related facility, including loading docks, parking areas, and other areas where shipments of used oil are held for more than 24 hours and less than 35 days during the normal course of transportation. Transfer facilities that store used oil for more than 35 days are subject to Part 279, Subpart F **(70)** standards for used oil processors and re-

refiners. Facilities that store used oil for less than 24 hours do not meet the definition of used oil transfer facility and would not be subject to these requirements. Transfer facilities and transporters can store used oil in the same units as generators, but must supply secondary containment for the storage units in accordance with §279.45. **(70)** If a release occurs during the course of transportation, the transporter is required to take "appropriate and immediate action to protect human health and the environment" (§279.43(c)). **(70)**

Rebuttable Presumption: The used oil management standards require transporters to apply the rebuttable presumption to all shipments of used oil. They must determine the total halogen content of used oil prior to accepting a shipment for transport. Field testing on every pick-up, or having the generator provide the required documentation, is necessary to ensure this issue as described above can be addressed. If the presumption of mixing can be rebutted by documenting the source of the halogens, then the used oil can be managed as used oil. Otherwise, the used oil contaminated with halogens is regulated as hazardous waste under applicable solid and hazardous waste requirements.

Record Keeping

Used Oil Transporters: Transporter reporting requirements as identified in 40 CFR §279.41. **(70)** (a) and (b) below identify restrictions on transporters who are not also processors or re-refiners and would not be subject to reporting requirements if they comply with:

a. Used Oil Transporters may consolidate or aggregate loads of used oil for purposes of transportation. However, except as provided in paragraph (b) of this section, used oil transporters may not process used oil unless they also comply with the requirements for processors/re-refiners in subpart F of this part.
b. Transporters may conduct incidental processing operations that occur in the normal course of used oil transportation (e.g., settling and water separation), but that are not designed to produce (or make more amenable for production of) used oil derived products unless they also comply with the processor/re-refiner requirements in subpart F of this part.
c. If you meet the exception as a transporter only your required documentation is identified in 40 CFR §279.46 **(26)** Tracking: **(70)**
    1. Acceptance. Used oil transporters must keep a record of each used oil shipment accepted for transport. Records for each shipment must include:

a. The name and address of the generator, transporter, or processor/re-refiner who provided the used oil for transport;

2. The EPA identification number (if applicable) of the generator, transporter, or processor/re-refiner who provided the used oil for transport;
    i. The quantity of used oil accepted;
    ii. The date of acceptance; and

d. Except as provided in paragraph (a) of this section, the signature, dated upon receipt of the used oil, of a representative of the generator, transporter, or processor/re-refiner who provided the used oil for transport.
  1. Intermediate rail transporters are not required to sign the record of acceptance.
  2. Deliveries. Used oil transporters must keep a record of each shipment of used oil that is delivered to another used oil transporter, or to a used oil burner, processor/re-refiner, or disposal facility. Records of each delivery must include:
      i. The name and address of the receiving facility or transporter;
      ii. The EPA Identification number of the receiving facility or transporter; and
      iii. The quantity of used oil delivered.
  3. The date of delivery; Except as provided in paragraph (a) of this section, the signature, dated upon receipt of the used oil, of a representative of the receiving facility or transporter.
  4. Intermediate rail transporters are not required to sign the record of delivery.
  5. Exports of used oil. Used oil transporters must maintain the records described in paragraphs (b)(1) through (b)(4) of this section for each shipment of used oil exported to any foreign country.
  6. Record retention. The records described above must be maintained for at least 3 years and be available for review by the TCEQ upon request.

Typically transporters who store used oil rarely operated under this exception and are identified as a transfer station at a minimum subjected to 40 CFR §279.45 (a) **(70)** identified below the applicability states that this section applies to used oil transfer facilities.

Used Oil Transfer Facilities Exception: Used oil transfer facilities are transportation-related facilities, including loading docks, parking areas, storage areas, and other areas, where shipments of used oil are held for more than 24 hours during the normal course of transportation and not longer than 35 days. Unless you meet a transporter exception and the transfer exceptions you are still subject to the 9. (A) section below on the record keeping requirements as a processor at a minimum.

Used Oil Processors and Re-Refiners: **(67)** Processing means any chemical or physical operation designed to produce or facilitate the production of fuel oils, lubricants, or other used oil-derived products from used oil. These operations include, but are not limited to, blending used oil with virgin petroleum products, blending used oils to meet fuel specification, filtration, simple distillation, chemical or physical separation, and re-refining. Processing, however, does not include generators processing used oil generated onsite for use on-site. If transporters or transfer facilities filter oil from oil-bearing transformers before returning it to its original use, they would not be subject to processing standards.

Re-Refining Versus Processing

Re-refining is not the same as processing even though they may be subject to many of the same requirements. The major difference is that re-refined used oil has sufficient quality to be used again as a lubricating oil. By contrast, processing produces a used oil of lower quality, which can be used as a fuel for producing electricity or other purposes. Processed used oil can also be blended for marine fuel or other use. It can even be mixed with asphalts.

Storage: Processors and re-refiners are subject to the same storage standards, including secondary containment, as transporters or burners (§279.54) **(67)**. Processors and re-refiners, however, have no time or quantity limitations on storing used oil. There is one major difference in the storage requirements for processors and re-refiners. Processing facilities must comply with stringent closure requirements (§279.54(h)).**(67)** Both aboveground tanks and containers need to have all used oil residues removed and any units still containing used oil or residues must be taken off-site. The owner and operator must decontaminate or remove used oil residues from all buildings, structures, and soils at the site. If the owner and operator cannot demonstrate the absence of contaminated soils at the time of closure, then the facility must be closed in accordance with hazardous waste landfill standards (§279.54(h)(1)(ii)).**(67)** Because the contaminated equipment will ultimately be disposed of, hazardous waste regulations would apply if the equipment exhibits a characteristic of hazardous waste.

Rebuttable Presumption

The rebuttable presumption requirements for processors and re-refiners are similar to the transporter requirements. Processors must determine the total halogen content of the used oil prior to accepting the used oil shipment in accordance with §279.53. If the presumption cannot be rebutted, then the shipment is considered to be hazardous waste and must to be managed by a hazardous waste TSDF.

General Facility Standards and Record Keeping

Facilities that conduct processing are subject to general land office , similar to interim status TSDFs (§279.52). **(67)** These requirements include general facility maintenance, fire and explosion prevention, emergency communication systems, and release response plans. Every processing facility must have a contingency plan to minimize the hazards to human health and the environment in the event of a fire, explosion, or release of used oil. The above statement identifies the requirement of Spill Prevention, Control and Countermeasures (40 CFR part 112).

Record Keeping: Processors and re-refiners of used oil are also subject to the following tracking, planning, and record-keeping requirements: The owner and operator must make a record of each shipment of used oil that the facility accepts for processing in accordance with §279.56 **(67),** the owner and operator must prepare an analysis plan that describes the procedures that will be used in rebutting the presumption and determining if the used oil meets specification requirements, and the owner and operator must submit a biennial report per the TAC §324.12. **(115)** that states the biennial report required by 40 CFR §279.57(b) **(67)** covering each odd numbered year must be provided to the commission by December 1 of the odd-numbered year if all used oil operations have been completed for that year. If not, you must submit the report by January 25 of the following even-numbered year. The information must be entered on a commission form. Mail the report to the Texas Commission on Environmental Quality, Used Oil Recycling Program.

Used Oil Fuel Marketers

Per 40 CFR 279.11 **(58),** a used oil fuel marketer is any person who either (1) directs a shipment of off-specification used oil from that facility to a used oil burner, or (2) first claims that the used oil going to be burned for energy recovery meets the specification requirements. Used oil marketers can be divided into two categories:

- Those that market off-specification used oil, and
- Those that market oil that meets specification.

For each category, different regulations apply under Part 279, Subpart H. The used oil fuel marketer requirements are applicable to anyone, including a generator, transporter, processor, or burner of used oil, who engages in marketing activities or who first claims that the used oil meets the specification criteria. Under the present definition of marketer, it is impossible for someone to only be a marketer without engaging in any other used oil management practices. For example, a used oil generator who first directs a shipment of off-specification used oil to a burner is a marketer as well as a generator and must comply with the applicable requirements of Part 279, Subparts C and H.

Off-Specification Used Oil Marketers

Used oil marketers handling off-specification oil are subject to the prohibitions found in §279.71 **(71),** as well as analysis of used oil fuel and notification to EPA of location and description of used oil management activities.

Record Keeping: It is necessary to comply with tracking requirements, including their EPA ID number, the name and address of both the marketer and the receiving facility, the quantity delivered, and the date of shipment. Prior to initiating the first shipment to a burner or other marketer, a marketer must acquire a one-time written and signed certification that the burner or marketer has notified EPA of all used oil management activities and, if a burner, that the used oil will only be burned in one of the allowable devices listed in §279.12(c) **(59)**. A marketer must keep copies of all invoices for 3 years from the date received or prepared. In addition, copies of all off-specification certifications from other burners or marketers must be kept for 3 years from the date of the last transaction with that person or facility. See 40 CFR §279.74 Tracking below for more detail associated with record keeping.

On-Specification Used Oil Marketers

Used oil marketers handling specification used oil must analyze the used oil fuel to show that it meets specification, as well as notify EPA if the marketer is the first one to claim the oil meets specification.

Record Keeping: An operating log must be kept for each shipment, including the name and address of the receiving facility, quantity, delivery date, and cross-reference to the record of analysis to show it meets specification. Records of all waste analysis documentation and operating logs must be kept for three years. See 40 CFR §279.74 Tracking below for more detail associated with record keeping.

40 CFR §279.74 Tracking (71): Any used oil marketer who directs a shipment of off-specification used oil to a burner must keep a record of each shipment of used oil to a used oil burner. These records may take the form of a log, invoice, manifest, bill of lading, or other shipping documents. Records for each shipment must include the following information:

1. The name and address of the transporter who delivers the used oil to the burner,
2. The name and address of the burner who will receive the used oil,
3. The EPA identification number of the transporter who delivers the used oil to the burner,
4. The EPA identification number of the burner,
5. The quantity of used oil shipped, and
6. The date of shipment.

On-Specification Used Oil Delivery

A generator, transporter, processor/re-refiner, or burner who first claims that used oil that is to be burned for energy recovery meets the fuel specifications under §279.11 must keep a record of each shipment of used oil to the facility to which it delivers the used oil. Records for each shipment must include the following information:

1. The name and address of the facility receiving the shipment,
2. The quantity of used oil fuel delivered,
3. The date of shipment or delivery, and

4. A cross-reference to the record of used oil analysis or other information used to make the determination that the oil meets the specification as required under §279.72(a).**(71)**

Record Retention: The records described in paragraphs (1) and (4) of this section must be maintained for at least 3 years.

USED OIL AGGREGATION POINTS

This is not a typical site, but is listed for reference and provides a possible option for generators as if you operated an aggregation point with its restrictions on maximum shipments of 55 gallons from facilities you own and even if you accept used oil from households do-it-yourselfers (DIYs) you must comply with only the generator standards not the used oil handlers requirements. For your reference:

40 CFR §279.32 Used oil aggregation points owned by the generator: **(69)**

a. *Applicability.* This section applies to owners or operators of all used oil aggregation points. A used oil aggregation point is any site or facility that accepts, aggregates, and/or stores used oil collected only from other used oil generation sites owned or operated by the owner or operator of the aggregation point, from which used oil is transported to the aggregation point in shipments of no more than 55 gallons under the provisions of §279.24(b). (68) Used oil aggregation points may also accept used oil from household do-it-yourselfers.
b. *Used oil aggregation point requirements.* Owners or operators of all used oil aggregation points must comply with the generator standards in subpart C of this part.

USED OIL COLLECTION CENTERS

A collection center is a facility that accepts, aggregates, or stores used oil received from generators, such as service stations, governments, or businesses (in shipments of not more than 55 gallons of used oil at a time). A center may accept used oil from household do-it-yourselfers (DIYs). Additional requirements found in TAC §324.7 are below. **(115)**

Posting Requirement: Facilities accepting household DIY used oil must post and maintain a durable and readable sign showing that the facility is a public or household DIY used oil collection center and showing the hours used oil is collected. *(Example signs are provided in DIY permit section of this guide)*

Registration: Used oil collection centers must register with the TCEQ's Used Oil Program within 30 days of starting operation. Collection centers must re-register no later than January 25 of *odd-numbered years.*

Rebuttable Presumption: Household used oil is not subject to the rebuttable presumption (a requirement to prove used oil is not hazardous) unless mixed with non-household used oil.

Reporting and Documentation: Also, no later than January 25th of *every year,* used oil collection centers must report to TCEQ the amounts of used oil collected during the previous year. The TCEQ will send a reporting form to registered collection centers by December 1 of each year. There is no minimum time period for record retention beyond insuring that data is collected for the annual reporting. It is recommended that record retention be for 3 years at a minimum.

DIY collection centers or other generators, transfer facilities, process facilities who accept DIY used oil are recommended even though not required by the EPA Guidance Manual **(138)** to keep a log for DIY collections where each DIY would sign with an estimated amount they provided that can satisfy the Comprehensive Environmental Response, Compensation, and Liability Act (CERCLA) exemption for service station dealers. Note this exemption does not apply to all DIY collectors unless the DIY used oil is managed completely separated from all other used oil managed. The data tracking above also will help evaluate the success of any DIY collection program.

Off-Site Shipments

Federal regulations 40 CFR §279.24 **(68)** and Texas TAC §324.7 **(115)** require generators must ensure that their used oil is transported only by transporters who have obtained EPA identification numbers unless you the generator are transporting your own oil the used oil in a vehicle owned by the generator or owned by an employee of the generator and no more than 55 gallons of used oil at any time. Another exception to the EPA number requirement is if a Tolling which is a contract arrangement that is in place where a generator may arrange for used oil to be transported by a transporter without an EPA identification number if the used oil is reclaimed under a contractual agreement pursuant to which reclaimed oil is returned by the processor/re-refiner to the generator for use as a lubricant, cutting oil, or coolant. The tolling arrangement must indicate: (1) The type of used oil and the frequency of shipments; (2) That the vehicle used to transport the used oil to the processing/re-refining facility and to deliver recycled used oil back to the generator is owned and operated by the used oil processor/re-refiner; and (3) That reclaimed oil will be returned to the generator.

Do-It-Yourselfer Used Oil Collection Centers

Do-it-yourselfer (DIY) is a used oil collection center is any site or facility that accepts/aggregates and stores used oil collected only from household do-it-yourselfers and must comply with all same requirements as a Used Oil Collection Center identified above.

Used Oil Generators

40 CFR §279.1 **(56)** identifies a used oil generator as any person, by site, whose act or process produces used oil or whose act first causes used oil to become subject to regulation. Used oil generators ranging from multinational corporations to local service stations, used oil generators are subject to limited standards under Part 279, Subpart C. Summary of compliance requirements are provided below some of which have been identified above but are worth repeating at this time.

Exceptions: Certain used oil generators are not subject to the regulations in Subpart C of Part 279.

1. Household do-it-yourselfers (DIYs) — individuals who generate used oil from the maintenance of their personal vehicles or other household sources;
2. Vessels at port or at sea;
3. Generators after mixing diesel fuel and used oil to be used in their own vehicles as a fuel; and
4. Farmers who, in a month, generate no more than 25 gallons of used oil from farm vehicles or machinery (§279.20(a)).

Used Oil Storage

The primary requirements placed on generators involve good housekeeping practices for storing used oil on site. Referenced requirements are from 40 CFR §279.22 **(61).**

Used oil generators are subject to all applicable Spill Prevention, Control and Countermeasures (40 CFR part 112) in addition to the requirements of this Subpart. Used oil generators are also subject to the Underground Storage Tank (40 CFR part 280) standards for used oil stored in underground tanks whether or not the used oil exhibits any characteristics of hazardous waste, in addition to the requirements of this subpart.

1. *Storage units.* Used oil generators shall not store used oil in units other than tanks, containers, or units subject to regulation under parts 264 or 265 of this chapter.
2. *Condition of units.* Containers and aboveground tanks used to store used oil at generator facilities must be: In good condition (no severe rusting, apparent structural defects or deterioration) and not leaking (no visible leaks).
3. *Labels:* Containers and aboveground tanks used to store used oil at Generator facilities must be labeled or marked clearly with the words "Used Oil." Fill pipes used to transfer used oil into underground storage tanks at generator facilities must be labeled or marked clearly with the words "Used Oil."
4. *Accumulation:* Generators of used oil have no time or quantity limitations on storing used oil because the Agency feels that used oil is a marketable commodity; therefore, there is an incentive for the generator to send used oil off-site for recycling.
5. *Release:* In the event of a release from one of these units to the environment, the generator must comply with the response measures set forth in §279.22(d) **(61)**, which include stopping and containing the release, properly managing any used oil or contaminated materials, and repairing or replacing the leaking container or tank.

Used Oil Collected by Another Than You as the Site Owner or Operator

TAC §324.6 **(115)** A person or entity that services equipment involving removal of used oil or changes used oil at a customer's home or business and transports the used oil from the site in quantities less than or equal to 55 gallons may choose to be the generator. If the service company removing the used oil from equipment does not assume generator responsibility, the site owner or operator will remain the generator.

Used Oil as Burner Fuel for Space Heaters

Per 40 CFR § 279.23 **(68)** On-site burning in space heaters. Generators may burn used oil in used oil-fired space heaters provided that:

1. The heater burns only used oil that the owner or operator generates or used oil received from household do-it-yourself used oil generators;
2. The heater is designed to have a maximum capacity of not more than 0.5 million Btu per hour; and
3. The combustion gases from the heater are vented to the ambient air.

Processing by Generators

There are five activities that a generator may perform and not be considered a processor as long as the used oil was generated on-site and it is not being sent directly off-site to a burner.

1. Filter, clean, or recondition used oil to be reused by the generator.
2. Separate used oil generated on-site from wastewater to meet Clean Water Act regulations prior to discharge of the wastewater.
3. Capture used oil droplets in the air by a mist collector to make in-plant air amenable for recirculation.
4. Drain or remove used oil from materials containing or contaminated with used oil.
5. Filter, separate, or recondition used oil before burning it in a space heater on-site.

Off-Site Shipments: Requirements are the same as identified above.

Used Oil Filters: In the September 23, 1991, Federal Register , EPA discussed exempting used oil filters that had been properly drained and crushed from the definition of hazardous waste (56 FR 48000) **(144)**. The Federal Agency codified this exclusion in §261.4(b)(13) **(146)** (57 FR 21531; May 20, 1992). The TCEQ identifies management of used oil filters in TAC 30 §328 **(116)** which will be the primary reference for this section along with other guidance provided in the EPA guidance and manuals identified above.

a) **TAC §328.22. Applicability:** (a) This subchapter applies to a used oil filter only if the filter has been determined to be nonhazardous or exempt from being hazardous waste due to draining of used oil per Title 40 Code of Federal Regulations,

§261.4(b)**(13)** (as amended through August 24, 1998), and the filter has not been: (1) separated into its component parts; or (2) burned for: (A) steel recovery; or (B) energy recovery. (b) **This subchapter does not apply to**: (1) an industrial generator that is: (A) registered with the agency as an industrial or hazardous waste facility; or (B) under the waste management authority of a state agency other than the commission. (2) a do-it-yourselfer. Adopted December 1, 1999 Effective December 23, 1999.

b) **TAC §328.21(11) Used oil filter definition: (116)** Any device that is an integral part of an oil flow system, the primary purpose of which is to remove contaminants from flowing oil contained in the system, and that as a result of use has become contaminated and unsuitable for its original purpose, is removed from service, and contains entrapped used oil. The term does not include a filter attached to the equipment containing the oil flow system. This term continues to apply regardless of prior processing until, but not after, the filter has been burned for steel recovery or energy recovery or it is separated into its component parts. Adopted December 1, 1999 Effective December 23, 1999.
c) **TAC §328.21(5) Generator (116)** - Person whose activities produce used oil filters. The term does not include do-it-yourselfers.
d) **TAC §328.21(6) Process (116)** - To prepare a used oil filter for recycling, steel recovery, energy recovery, or proper disposal.
e) **TAC §328.21**(7) **Processor (116)** - A person that processes used oil filters generated by another person. The term does not include a generator that consolidates, drains, or crushes used oil filters for off-site recycling or disposal.
f) **TAC §328.21**(8) **Store (116)** - To hold in a location for any period.
g) **TAC §328.21**(9) **Storage facility (116)** - A location that stores used oil filters before transportation, processing, recycling, or disposal of the filters. The term does not include the location of a generator.
h) **TAC §328.21**(10) **Transporter (116)** - A person that transports used oil filters to a location for storage, processing, recycling, or disposal.
i) **TAC §328.24. Registration and Reporting: (116)** (a) A transporter, storage facility, or processor may not store, process, recycle, or dispose of used oil filters unless the person is registered with the agency. These persons must register using an agency form and have a valid registration prior to operation.

1. Unless the person is registered with the agency as a storage facility, a person may not store used oil filters: (1) that in the aggregate have a volume greater than six 55-gallon drums; or (2) in more than one bulk filter container.
2. Transporters, storage facilities, and processors must renew their registrations by January 25th of every even-numbered year, and must report by January 25th of every even-numbered year, in accordance with paragraphs (a) - (b) of this subsection.
   a. The report must include the number of used oil filters the person transported, stored, or processed in the preceding two calendar years.
   b. Registration renewal and reporting information must be entered on an agency form.
   c. The registration and reporting information must be mailed to the TCEQ.
   d. A transporter, storage facility, or processor must renew its registration by January 25th of every even-numbered year.

General Requirements, Transportation, and Transporters:

Financial Responsibility Requirement: A transporter, storage facility, or processor is required to provide evidence of financial responsibility with registration as the executive director deems necessary to assure that the facility has sufficient assets to provide for proper closure. Financial assurance for closure may be demonstrated by using one or more of the following mechanisms: trust funds, surety bonds guaranteeing payment or performance, letters of credit, insurance, or financial test and corporate guarantee. These mechanisms shall be prepared on forms approved by the executive director. *(Financial Responsibility Requirements are covered in Financial Assurance section Part XIV in this guide.)*

Transportation: Persons transporting used oil filters may transport up to two 55-gallon containers, or the volumetric equivalent, without registering as a transporter.

Storage: The following requirements must be observed:

1. A person may not store, process, or dispose of a used oil filter in a manner that results in the discharge of oil into soil or water.
2. A person may not knowingly place on land a used oil filter that contains used oil

unless the used oil filter is in a container.

3. A bulk filter container used to store used oil filters: (1) must not leak; and (2) must be securely closed, waterproof, and in good condition.
4. A used oil filter may not be intentionally or knowingly placed in or accepted for disposal in a landfill permitted by the commission.
5. A processor that stores used oil filters in a container shall label each container clearly with the phrase "Used Oil Filters."

Limitations on Storage:

1. A storage facility may not store a used oil filter for more than 120 days.
2. A transporter may not store a used oil filter for more than 10 days.
3. A processor may not store a used oil filter for more than 30 days before it is processed.

Variances: The executive director may grant an individual variance to allow:

1. A generator to store used oil filters in a greater aggregate volume than the volume prescribed in §328.24(b) **(116)** of this title (relating to Registration and Reporting); or
2. A person to store used oil filters for a period longer than the period prescribed for that person in §328.26 **(116)** of this title (relating to Limitations on Storage).

Factors to be considered in determining whether a variance should be granted include but are not limited to, the risk to human health and the environment that is posed by the requested variance. The burden of justifying the need for a variance is on the requestor, and the requestor must submit the information in writing to clearly indicate the issues involved, the reason(s) for the request, and both positive and negative impacts that may result from the granting of the variance. Prior approval of the variance must be obtained before any change is authorized. If a variance is denied, the executive director will provide an explanation of the reason(s) for the denial in a written response to the requestor.

The executive director may not grant a variance under this section for a period longer than 2 years.

Shipment Records:

1. Each shipment of used oil filters **must be accompanied** by a bill of lading. The bill of lading must demonstrate a transfer of custody of the used oil filters from the shipping facility to the registered transporter, and from the transporter to the registered storage facility or processor.
2. The bill of lading must contain the date of such transfer, the name and physical address of the shipping facility, the name and address of the receiving facility, and the name and address of the transporter, the quantity of used oil filters removed and any other information which the executive director may deem necessary to protect the environmental quality of the State of Texas. The shipping facility personnel must verify the information within the bill of lading, and demonstrate concurrence by the signature of an authorized representative.
3. A copy of the bill of lading for each shipment of used oil filters must be maintained by the generator of the filters, transporter of the filters, storage facility at which the filters were stored, and processor of the filters for at least three years after the date the filters were transported, stored, or processed.
4. The copies of bills of lading must be made available for agency personnel to inspect at any reasonable time.

Spill Prevention and Control – SPCC Requirement:

Each registered storage facility and each facility of a registered processor shall develop a plan to prevent spills and respond to spills in accordance with the federal spill prevention, control, and countermeasure requirements provided by Title 40 Code of Federal Regulations, Part 112 (as amended through July 1, 1998).

Used Oil Absorbents:

Absorbents management is extremely challenging area of used oil management due to the detail management and inspection requirements associated with compliance. Due the nature of the toxins associated with the normal generation of used oil that are typically found in absorbents they can significantly pollute the environment if not managed correctly. The guidance provided in this section is based on TCEQ and EPA guidance and manuals identified above. The federal regulation and Texas TAC 30 basis for management practices 40 CFR §279.10(c) **(57)** (c) *Materials containing or otherwise contaminated with used oil.* (1) Except as provided in paragraph (c)(2) of this section, materials containing or otherwise contaminated with used oil from which the used oil has been properly drained or removed to the extent possible such that no visible signs of free-flowing oil remain in or on the material: The exception (c)(2) identified above states "Materials containing or otherwise contaminated with used oil that are burned for energy recovery are subject to regulation as used oil under this part."

Absorbents:

When properly selected and used, absorbents can soak up and slow the movement of used oil. Many factors affect the performance and cost of using absorbents for cleaning up used oil spills, including safety, capacity, characteristics, applicability, reuse, and disposal. There are three general classes of absorbents are synthetic pads, clay (commonly called "kitty litter"), and cellulose materials.

Synthetic Pads:

Pads made of polypropylene, are water repellant so that moisture will not affect their usefulness. Such pads are not biodegradable. However, they can be disposed of by incineration, and the pads have low ash content. Synthetic pads can also be burned for energy recovery, which is preferable to incineration or land disposal. Some absorbent pads are made to be reusable, which makes them more desirable for recycling and saves on disposal costs when the used oil has been properly removed.

Clay, Commonly Called Kitty Litter:

Kitty litter is more like an adsorbent (where the oil sticks to the surface) rather than an absorbent material (where the oil soaks into the material). Clay is heavy, highly abrasive, and not biodegradable. It also has a high ash content and low heating value—qualities that make it the least desirable of the three classes to incinerate or use in fuel blending.

Cellulose Materials:

Cellulose materials are biodegradable organic absorbents, such as peat moss, corn cobs, or recycled paper products. Cellulose materials can be burned for energy recovery.

Soil as an Absorbent:

Based on 40 CFR §279.10(c) from EPA Guidance **(138)** regarding disposition of used oil containing soil is regulated as used oil until the used oil is removed and can be burned for energy recovery if it meets the over 5,000 Btu/lb. requirement otherwise it must be managed as a solid waste either hazardous waste or in Texas petroleum-contaminated materials are "special waste" when disposed of in municipal solid waste landfills (MSWLFs). **(149)**

Management Practices:

Absorbent materials with signs of free-flowing oil are managed as used oil. If there are no visible signs of free-flowing oil, these materials are not regulated as used oil—unless they are burned for energy recovery. If they are to be disposed of in a landfill, absorbent materials are solid waste and subject to a hazardous waste or special waste determination.

Storage:

Ensure the same storage practices are utilized as you would for any used oil material.

Disposition Options:

1. Recycling: There are several options for recycling used oil absorbents. (1) Wring the oil out of the absorbent pad, reclaim the used oil, and reuse the pads. (2) Send the used oil absorbent material to a facility that burns it for energy recovery. Absorbent materials recycled by burning for energy recovery are regulated as used oil and must meet the used oil specifications identified above.
2. Incineration: Incineration offers the greatest control of toxic organic contamination, combined with very low human health risks. Hydrocarbons, in particular, incinerate well because of their organic content and energy value. Hazardous waste incinerators are regulated under Title 40 CFR Parts 264 or 265.
3. Bioremediation: This method involves applying organisms or oxygen and mineral nutrients, such as phosphate and nitrogen, to the contaminated used oil absorbent material. Under a controlled environment, bioremediation can be a cost-effective alternative to landfill disposal. This method can (1) reduce liability and expenses associated with using landfills (2) reduce or eliminate levels of contaminants such as TPH (total petroleum hydrocarbons) and BTEX (benzene, toluene, ethylben-

zene, and xylenes). Bioremediation of absorbents that do not contain free-flowing oil is not subject to the used oil management standards.

4. Landfill Disposal: Hopefully your last option is disposal in municipal solid waste landfills (MSWLFs) that requires the absorbent materials to be sampled, tested, and certified as solid waste or special waste before disposal.

PART IV: USED OIL FACILITY EXAMPLE CONFIGURATION

Facility Reference: Provided in the descriptions below are the site and area descriptions associated with the site plot and descriptions as part of a one page reference of an example Used Oil Facility identifying typical operation areas. Note that other than filter heating and crushing operations are limited to collection and storage of used oil and antifreeze. Many of the activities are addressed within this guide related to compliance requirements but not all and is provided only as an example of typical activities.

Site Area Description

Site Operations: The ABC Used Oil Facility (ABC) is a Used Oil Storage Facility, Processor, Filter Handler and Transporter of used oil. Activities include management of used oil collected from third parties, oily water, used oil filters, antifreeze and oil absorbents for recycle. Facility recycles 1,500,000 gallons of used oil, 300,000 gallons of oily water, 16,000 55 gallon drums of used oil filters, 300,000 gallons of antifreeze, 4,000 pounds of oil absorbents per year. Transports liquids from customer sites using tanker trucks and box trucks for totes and drums of filters, oil absorbents and antifreeze.

General Site Construction: The approximately 2-acre ABC property is developed with three buildings. The buildings consist of; (1) the main office building that contains a conference room, restrooms, and change rooms is located on the northeastern portion of the property; (2) the drum management and facility support building that contains empty drum storage, drum crusher, parts and equipment storage area, lab, shop and field supervisor office, and employee breakroom is located on the north central portion of the property; (3) the filter process building containing the drum washing area, filter heater processing unit, filter crusher, and absorbent management area is located at the southwest corner of the property; additional operating areas include, tank farm with seven aboveground storage tanks used for used oil, oily water, and antifreeze storage located in the northwest corner of the property. A hazardous waste storage area. A truck loading and off-loading pad is located in the west portion of the property. A cement pad for storage of totes, empty used oil and antifreeze tanks, and filter bins is located in the west end of property. A roll-off storage cement pad is located on the south property area. A truck parking and employee parking located at the south of property. A water retention pond is located in the southeast portion of the property. The remainder of the subject property consists of a perimeter security metal fencing with electric gate access, perimeter berm and paved or gravel areas.

Area A: Tank Farm is a 2,040 sq. ft. area cement construction 8" thick with rebar enforcement and a 20 mil HDPE liner above 2' compacted sand and compacted soil and with 2' 6" x 8" containment wall. Tank Farm contains six Aboveground Storage Tanks (AST) which are 16,300 gallon tanks, one lateral 10,000 gallon tank, pump house with three pumps, and a grated sump 4' x 6' x 3' with submerged pump to pump water to oily water tank.

Area B: (2) Building is 2,400 sq. ft. enclosed structure with cargo access door 8' wide x 10' high. Building is used for empty 55 gallon drum storage, drum crusher, miscellaneous storage area. Three areas with access from outside consisting of a small lab used for limited used oil related testing associated with used oil acceptance and transportation off-site, shop area with field supervisor office, and employee breakroom.

Area C: (1) Building is 1,936 sq. ft. containing offices, office storage, conference / training room, change rooms with showers, and restrooms.

Area D: Truck loading and offloading is a 1,560 sq. ft. 6" thick cement area with a 20 mil geomembrane liner of High Density Polyethylene (HDPE) with 4" berm containment. Area also contains a 4' wide x 6' length x 3 ft. depth sump with submerged pump for transferring liquid to oily water tank. Additional liquids control provided by an 11' length x 2' width x 3" depth grated strip drain in front of manifold providing connection to pump house and tank lines from truck lines connected to manifold.

Area E: Container storage area 612 sq. ft. compacted soil with gravel cover used for Empty Liquid Totes (1) 275-330 gal., (2) Empty 500 gal. used oil customer Steel Tanks used by customers on their sites for collection prior to pump out, (3) Empty 95 gal. roll-away Filter Totes and 275 gal. Filter Bins

Area F: (3) Building is 7,500 sq. ft. Processing area, floor is cement construction 4" thick with rebar enforcement and a 20 mil HDPE liner above 2' compacted sand and compacted soil subsurface with 6" x 8" containment inside berm at the base of all walls. Building contains high temperature filter furnace, filter crusher cuber, roll-off for crusher cubes, 216 sq. ft. 3" deep grated drum wash pit, 4' wide x 8' length x 3' deep grated sump with submerged transfer pump, Tank 8 - 6000 gal. vertical steel oily water tank, absorbent material drums collection area, outside north wall 5,000 gal. lateral used oil tank in 14' x 28' cement containment with 24" walls, and 24' x 12' dock door.

Area G: 2, 100 sq. ft. cement pad 4" thick with 6 mil liner above compacted soil with 6" berm and access ramp used for roll-off storage.

Area H: Truck parking compacted area covered with gravel.

Area I: 12' x 15' cement pad 2" thick with 6 mil liner above compacted soil with 6" berm and chain-link fencing with locked gate access used for hazardous waste storage of drums and small containers.

Area J: Evaporation Pond at southeast of property downgradient from all surface water runoff to collect all contact water from site. Pond has 40 mil Linear Low Density Polyethylene (LLDPE) that has high elongation, tear resistance and burst strength properties. Also contains carbon black and UV stabilizers to enhance longevity and outdoor performance as it covers this 28' x 40' concave pond area to a depth of 3' over a 2' clay compaction that includes an underdrain with a leak detection and collection system.

Site Plot 1 – Example Facility Plot

Site Plot 1 – *Example Facility Plot*

Site Operations

The ABC Used Oil Facility (ABC) is a Used Oil Storage Facility, Processor, Filter Handler and Transporter of used oil. Activities include management of used oil collected from 3rd parties, oily water, used oil filters, antifreeze and oil absorbents for recycle. Facility recycles 1,500,000 gallons of used oil, 300,000 gallons of oily water, 16,000 55 gallon drums of used oil filters, 300,000 gallons of antifreeze, 4,000 pounds of oil absorbents per year. Transports liquids from customer sites using tanker trucks and box trucks for totes and drums of filters, oil absorbents and antifreeze.

General Site Construction

The approximately 2 acre ABC property is developed with three buildings. The buildings consist of; (1) the main office building that contains a conference room, restrooms, and change rooms is located on the northeastern portion of the property; (2) the drum management and facility support building that contains empty drum storage, drum crusher, parts and equipment storage area, lab, shop and field supervisor office, and employee breakroom is located on the north central portion of the property; (3) the filter process building containing the drum washing area, filter heater processing unit, filter crusher, and absorbent management area is located at the southwest corner of the property; additional operating areas include, tank farm with seven aboveground storage tanks used for used oil, oily water, and antifreeze storage located in the northwest corner of the property. A hazardous waste storage area. A truck loading and off-loading pad is located in the west portion of the property. A cement pad for storage of totes, empty used oil and antifreeze tanks, and filter bins is located in the west end of property. A roll-off storage cement pad is located on the south property area. A truck parking and employee parking located at the south of property. A water retention pond is located in the southeast portion of the property. The remainder of the subject property consists of a perimeter security metal fencing with electric gate access, perimeter berm and paved or gravel areas.

AREA A

Tank Farm a 2,040 sq. ft. area cement construction 8" thick with rebar enforcement and a 20 mil HDPE liner above 2' compacted sand and compacted soil and with 2' 6" x 8" containment wall. Tank Farm contains six Aboveground Storage Tanks (AST) which are 16,300 gallon tanks, one lateral 10,000 gallon tank, pump house with three pumps, and a grated sump 4' x 5' x 3' with submerged pump to pump water to oily water tank.

AREA F

(3) Building is 7,500 sq. ft. Processing area, floor is cement construction 4" thick with rebar enfarcement and a 20 mil HDPE liner above 2' compacted sand and compacted soil subsurface with 6" x 8" containment inside berm at the base of all walls. Building contains high temperature filter furnace, filter crusher cuber, roll-off for crusher cubes, 216 sq. ft. 3" deep grated drum wash pit, 4' wide x 8' length x 3' deep grated sump with submerged transfer pump, Tank 8 - 6000 gal. vertical steel oily water tank, absorbent material drums collection area, outside north wall 5,000 gal. lateral used oil tank in 14' x 28' cement containment with 24" walls, and 24' x 12' dock door.

AREA G

2, 100 sq. ft. cement pad 4" thick with 6 mil liner above compacted soil with 6" berm and access ramp used for roll-off storage.

AREA H

Truck parking compacted area covered with gravel.

AREA I

12' x 15' cement pad 2" thick with 6 mil liner above compacted soil with 6" berm and chain-link fencing with locked gate access used for hazardous waste storage of drums and small containers.

AREA J

Evaporation Pond at southeast of property downgradient from all surface water runoff to collect all contact water from site. Pond has 40 mil Linear Low Density Polyethylene (LLDPE) that has high elongation, tear resistance and burst strength properties. Also contains carbon black and UV stabilizers to enhance longevity and outdoor performance as it covers this 28' x 40' concave pond area to a depth of 3' over a 2' clay compaction that includes an underdrain with a leak detection and collection system.

AREA B

(2) Building is 2,400 sq. ft. enclosed structure with cargo access door 8' wide x 10' high. Building is used for empty 55 gallon drum storage, drum crusher, miscellaneous storage area. Three areas with access from outside consisting of a small lab used for limited used oil related testing associated with used oil acceptance and transportation off-site, shop area with field supervisor office, and employee breakroom.

AREA C

(1) Building is 1,936 sq. ft. containing offices, office storage, conference / training room, change rooms with showers, and restrooms.

AREA D

Truck loading and offloading is a 1,560 sq. ft. 6" thick cement area with a 20 mil geomembrane liner of High Density Polyethylene (HDPE) with 4" bermed containment. Area also contains a 4' wide x 6' length x 3 ft. depth sump with submerged pump for transferring liquid to oily water tank. Additional liquids control provided by an 11' length x 2' width x 3" depth grated strip drain in front of manifold providing connection to pump house and tank lines from truck lines connected to manifold.

AREA E

Container storage area 612 sq. ft. compacted soil with gravel cover used for Empty Liquid Totes (1) 275-330 gal., (2) Empty 500 gal. used oil customer Steel Tanks used by customers on their sites for collection prior to pump out, (3) Empty 95 gal. rollaway Filter Totes and 275 gal. Filter Bins

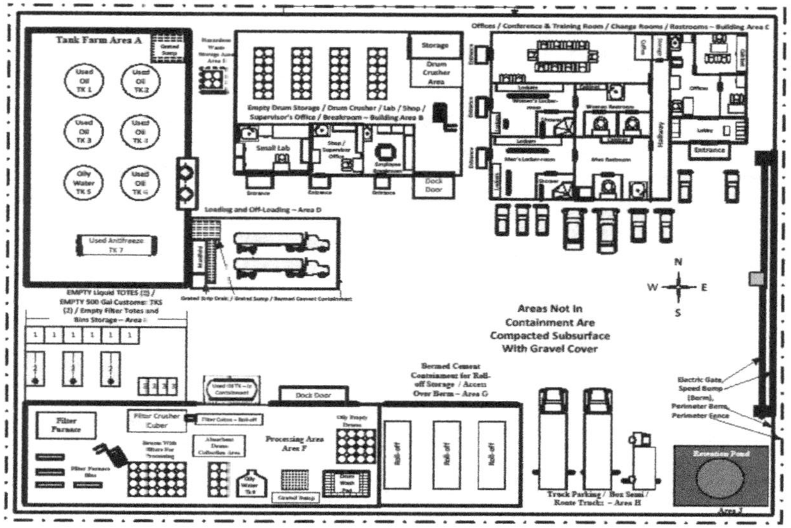

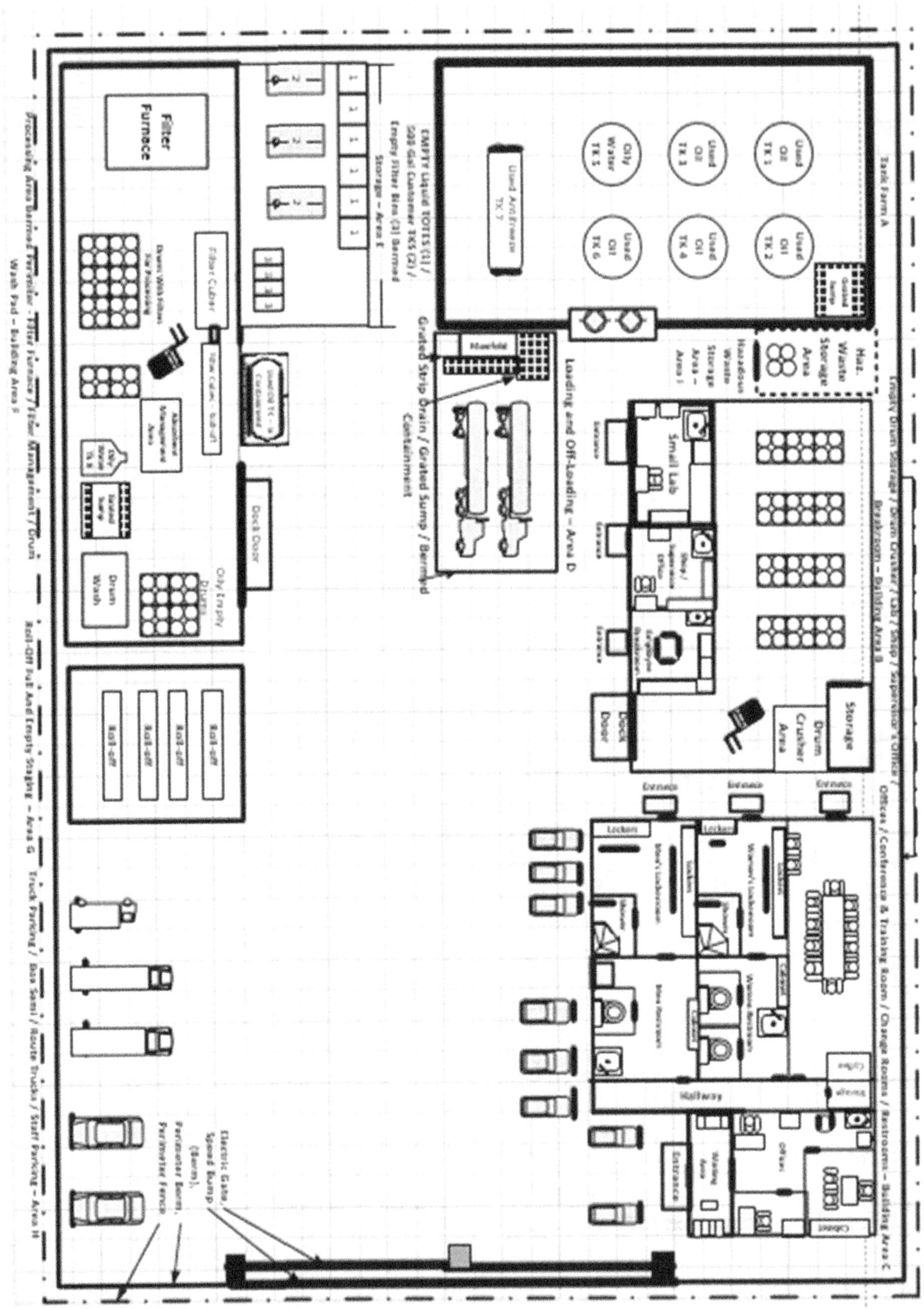
Tank Farm A
Used Oil TK 1
Used Oil TK 2
Used Oil TK 3
Used Oil TK 4
Oily Water TK 5
Used Oil TK 6
Used Anti-Freeze TK 7
Filter Furnace
Filter Crusher
Dock Door
Drum Wash
Roll-off
Roll-off
Roll-off
Roll-off
Loading and Off-Loading – Area D
Grated Strip Drain / Grated Sump / Bermed Containment
Small Lab
Storage
Drum Crusher Area
Dock Door
Hallway
Entrance

PART V: TEXAS COMMISSION ON ENVIRONMENTAL QUALITY (TCEQ)

The Texas Commission on Environmental Quality (TCEQ)(1) is the environmental agency for the state of Texas and your primary contact for all Used Oil compliance issues. The commission's headquarters are located at 12100 Park 35 Circle in Austin. The fourth largest environmental agency in the United States (and the third largest state environmental agency, behind the US Environmental Protection Agency, the California EPA, and the New York DEC), it employs approximately 2,780 employees, has 16 regional offices, and has a $456 million operating budget for the 2017 fiscal year. **(106)**

TCEQ Organization History (150)

The history of natural resource protection by the State of Texas is one of gradual evolution from protecting the right of access to natural resources (principally surface water) to a broader role in protecting public health and conserving natural resources for future generations of Texans.

Natural resource programs were established in Texas at the turn of the 20th century, motivated initially by concerns over the management of water resources and water rights. In parallel with developments in the rest of the nation, and at the federal level, state natural-resource efforts broadened at mid-century to include the protection of air and water resources, and later to the regulation of generating hazardous and non-hazardous waste which is where the agency impacts how you do used oil management.

During the 1990s, the Texas Legislature moved to make natural-resource protection more efficient by consolidating programs. This trend culminated in the creation of the Texas Natural Resource Conservation Commission in the fall of 1993 as a comprehensive environmental protection agency.

Sunset legislation passed by the Texas Legislature in 2001 continued the agency and changed its name to the Texas Commission on Environmental Quality. In 2011, sunset legislation continued the TCEQ through 2023.

Flowchart 2, from the TCEQ website, shows the history of the TCEQ and its predecessor agencies to give you an understanding of how the TCEQ became the agency it is today with the consolidations that were part of the legislative development.

A brief history of the TCEQ starts with The Texas Board of Water Engineers, established in 1913, became the Texas Water Commission in 1962, which then became the Texas Water Rights Commission in 1965. The Texas Water Pollution Control Advisory Council, established in 1953, became the Texas Water Pollution Control Board in 1961, which then became the Texas Water Quality Board in 1967. The Texas Water Development Board was established in 1957 and the Texas Air Control Board was established in 1965.

The Texas Water Rights Commission, the Texas Water Development Board, and the Texas Water Quality Board were combined in 1977 to form the Texas Department of Water Resources. This agency was split into the Texas Water Commission and the Texas Water Development Board in 1985. The Texas Water Commission and the Texas Air Control Board were combined to form the Texas Natural Resource Conservation Commission in 1993. The TNRCC's name was changed to the Texas Commission on Environmental Quality in 2002. Sunset legislation passed by the Texas Legislature in 2001 continued the agency until 2013 and changed its name to the Texas Commission on Environmental Quality. During the special session of the 81st Legislature (2009), legislation was adopted amending the 2013 date to 2011, at which time the agency was continued for an additional 12 years and subject to review in 2023.

Flowchart 2 – History of the TCEQ. Retrieved From TCEQ

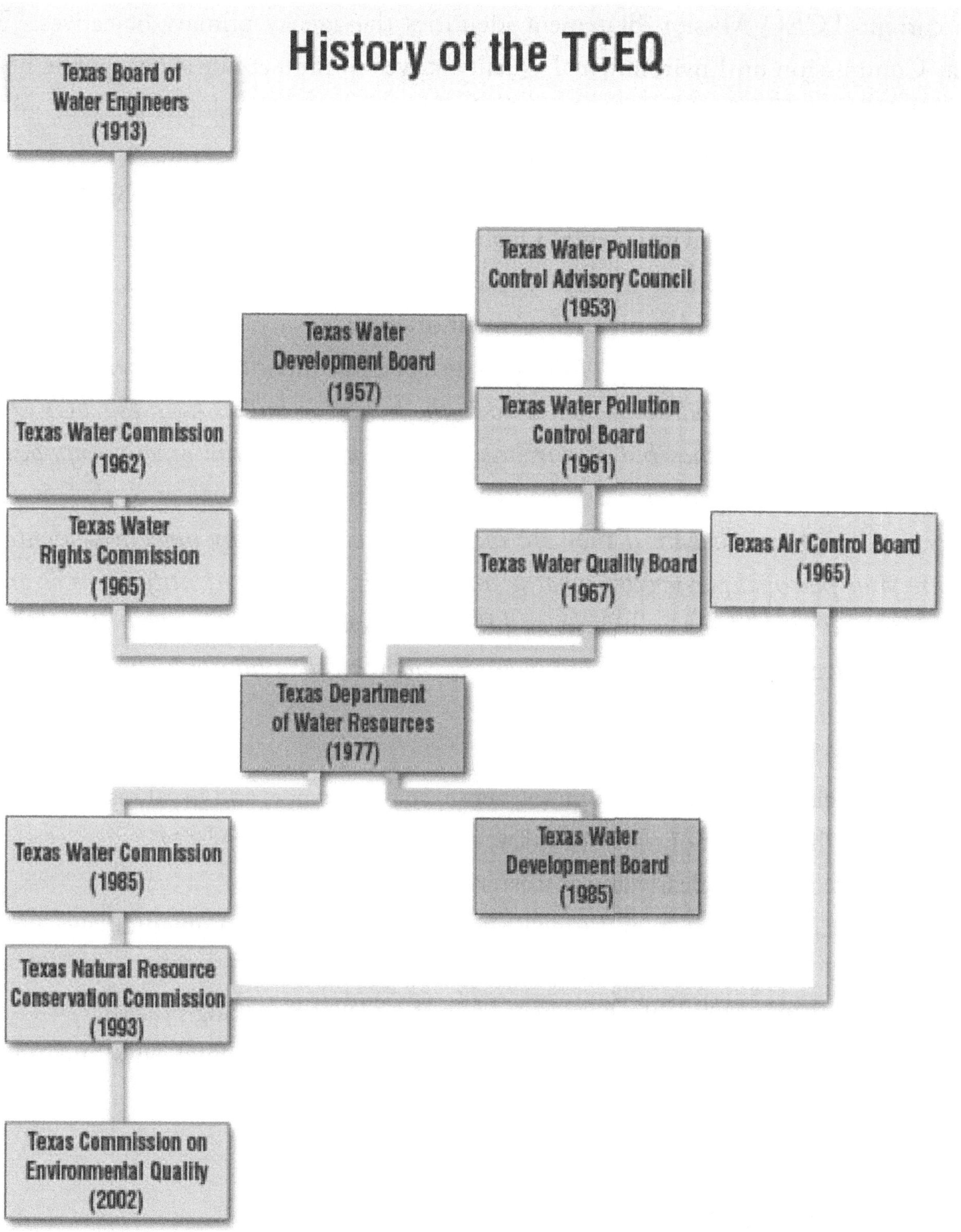

https://www.tceq.texas.gov/agency/about-the-tceq

TCEQ Mission Statement and Present Organization (151)

The current TCEQ Mission Statement identifies the agency primary objectives: "The Texas Commission on Environmental Quality strives to protect our state's public health and natural resources consistent with sustainable economic development. Our goal is clean air, clean water, and the safe management of waste."

The current TCEQ Mission Statement identifies the agency primary objectives: "The Texas Commission on Environmental Quality strives to protect our state's public health and natural resources consistent with sustainable economic development. Our goal is clean air, clean water, and the safe management of waste."

Commentary: Overall agency philosophy perspectives identifying how they intend to administer their responsibilities is identified in the following bullets from the TCEQ that identify positive approaches but as any organization or agency the practical application are significantly challenging and as Used Oil Handlers that are subject to their requirements one can always debate if they are meeting their philosophy objectives identified. Despite your personal perspective what is important is that they identify them as an organization and we all hope as citizens and those subject to compliance that they are continuing to strive to meet their mission requirements with the philosophy perspectives as a firm foundation.

- Base decisions on the law, common sense, sound science, and fiscal responsibility;
- Ensure that regulations are necessary, effective, and current;
- Apply regulations clearly and consistently;
- Ensure consistent, just, and timely enforcement when environmental laws are violated;
- Ensure meaningful public participation in the decision-making process;
- Promote and foster voluntary compliance with environmental laws and provide flexibility in achieving environmental goals; and
- Hire, develop, and retain a high-quality, diverse workforce.

Current Organization (151)

Flowchart 3 below shows the current TCEQ Organization; this is followed by brief descriptions of the overall responsibilities for areas in which Used Oil Handlers are typically involved. Specific contact information for agency staff can be found later in this chapter.

Flowchart 3 – TCEQ Organization Chart.

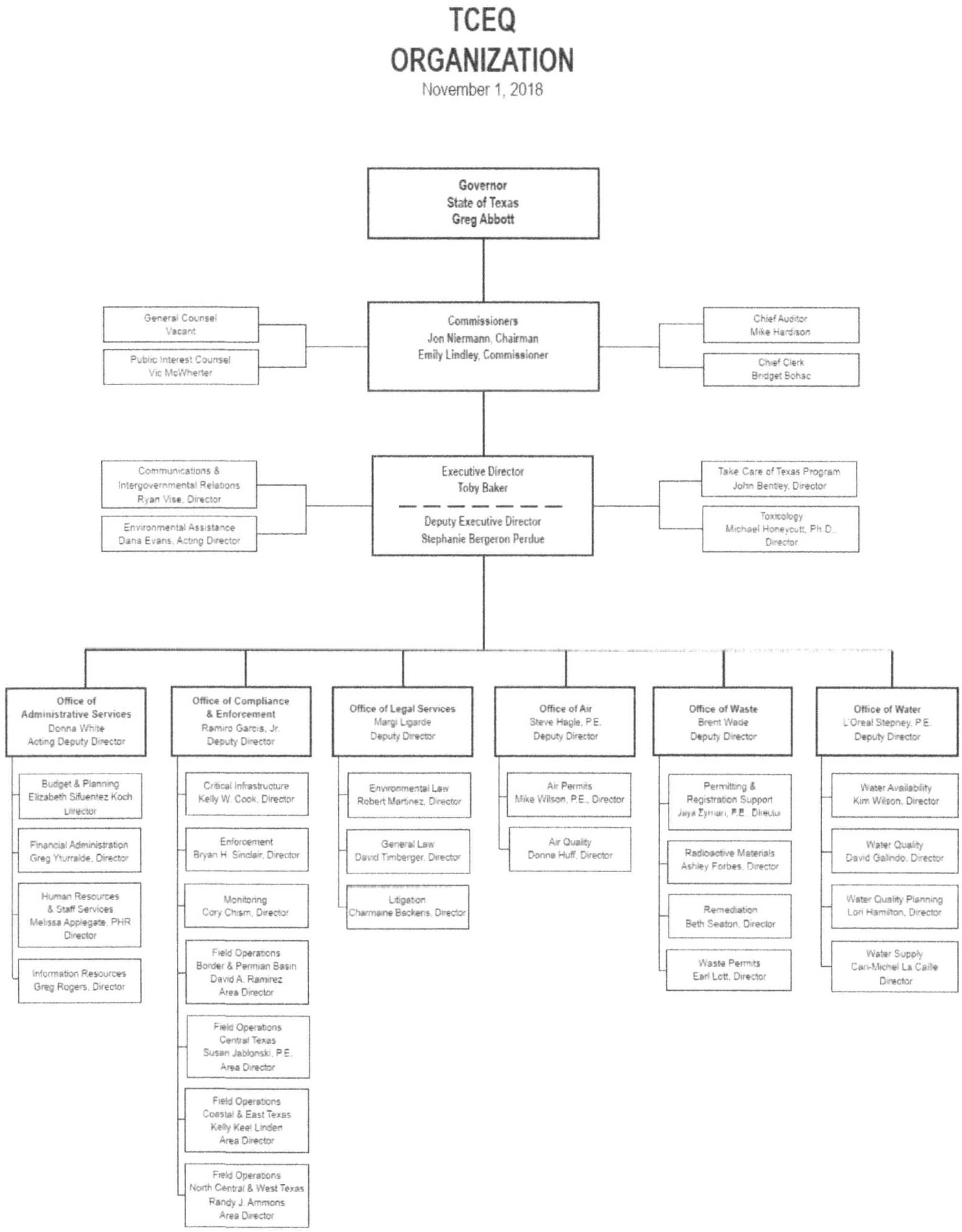

Retrieved From https://www.tceq.texas.gov.

TCEQ Organization as of March 1, 2017 (151)

Comments on various group functions are provided, but always refer to the TCEQ website for more detailed information on each group and their functions.

The TCEQ has three full-time commissioners, who are appointed by the governor to establish overall agency direction and policy, and to make final determinations on contested permitting and enforcement matters. The commissioners are appointed for 6-year terms with the advice and consent of the Texas Senate. A commissioner may not serve more than two 6-year terms, and the terms are staggered so that a different member's term expires every 2 years. The governor also names the chairman of the commission.

- Greg Abbott, current governor of Texas
- Commissioners Bryan W. Shaw, Ph.D., P.E., Chairman; Toby Baker; and Jon Niermann

The Environmental Assistance Division (EAD) is perhaps the most important support group as a Used Oil Handler, as their focus is on technical assistance, education, and pollution prevention programs to encourage environmental improvements. The EAD steers many of these programs in a direction that focuses on agency priorities and aligns with agency regulatory systems. Brian Christian is the current director of the EAD.

In fiscal 2015 and 2016, the division responded to 13,986 requests for assistance from small businesses and local governments. Of those, 663 received one-on-one assistance at their business site or facility.

In fiscal 2015, more than 180 small businesses and local governments took advantage of the EAD's Site Visit Program, which allowed them a site visit, during which a contractor of the TCEQ used a checklist to identify problems with environmental compliance. After the visit, the businesses and facilities received recommendations about actions they could take to resolve those problems. In fiscal 2015, 48 participants resolved the issues that were identified. Recognizing that having a TCEQ representative on your site sounds intimidating the reality from this author's personal experience is just the opposite and they are really there to help you but you must be committed to implementing the recommendations to get the most out of the help.

The Small Business and Local Government Assistance (SBLGA) guide **(5)** provides an intent of the program: "Federal and state laws require the TCEQ to provide compliance assistance to small businesses. It is confidential, except when there is an imminent threat to the environment. By keeping assistance confidential, and separate from enforcement, the agency encourages entities to seek assistance and achieve compliance. The SBEA offers compliance assistance through: direct on-site assistance; a toll-free hot line answered by its staff; active participation on agency rule, standard permit, and general permit teams; regulatory guidance development; and advisory committees."

The TCEQ's Environmental Assistance, Small Business and Local Government Assistance Section (SBLGA) **(141)** is a significant resource within the TCEQ. If you keep only one number available for TCEQ contacts, this is the one to keep. Additional information on the SBLGA support can be found at https://www.tceq.texas.gov/assistance and TexasEnviroHelp.org.

Small Business Assistance (800-447-2827) is a hotline for small businesses to obtain information, advice, and confidential assistance from the TCEQ. Business hours are Monday-Friday from 8 a.m. to 12 p.m. and 1 p.m. to 5 p.m. This is a number you should know and not hesitate to call if you have a question.

The TCEQ website provides several typical areas of support, listed below; according to their website, "the TCEQ's Small Business and Local Government Assistance Section offers free, confidential help to small businesses." You can ask any question about your business or any other environmental issue; they also provide a free site visit and compliance review, explain what and how corrections can be done, and will work with you to achieve compliance. Of special note is that, even if something of significance is identified (and only if an imminent threat to the environment) no enforcement will take place as long as corrective actions are completed in a timely manner. From the TCEQ "Guidance to Agency Programs" **(141)**, "It is confidential, except when there is an imminent threat to the environment. By keeping assistance confidential, and separate from enforcement, the agency encourages entities to seek assistance and achieve compliance." This writer has personal experience with this group over many years and the staff have always been helpful, accurate, and timely with any assistance. Remember, when seeking help or information available through the site Used Oil is under Waste and Recycling support, but they offer help on any environmental issue such as the Clean Water Act (CWA), Clean Air Act (CAA), reporting, petroleum storage tanks, and waste management.

The following are additional support resources that may be utilized.

Recycling:

Recycling: Am I Regulated?

Resource Exchange Network for Eliminating Waste (RENEW) Listings

The Used Oil Recycling Handbook -Guidance for Used Oil Handlers.

Other Waste Topics

Special Waste Regulations in Texas

Petroleum Storage Tanks (PST): Owners and operators of PSTs can find regulatory guidance and forms for operators of PSTs on this webpage.

Common Questions on Empty Waste Containers

Used Oil, Batteries, Antifreeze, or Used Tires

Waste from Grease Traps, Grit Traps, and Septic Tanks: Questions and Answers

Specific Information for Small Businesses – Information and resources by industry

The following are excerpts from the Guidance to Agency Programs document identified above regarding the small business assistance and the SBLGA as a whole functions and history.

"The Small Business and Environmental Assistance Division (SBEA) provides confidential compliance assistance on air, water, and waste regulations to small businesses and small local governments; works with regulated entities to implement pollution prevention and innovative environmental programs; offers technical assistance for recycling; and educates the public and the regulated community on rules and environmental issues. The SBEA's major activities are described below.

Small Business and Local Government Assistance (SBLGA): Federal and state laws require the TCEQ to provide compliance assistance to small businesses. The commission also offers that service to small local governments. It is confidential, except when there is an imminent threat to the environment. By keeping assistance confidential, and separate from enforcement, the agency encourages entities to seek assistance and achieve compliance. The SBEA offers compliance assistance through direct on-site assistance; a toll-free hotline answered by its staff; active participation on agency rule, standard permit, and general permit teams; regulatory guidance development; and advisory committees.

The SBEA was created in 1999 by merging multiple assistance programs: the former Office of Pollution Prevention and Recycling, the Small Business Assistance Program, and the Local Government Assistance Program. At the same time of this merger, the agency deployed more assistance resources to the TCEQ regional offices. The SBEA is located in the Executive Director's Office. Consolidation resulted in greater efficiency and uniformity in the delivery of assistance.

Section 507 of the 1990 Federal Clean Air Act Amendments require all states to implement a program to help small businesses comply with all facets of the Act, employ an ombudsman to represent small-business interests before the state environmental regulatory agency, and convene a Compliance Advisory Panel (CAP) comprised of individuals that advise the agency on the concerns and interests of small businesses. The state analog to that statute is Texas Water Code, Section 5.135, which goes further to require that compliance assistance be provided across *all* environmental media (not only air). Hence the multimedia approach of the compliance program.

Congress also passed a comprehensive pollution prevention statute in 1990, which was codified in 42 CFR 133 (Pollution Prevention). At the state level, pollution prevention, recycling, and educational activities are driven by multiple statutes found in Texas Health and Safety Code, Chapter 361 and Texas Water Code Chapter 5."

For detailed discussion on specific activities the SBLGA provides support for or has responsibility to implement, please see Appendix B.

PART VI: USED OIL OPERATING PERMITS IN TEXAS

USED OIL HANDLERS REQUIRING PERMITS

Used oil and related permits are critical to your operations compliance and are required to be in place prior to any related operations. Both the TCEQ and Federal requirements must be met and acknowledged by the TCEQ by granting the permit associated with your intended activity. This section describes the type and applicability to the example facility in this guide that represents typical facility operations but as always if the applicability does not match your specific operations you need to call the TCEQ to get further assistance.

As section VII below indicates, your notice of the related permit activities is given to the TCEQ utilizing the TCEQ Core Data Form TCEQ-10062, TCEQ 10400, and EPA Form 8700-12 as applicable below. In deciding which permit is required, you can utilize the Decision Tree Flow Chart for Applicability Determination provided below. If you have any questions on your applicability or if you need to ensure your understanding is correct, review the permit type regulatory reference provided following the flowchart. Please note that just meeting the applicability requirement does not complete your compliance requirement; you must also ensure the other requirements are in place and that you understand your reporting requirements for various permit types. The additional compliance requirements associated with the example facility and permits it would typically hold are addressed in other sections of this guide.

USED OIL HANDLER (UOH) PERMIT

First, you must establish that you meet the definition of a Used Oil Handler as identified in the Texas Administrative Code (TAC) §324.2, which defines a Used Oil Handler as "A transporter or an owner or operator of a used oil transfer, processing, re-refiner, or off-specification used oil burning facility." (115)

As noted above, the TCEQ also requires you to comply with the federal rules that are found in 40 Code of Federal Regulations (CFR) Part 279 as identified in the TAC §324.1 **(115)** indicating that the TCEQ has adopted the CFR 40 Part 279 regulations by reference. **(115)**

The following TCEQ Used Oil Handler permit types are identified below: Transfer Facility, Transporter, Used Oil Filter Handler, Processor or Re-Refiner, and Marketers of Used Oil Fuel.

Table 4 – TCEQ Which Requirements of the Used Oil Management Standards Apply to You? TCEQ UOH Guide (2005)

Requirements	Generator	Collection Center	Transporter/ Transfer Facility	Processor/ Re-refiner	Marketer	Off-Spec. Burner
Storage Management	Yes	Yes	Yes	Yes	Yes	Yes
Release Response	Yes	Yes	Yes	Yes	Yes	Yes
Secondary Containment	Recommended	Recommended	Yes	Yes	Yes1	Yes
Notification to EPA	No	No	Yes	Yes	Yes	Yes
Tracking	No	No	Yes	Yes	Yes	Yes
Financial Assurance	No	No	Yes	Yes	Yes1	Yes
Rebuttable Presumption	Yes, if mixed with hazardous waste (HW)	Yes, if mixed with HW	Yes	Yes	Yes1	Yes
Register with 2 the TCEQ	No	Every two years	One-time	One-time	One-time	One-time
Report to the TCEQ2	No	Annually	No	Every Two Years	No	No

Note: This table serves only as a guideline. For more detailed information on regulations that may apply to you, please refer to state and federal used oil regulations.

1. *This requirement does not apply to a marketer that is also a generator or collection center. Generators must still meet the requirements listed for them in*

the table's second column above.

2. *Registration and reports must be submitted to the TCEQ Used Oil Program.*

PROHIBITIONS

In determining your UOH status, it is important to make sure you understand what you cannot do as established by the TAC identifying prohibitions in TAC §324.4 **(115)**; this includes compliance with 40 CFR Subpart B, §279.12. **(59)** Identified below are the prohibitions identified by both these regulations:

1. TAC §324.4 Prohibitions:
    a. A person must not collect, transport, store, burn, market, recycle, process, use, discharge, or dispose of used oil in any manner that endangers the public health or welfare of the environment.
    b. A person commits an offense if the person:
        1) intentionally discharges used oil into a sewer, drainage system, septic tank, surface water or groundwater, watercourse, or marine water;
        2) knowingly puts used oil in waste that is to be disposed of in landfills or
        3) directly disposes of used oil on land;
        4) knowingly transports, treats, stores, disposes of, recycles, markets, burns, processes, re-refines used oil within the state without first complying with the registration requirements of this rule and/or in violation of rules for the management of used oil;
        5) intentionally applies used oil to roads or land for dust suppression, weed abatement, or other similar uses;
        6) violates an order of the commission to cease and desist any activity prohibited by this section or any rule applicable to a prohibited activity;
        7) or intentionally makes any false representation in any document used for program compliance.

An exception to paragraph (2) of this section is if a person knowingly disposes into the environment any used oil that has not been separated by the generator from other solid wastes.

Another exception to paragraph (2) of this section is if the mixing or commingling of used oil with waste to be disposed of in landfills is the unavoidable result of the mechanical shredding of motor vehicles, appliances or other metals.

1. 40 CFR § 279.12 Prohibitions: (59)
    a. Surface impoundment prohibition. Used oil shall not be managed in surface impoundments or waste piles unless the units are subject to regulation under parts 264 or 265 of this chapter.(53)
    b. Use as a dust suppressant. The use of used oil as a dust suppressant is prohibited, except when such activity takes place in one of the states listed in §279.82(c).
    c. Burning in particular units. Off specification used oil fuel may be burned for energy recovery in only the following devices:
        1) Industrial furnaces identified in §260.10 (159) of this chapter; 2. Boilers, as defined in §260.10 of this chapter that are identified as follows:
            a. Industrial boilers located on the site of a facility engaged in a manufacturing process where substances are transformed into new products, including the component parts of products, by mechanical or chemical processes;
            b. Utility boilers used to produce electric power, steam, heated or cooled air, or other gases or fluids for sale; or
            c. Used oil-fired space heaters provided that the burner meets the provisions of § 279.23. (62)
            d. Hazardous waste incinerators subject to regulation under subpart O of parts 264 or 265 of this chapter.

Transfer Facility

As identified in the TCEQ Used Oil Recycling Handbook (TUORH) **(109)**, a Used Oil Transfer Facility is any transportation-related facility (including loading docks and parking and storage areas) where shipments of used oil are held for more than 24 hours and not longer than 35 days during the normal course of transportation.

The CFR 40 279.1 definition goes into more detail as it states "*Used oil transfer facility* means any transportation related facility including loading docks, parking areas, storage areas and other areas where shipments of used oil are held for more than 24 hours and not longer than 35 days during the normal course of transportation or prior to an activity performed pursuant to §279.20(b) **(66)**" that states "*Other applicable provisions.* Used oil generators who conduct the following activities are subject to the requirements of other applicable provisions of this part as indicated in paragraphs (b)(1) through (5) of this section: (1) Generators who transport used oil, except under the self-transport provisions of §279.24 (a) and (b), must also comply with subpart E of this part. (2) (i) Except as provided in paragraph (b)(2)(ii) of this section, generators who process or re-refine used oil must also comply with subpart F of this part. (ii) Generators who perform the following activities are not processors provided that the used oil is generated on-site and is not being sent off-site to a burner of on- or off-specification used oil fuel." Along with the activities above you must comply with 40 CFR Subpart F.

Transfer facilities must also comply with TAC §324.11 that requires that if you have Underground storage tanks (USTs) containing used oil are subject to Chapter 334 **(112)** of this title (relating to Underground and Aboveground Storage Tanks) and 40 CFR Part 279.

Transporter

In addition, according to the TUORH a Used Oil Transporter is identified as any person who transports more than 55 gallons of used oil off site or collects used oil from generators to transport it.

The CFR 40 279.1 definition again goes into more detail as it states *"Used oil transporter* means any person who transports used oil, any person who collects used oil from more than one generator and transports the collected oil, and owners and operators of used oil transfer facilities. Used oil transporters may consolidate or aggregate loads of used oil for purposes of transportation but, with the following exception, may not process used oil. Transporters may conduct incidental processing operations that occur in the normal course of used oil transportation (e.g., settling and water separation), but that are not designed to produce (or make more amenable for production of) used oil derived products or used oil fuel."

Transporters must also comply with TAC §324.11 **(160)**, which requires transporters to register their used oil activities if they have not previously registered their specific used oil activities with the commission and the United States Environmental Protection

Agency (EPA). Transporters must register, through the commission, using EPA Form 8700-12 and a commission form. Mail registration forms to the Texas Commission on Environmental Quality, Used Oil Recycling Program.

It should be noted that along with the TAC and CFR 40 requirements, all transporters must also comply with all applicable Department of Transportation **(9)** regulations as they apply to the truck operations, permits, and driver training and licensing. However, this guide will not address those issues and you will need to investigate all required compliance requirements on your own.

Used Oil Filter Handler and Used Oil Filter Storage Facility

In addition to other registrations as a UOH if you collect and manage or process used oil filters you must file for a permit. The TCEQ defines a used oil filter in TAC §328.21**(161)** as "Any device that is an integral part of an oil flow system, the primary purpose of which is to remove contaminants from flowing oil contained in the system, and that as a result of use has become contaminated and unsuitable for its original purpose, is removed from service, and contains entrapped used oil. The term does not include a filter attached to the equipment containing the oil flow system. This term continues to apply regardless of prior processing until, but not after, the filter has been burned for steel recovery or energy recovery or it is separated into its component parts."

Also identified in TAC §328.21 **(161)** sections 6-10 (listed below) are other activities that make you a Used Oil Filter Handler. If you store the filters for any period, you must also register as a Used Oil Filter Storage Facility. Therefore, unless you just pick up used oil filters and transfer them to someone else's facility, you are also a filter storage facility and must register as one. TAC provides the following definitions:

1) "Process—To prepare a used oil filter for recycling, steel recovery, energy recovery, or proper disposal.
2) Processor—A person that processes used oil filters generated by another person. The Term does not include a generator that consolidates, drains, or crushes used oil filters for off-site recycling or disposal.
3) Store—To hold in a location for any period.
4) Storage facility—A location that stores used oil filters before transportation, pro-

cessing, recycling, or disposal of the filters. The term does not include the location of a generator.

5) Transporter—A person that transports used oil filters to a location for storage, processing, recycling, or disposal."

Other requirements for the management and processing of used oil filters will be discussed in the guide section on filter management related to filter processing area associated with TAC Title 30, Part 1, Chapter 328 Subchapter D Used Oil Filter Management And Recycling. **(161)**

Processor or Re-Refiner

According to the TUORH, a used oil processor or re-refiner is any person or facility that makes used oil more amenable for production of fuel oils, lubricants, or other products derived from used oil. Further definition for processing is provided in CFR 40 279.1:

"Processing means chemical or physical operations designed to produce from used oil, or to make used oil more amenable for production of, fuel oils, lubricants, or other used oil-derived product. Processing includes, but is not limited to: blending used oil with virgin petroleum products, blending used oils to meet the fuel specification, filtration, simple distillation, chemical or physical separation, and re-refining."

> ***Commentary:*** *Typically any used oil storage facility (as in the case of our example) would also be a processor by one or more of the process examples above requiring a permit.*

Marketers of Used Oil Fuel

According to the TUORH, a used oil marketer is anyone who sends a shipment of off specification used oil from his facility to a used oil burner or first claims that used oil that is to be burned for energy recovery meets the used oil fuel specification set forth in 40 CFR 279.11.

Further definition of what a Marketer is can be found in CFR 40 Subpart H §279.70 **(71)** Applicability:

(1) "Any person who conducts either of the following activities is subject to the re-

quirements of this subpart:

a. Directs a shipment of off-specification used oil from their facility to a used oil burner; or
b. (2) First claims that used oil that is to be burned for energy recovery meets the used oil fuel specifications set forth in §279.11.

(2) The following persons are not marketers subject to this subpart:

a. Used oil generators, and transporters who transport used oil received only from generators, unless the generator or transporter directs a shipment of off-specification used oil from their facility to a used oil burner. However, processors/re-refiners who burn some used oil fuel for purposes of processing are considered to be burning incidentally to processing. Thus, generators and transporters who direct shipments of off-specification used oil to processor/re-refiners who incidentally burn used oil are not marketers subject to this Subpart;
b. Persons who direct shipments of on-specification used oil and who are not the first person to claim the oil meets the used oil fuel specifications of §279.11.

(3) Any person subject to the requirements of this Subpart must also comply with one of the following:

a. Subpart C of this part—Standards for Used Oil Generators;
b. Subpart E of this part—Standards for Used Oil Transporters and Transfer Facilities;
c. Subpart F of this part—Standards for Used Oil Processors and Re- refiners; or
d. Subpart G of this part—Standards for Used Oil Burners who Burn Off- Specification Used Oil for Energy Recovery."

Collection Center for DIY

Obtaining a Collection Center permit to collect used oil from households as a "do-it-yourselfer" (DIY) as a UOH requires notice to the TCEQ similar to other activities and is encouraged but not required. Our example facility does have a Collection Center permit and specific management requirements will be covered in this guide.

TAC Title 30 §324.7 identifies a DIY and Collection Center as referenced to the federal regulations. "The commission incorporates by reference rules for owners or operators of all "do-it-yourselfer used oil collection centers" and "used oil collection centers" (as defined in 40 Code of Federal Regulations (CFR) §279.1) in 40 CFR Part 279, particularly

Subpart D." **(69)**

CFR 40 Part 279 §279.1 defines DIY, Household and Used Oil Collection Center as follows:

"Household "do-it-yourselfer" used oil means oil that is derived from households, such as used oil generated by individuals who generate used oil through the maintenance of their personal vehicles.

Household "do-it-yourselfer" used oil generator means an individual who generates household "do-it-yourselfer" used oil.

Used oil collection center means any site or facility that is registered /licensed /permitted /recognized by a state/county/municipal government to manage used oil and accepts /aggregates and stores used oil collected from used oil generators regulated under subpart C of this part who bring used oil to the collection center in shipments of no more than 55 gallons under the provisions of §279.24. Used oil collection centers may also accept used oil from household do-it yourselfers."

When you decide to be a DIY Collection Center be able to collect the DIY used oil in a separate container as you must be sure your collection efforts from sources—households in this situation—don't contaminate your larger tanks and it is a best practice to test the collection container as you would any third-party material prior to bringing it into your facility.

Suspension or Revocation of Registration

With any permit issued, you must ensure you follow all rules associated with your permit. To help you understand what could cause a suspension or revocation of that permit, which would mean you would have to stop your operations in that area or possibly suffer significant action by an agency, the TCEQ in TAC §324.21 identifies the following actions by the permit holder than would trigger this type of action. Section (7)(d) below identifies an appeal process but it could take time and costs could be significant. In every potential enforcement situation where things may have gone wrong it is always advisable to let the TCEQ know as soon as you do to possibly resolve the issue quickly. As noted in the listing of issues, you will notice subjective statements such as in **(115)** "failure to reasonably perform" or (7)(b) "However, depending upon the seriousness of the offense(s), the time of suspension may be increased or decreased." From an enforcement perspective, which will be discussed further in this guide in the enforcement section, the

TCEQ has latitude to alter the degree of enforcement but your reliance on any level of forgiveness could very well be to your detriment. As in all violation situations you should consider seeking legal and / or an environmental compliance specialist help.

TAC §324.21 identifies actions where the commission may suspend or revoke a registration as stated:

1. "The commission may suspend or revoke a registration for:
    a. failure to maintain complete and accurate records;
    b. alteration of any record, except justified and documented corrections;
    c. delivery of used oil to an unregistered entity;
    d. failure to comply with this rule or an order issued by the commission;
    e. failure to submit required reports;
    f. failure to maintain financial assurance; or
    g. failure to reasonably perform the used oil activities for which the registration was issued.
2. A registration shall be suspended for a period of one year. However, depending upon the seriousness of the offense(s), the time of suspension may be increased or decreased. A registration is revoked automatically on a second suspension. If the registration is suspended or revoked, a facility must not possess or accept used oil regulated under this rule.
3. The holder of a revoked used oil registration may reapply for registration after a period of at least one year. If a registration is revoked by the commission a second time, the revocation shall be permanent.
4. The procedure for an appeal of a suspension or revocation of registration is:
    a. An opportunity for a formal hearing on the suspension or revocation of registration may be requested in writing by certified mail, return receipt requested. The request must be postmarked within 20 days after a notice of proposed suspension or revocation of registration has been sent to the last known address of the applicant.
    b. An opportunity for a formal hearing may be requested in writing by the applicant by certified mail, return receipt requested. The request must be postmarked within 20 days after a notice of denial of registration has been sent to the last known address of the applicant. If the registration is denied,

a person must not possess used oil regulated under this rule.

c. The formal hearing under this paragraph shall be in accordance with the requirements of §305.68 of this title (relating to Action and Notice of Petition for Revocation or Suspension)."

PART VII: USED OIL HANDLER REQUIRED TCEQ PERMIT FILINGS

Permit Filings

TCEQ requires you to file the following permits prior to doing business related to the permit you are filing. From the TCEQ site (135) below is the basic form filing requirement, this guide will address all but "the burners of off-specification used oil" filing. The Core Data Form is significant because it provides the TCEQ with all your business basic information and you want it to be as complete as possible.

"Used oil handlers are used oil transporters, transfer facilities, processors/re-refiners, and burners of off-specification used oil. If you are a used oil handler you must register one time through the TCEQ using the EPA's Notification of Regulated Waste Activity Form (EPA8700-12) and the TCEQ's Registration for Used Oil Handler and/or Used Oil Filter Handlers Form (TCEQ-10062). This form must be submitted with the Core Data Form (TCEQ-10400)."

The following process is taken directly from the TCEQ Core Data Form Instructions (163) with comments to amplify specific requirements. As always on any of the forms discussed below you should call the TCEQ directly, the phone number at the top of the form (512-239-5175) with any questions prior to filing the form. A copy of each form is provided below the individual instructions. Also, a PDF auto fill in form is available at the TCEQ website available from

https://www.tceq.texas.gov/permitting/central_registry/guidance.html.

The PDF also has active links to different Texas and TCEQ agency sites referenced in the form.

What is the Core Data Form? The Core Data Form (CDF) collects core data about the people, businesses, and organizations that are regulated by TCEQ. "Core Data" refers to names, locations, Authorization IDs, business entity types, and other information that is shared across the Agency.

> ***Commentary:*** *Sections of the Core Data Form data is also part of the public record and contained in the Central Registry (CR) database that is maintained by the TCEQ for inquiry by anyone and as indicated the information is shared across the agency as needed.*

WHEN DO I COMPLETE THIS FORM?

A Core Data Form should be submitted under the following circumstances:

1. Your core information is not yet in or is incomplete in the Central Registry (CR) database.
2. Your current information has changed and is not being displayed in the CR database.
3. It is requested by the Agency.

You can view/review your data in Central Registry at http://www15.tceq.texas.gov/crpub/.

Note: The CDF is a supplemental form used if you are requesting a new permit, change of ownership, or termination. Please contact Central Registry for assistance with the CDF at (512) 239-5175.

WHERE DO I SEND THIS FORM?

1. It should be sent to the address listed on the other TCEQ forms that the Core Data Form is accompanying.
2. It should be sent to the location requested by TCEQ staff.

What is a customer?

A customer is an individual or organization responsible for one or more Regulated Entities. Examples include owners, operators, or responsible parties of a site or an individual who holds a license. Note: The Customer Name submitted may be updated automatically based on what is current and active with the Texas Secretary of State (SOS) or Texas

Comptroller of Public Accounts (CPA).

Customer Reference Numbers (CNs) are used to identify Customers in Central Registry. Each Individual or Organization is issued a unique 11-digit identification number called a **CN** (e.g.CN123456789). Each Customer should have only one CN based upon their name and SOS/CPA information as applicable.

> ***Commentary:*** *Periodically you should inquiry the Central Registry (CR) to insure your data is still accurate as it is open to inquiry by anyone and you should insure its accuracy and if you find a discrepancy you need to correct it as soon as practical. The best way to start any correction process is to make a call first for instructions.*

What is a Regulated Entity?

A Regulated Entity (RE) is the activity, site, or thing that's being regulated by TCEQ. The Regulated Entity is located at the actual physical address where the regulated activity is occurring. Examples are a site, facility, or license. Note: The Regulated Entity Name submitted may be updated in order to meet TCEQ Agency Data Standards (removal of organizational endings such as Inc, LP, or LLC etc. from the Regulated Entity name).

Each site location is issued a unique 11-digit identification number called a RN (e.g., RN123456789). Each Regulated Entity should have only one RN number unless their activity is in a program with a stand-alone designation. Note: The CN and/or RN will not replace program-specific numbers (e.g., a permit, registration, account, etc.).

<u>Example:</u>

The table below provides examples of Customers who own or are responsible for the corresponding Regulated Entities.

Customer	Regulated Entity
Worlds Inc	Roads Landfill
Duke Energy Field Services LP	Englewood Compression Station
Exxon Mobil Corporation	Tiger Mart 791
Lubrizol Corporation	Lubrizol Bayport Plant
Wal-Mart Stores Inc	Wal-Mart 2468

Do I receive a new CN or RN Number each time I submit a CDF?

No. A new CN or RN number is assigned only if:

1. The Customer does not have a CN in Central Registry.

> ***Commentary:*** *It is very important that you use your correct CN or RN on any new application as it provides the link to all other data and if incorrect can create a lot of work getting data corrected. Note: TCEQ's qualifying the application of a single CN identified: A Customer may be involved with many sites and/or licenses but should only be issued a single CN in Central Registry based upon the Individual's Name or Organization's SOS/CPA information as applicable.*

2. The Regulated Entity does not have a RN in Central Registry.

> ***Commentary:*** *Again, note the qualifying statement for how the RN is managed. The RN becomes the historical link to all data about the site for its existence even if the CN changes.*

A Regulated Entity identifies where the regulated activity is occurring with the exception of portable Regulated Entities and licenses. There should be only one RN number issued for each location unless the activity is in a program with a stand-alone designation. The RN will stay with that site/equipment regardless of how many times the responsible Customer changes.

Line-By-Line Instructions: If you go to the TCEQ website

Section I: General Information

1. Reason for Submission:

- **New** – Check this box if the Core Data Form is being submitted in conjunction with a new Permit, Registration or Authorization application. Attach the completed Core Data Form to the program application.
- **Renewal** – Check this box if the Core Data Form is being submitted in conjunc-

tion with a renewal application. Attach the completed Core Data Form to the program renewal form.

- **Other** – Check this box if the Core Data form is being submitted for a reason other than those listed above. Briefly state why the form is being submitted. Attach the completed Core Data Form to the applicable program form (e.g. amendment, modification, etc.).

Commentary: *Program forms are forms to add a handler status such as collection center, filter handler etc;*

- **Examples:** Updating Customer Information and or Updating Regulated Entity Information. ABC Inc owns and operates Dallas Facility. Neither have CN or RN numbers in the Central Registry Database (search Central Registry first to confirm these numbers do not already exist).
  - ABC Inc. = Customer (owner & operator) WITHOUT A CN NUMBER
  - Dallas Facility = Regulated Entity WITHOUT AN RN NUMBER

- Process to get new RN and CN Numbers:
  - Use only one Core Data Form
  - Enter Worlds Inc's information in Section II
  - Enter Roads Landfill's information in Section III
  - Enter the preparer's information in Section IV
  - Always fill out Sections I, IV and V. (In this example, a CN and RN number have not been issued , so leave the CN and RN fields blank in Section I)
  - It may be necessary to submit a program application when getting new CN and RN numbers

2. **Customer Reference Number (CN):**

- Leave this field blank if this is the first time you or your company are completing this form. If you or your company has been issued a CN, enter it here.

3. **Regulated Entity Reference Number (RN):**

- Leave this field blank if no regulated activity has occurred or been reported at this site/location. If this location has been issued a RN, enter it here.

Section II: Customer Information

4. **General Customer Information:** Check ALL boxes that apply.

- **New Customer:**
  The Customer does not have a CN. Complete all items in Section II.
- **Update to Customer Information**
  The Customer has a CN but some of the information in Section II has changed and needs to be updated. Complete the items that have changed.
- **Change in Regulated Entity Ownership**
  The Customer for this Regulated Entity (listed in field 3) has changed (e.g. Facility has been sold or is now under operational control of another Customer). Complete all items in Section II and fill in the previous Customer's name and effective date.

Commentary: *If for some reason you don't know this information you can check the Central Registry.*

Note: Not all permits/authorizations are transferable to a new owner. Check with the program area(s) regarding their requirements for a change of ownership.

Commentary: *This is important to know as if not transferable you would need to re-apply causing more delay, in most cases you may operate in the interim but time is limited and you need to contact the agency for guidance.*

- **Change in Legal Name**
  A Customer name change will require verification with the SOS/CPA prior to it being processed by the Agency. Contact TX SOS at 512-463-5555 for more information related to your organization.

Commentary: *This is an area where you should seek legal advice if not already done to insure this is done correctly. More than compliance issues are impacted by any change in legal name.*

A new CN will be needed if the SOS/CPA Charter or Filing Number changes for a Business Entity. Contact Central Registry at 512-239-5175 for additional guidance.

ABC Incorporated is in Central Registry as the owner/operator of the Regulated Entity – Dallas Facility. A few months later ABC Incorporated goes through a reorganization which results in changes to the information the company originally submitted to TCEQ on the Core Data Form. The changes include a company name change to DEF Energy USA a company mailing address change, and a change in the Regulated Entity name to North Dallas Facility.

- Hyper Global Net Incorporated = Customer WITH A CN NUMBER Global Net Services = Regulated Entity WITH AN RN NUMBER
  - Process for updating CN/RN information
    - Fill out a single Core Data Form. Put the Customer's and Regulated Entity's CN and RN numbers in for Fields 2 and 3.
    - Mark "New Customer" and "Update to Customer Information" for Field 4 and fill in only the fields that are being updated for Section II. In this example it would be the Customer's new name and new mailing address.
    - Mark update to Regulated Entity information on Field 21 and fill out only the fields that are being updated in Section III. In this example it would be the Regulated Entity's new name.
    - Make sure to always fill out Section IV and V.

5. Effective Date: Enter the effective date of the update (e.g. date of sale, SOS filing date, etc.).

6. Customer Legal Name: Enter the complete legal name of the Customer.

- Verify the legal name of your organization by calling their help lines or by searching the SOS and CPA websites.

Commentary: *The reason for this TCEQ comment is that this is a common error and needs to be exact or your form could be rejected and returned.*

Secretary of State (SOS) 512-463-5555
https://direct.sos.state.tx.us/acct/acct-login.asp
Texas Comptroller (CPA) 1-800-252-1386
https://comptroller.texas.gov/

Note: The Customer Name submitted may be updated automatically based on what is current and active with the Texas Secretary of State (SOS) or Texas Comptroller of Public Accounts (CPA).

- **Previous Customer Name**
  Enter the prior name of the Customer if the Customer name is changing due to a change in the legal name.

OR

Enter the name of the previous owner if the Customer name is changing due to an ownership change for the Regulated Entity.

7. TX SOS/CPA Filing Number:

Search for your SOS Filing number at http://direct.sos.state.tx.us/acct/acct-login.asp or call at 512-463-5555.
Leave this field blank if the Customer has not filed with SOS/CPA as a Corporation, Limited Partnership or non-profit organization. If the Customer has filed with SOS/CPA, enter the Filing/Charter Number.

8. TX State Tax ID:

- Check the CPA website for a listing of your tax ID.
  https://mycpa.cpa.state.tx.us/coa/Index.html
- Leave this field blank if the Customer has not been issued a TX State Tax ID. If one has been issued, enter it here.

9. Federal Tax ID:

- All businesses, except some small sole proprietors and individuals should have a federal taxpayer identification number (FEIN). Enter it here. Do

NOT list hyphens, dashes, or a Social Security Number.

- For more information on Federal Tax IDs, use this web link: http://www.irs.gov/Businesses/Small-Businesses-&-Self-Employed/Do-You-Need-an-EIN.

10. DUNS Number:

- Leave this field blank if a Data Universal Numbering System (DUNS) number has not been issued by Dun and Bradstreet Corp (usually larger companies). If a DUNS Number has been issued, enter it here.

11. Type of Customer: Check only ONE box.

Corporation

The Customer meets all of the following:

- Is legally incorporated under the laws of any state or country
- Is recognized as a corporation by the Texas Secretary of State
- Has proper operating authority to operate in Texas

Examples of Corporations

- General corporations
- Subchapter S corporations
- Limited liability companies (LLC)
- Water supply corporations

Individual

An individual is a person who is regulated by TCEQ, but has not established a business.

Partnership

- General Partnership - A general partnership is created when two or more persons associate to carry on a business for profit. A partnership generally operates in accordance with a partnership agreement, but there is no requirement that the agreement be in writing and no state-filing requirement. Please submit a tax ID if you have a sales tax permit.
- Limited Partnership (LP & LLP) - A Texas limited partnership is a partnership formed by two or more persons, having one or more general partners and one or more limited partners. The limited partnership operates in accordance

with a partnership agreement, written or oral, of the partners as to the affairs of the limited partnership and the conduct of its business. While the partnership agreement is not filed for public record, the limited partnership must file a certificate of limited partnership with the Texas SOS. They provide a form for the certificate of limited partnership which meets minimum state law requirements.

- Government – City, County, Federal, or State - The Customer is either an agency of one of these levels of government or the governmental body itself (e.g. Blanco County, City of Houston, Texas Department of Transportation, etc.).
- Government – Other - The Customer is a utility district, water district, and tribal government, and college district, council of governments or river authority (e.g. Lower Colorado River Authority).
- Sole-Proprietorship - The Customer is an individual who owns or runs an unincorporated business with no distinction between the business and the owner.
- Other - The Customer does not fit any of the descriptions above (e.g. joint venture, trust, estate, etc.). Enter a short description in the space provided.
- Change in ownership of a Regulated Entity:

Commentary: *This area can be complicated so please insure changes are completed correctly for the benefit of the previous owner / operator and the new owner / operator. Call the TCEQ and be sure that you have covered all changes required based on what the previous entity was responsible for and what the current owner is now responsible for.*

(Contact the Program Areas before attempting to use this form to process a change of ownership, other forms may be required)

ABC Company owns and operates Dallas Facility. Each of these is in Central Registry and has a CN and RN number.

ABC Inc. = Customer with a CN with Dallas Facility = Regulated Entity with RN

A few months later DEF Recycle Inc. buys ABC Inc. and becomes its sole owner and operator. ABC Inc. is no longer associated with the site. DEF Recycle Inc. is NOT in the Central Registry Database yet.

DEF Recycle Inc. = Customer, WITHOUT A CN NUMBER

Dallas Facility = Regulated Entity, WITH AN RN NUMBER

Process for Change in Ownership

Fill out a single Core Data Form for DEF Recycle Inc. (new Customer) and Dallas Facility (Regulated Entity). In Section I, leave the CN blank because it has not been issued yet and use Dallas Facility RN Number. In Section II check "New Customer" and "Change in Regulated Entity Ownership" for field 4. Enter DEF Recycle Inc. as the Customer Legal Name in field 6. Enter ABC Company as the previous name and put the date they stopped being associated with the site.

> ***Commentary:*** *This date is very important as to compliance enforcement if issues come up.*

After the first Core Data Form has been submitted the prior owner (ABC Company) should submit a letter to all interested program areas at TCEQ stating that they are no longer the Customer for the Regulated Entity that they sold. (DEF Recycle Inc.)

> ***Commentary:*** *This is important as recommended by the TCEQ to make sure notice is given. Send the letter(s) registered to insure they have received it. The letter as stated by the TCEQ comment above should be brief and include the date you were no longer associated with the Regulated Entity.*

Note: Fill in the permit/registration that are changing hands in the TCEQ Programs and ID Numbers Field 39 if the change of ownership is for specific permits at a site and not the whole site.

Number of Employees:

- Enter the number of employees for the Customer's entire company, including subsidiaries. The number should include **ALL** locations, not just the location of the Regulated Entity listed on the CDF.

13. Independently Owned and Operated:

- Check "No" if the Customer is a subsidiary or part of a larger company. Otherwise, check "Yes".

14. **Customer Role:**
    - Check **ONE** box that best indicates the Customer's relationship to the Regulated Entity. Check "Other" if none of the boxes fit your situation and write in the relationship.

15. **Mailing Address:**

Commentary: *This is another critical entry and you need to make sure mail will get to this address and it is not always the case at a facility as another address needs to be used. Not getting some sort of notice or agency documents based on an address you used in a filing can be very problematic.*

Enter the mailing address where general correspondence will be received regarding Customer information (e.g. the Customer's Headquarters). This should be a deliverable address recognized by the United States Postal Service. Use this web link for address verification: www.usps.com.

Enter the mailing address for the overall/primary Customer contact for TCEQ. Do **NOT** enter specific program contacts (e.g. New Source Review Contact, Petroleum Storage Contact, Stormwater Contact, etc.).

If this address is outside the USA: Enter the mailing address and city if applicable. Enter all other mailing information under "Country" (Field 16).

Country Mailing Information:

- Leave this field blank if the address is in the USA. If not, enter the country name, mailing zone, non-ZIP mailing codes and non-USA Postal Service features.

17-20. E-mail, Phone and Fax Information:

- Enter the communication information that TCEQ will need to contact the Customer. Leave field 19 blank if your phone system does not use extensions or codes.

Section III: Regulated Entity Information

21. General Regulated Entity Information: Check **ONE** box

- New Regulated Entity - The Regulated Entity does not have a RN. Complete all items in Section III.
- Update to Regulated Entity Name - The known name of the Regulated Entity has changed and needs to be updated. This name will be used by all program areas.
- Update to Regulated Entity Information - The Regulated Entity has a RN but some of the information in Section III has changed and needs to be updated. Complete the items that have changed.

Note: If the street address of the RN is changing, contact Central Registry at 512-239-5175 to determine if a new RN number and/or new permits/authorizations are needed. This does not apply to location changes for portable Regulated Entities.

22. Regulated Entity Name:

- Enter the name the Regulated Entity will be known by. It will be used by all TCEQ programs when referring to the Regulated Entity. Make sure this name is consistent on all forms submitted to TCEQ.
- If the Regulated Entity name is the same as an individual, please print last name first, followed by the first name.

***Note:** The Regulated Entity Name submitted may be updated in order to meet TCEQ Agency Data Standards (removal of organizational endings such as Inc, LP, or LLC, etc. from the Regulated Entity name).*

23. Street Address of the Regulated Entity:

- Enter the physical street address of the Regulated Entity. Do not provide a P.O. Box. If there is no street address, complete fields 25 and 26.

Example: 1234 MAIN ST

HOMETOWN, TX 78711

24. **County:**
    - **Enter the** county where the Regulated Entity is located. **If there is more than one county,** enter the primary county in which the majority of the Regulated Entity is located. Include the other counties as applicable.

25. **Description to Physical Location:**
    - If the physical location has no street address, enter specific directions of how to reach the Regulated Entity.

Commentary: *This is important, if possible use a mapping system like Google or MAPQUEST to insure physical directions make sense, what may be simple to you or obvious may not be to anyone else trying to find your facility and frankly they have a right to come to the site unannounced. It is suggested that actually use the directions provided and drive them to verify. Example: FROM US 83, GO E ON COUNTY RD 101 GO 0.3 MI PAST FIRST TANK ON LEFT.*

26. **Nearest City, State, and Nearest ZIP:**
    - If the Regulated Entity is in a city, that city will be the nearest city. If the nearest city is in another state or country, please provide that information in this blank as well.
    - **State:** This will primarily be Texas.
        - **Nearest ZIP Code: Even Regulated Entities that are not within city limits are located within a zip code. That zip code is listed here, the zip code that this Regulated Entity is physically located within. If not known, give the zip code that is closest to this Regulated Entity, e.g. the zip for the Nearest City.**

27-28. **Latitude (N) and Longitude (W):**

- Enter the latitude and longitude for a fixed location. For larger sites use the center point of the operation. It can be entered in either decimal or degree format.

29-32 SIC and NAICS Codes:

- All Regulated Entities should have North American Industrial Classification System (NAICS) or Standard Industrial Classification (SIC) and codes. A Primary NAICS or SIC code is the code that best describes the business conducted at this Regulated Entity.
- For a list of SIC codes, use this web link: https://www.osha.gov/pls/imis/sicsearch.html
- For a list of NAICS codes, use this web link: http://www.census.gov/eos/www/naics/

33. What is the Primary Business of this Entity:

- Briefly describe the primary business of this Regulated Entity *(e.g.* ***Used Oil Transfer Facility, Used Oil Transporter,*** *Gas Compression, Fleet Refueling, Rubber Manufacturing, etc.). Do not repeat the SIC or NAICS description.*

34. Mailing Address: Use same guidance as in 15. Above.

- Enter the mailing address where general correspondence will be received regarding Regulated Entity information (e.g. the Main Office). This should be a deliverable address recognized by the United States Postal Service. Use this web link for address verification: www.usps.com.
- Enter the mailing address for the overall/primary Regulated Entity contact for TCEQ. Do **NOT** enter specific program contacts (e.g. New Source Review Contact, Petroleum Storage Contact, Stormwater Contact, etc.).

35-38. E-mail, Phone and Fax information:

- Enter the communication information that TCEQ will need to contact the Regulated Entity. Leave Field 37 blank if your phone system does not use extensions or codes.

39. TCEQ Programs and ID Numbers:

- Use these fields to indicate the Programs and Permit/Registrations that will be impacted by the updates noted on this form.
- For example: List the Programs and ID numbers linked to the Regulated Entity

that you want updated if the Regulated Entity's name is being changed. This identification will help ensure this form will go to the correct Program Areas and that the appropriate permits / registrations are updated.

- **If there are more Additional ID numbers than space available, list them on a separate sheet and attach it to the TCEQ Core Data Form.**

Section IV: Preparer Information:

Commentary: *This person should not only have direct knowledge of data entered but should be able to respond to any questions about the entry requirements related to the entity quickly if called by the TCEQ.*

40-45. Name, Title, Phone Numbers, E-mail.

- Enter the name of the person who prepared this form. This may not be the manager or authorized individual for this site, but is the person who actually prepared this form.

Section V: Authorized Signature

46. Name and Signature

- The signer must have signature authority and will certify that the information provided in this form is true and complete.
- TCEQ rules and statutes may restrict who can sign this form depending on the updates requested and the Program Areas involved. Refer to the reference regulation to determine signature authority.
- Find the rule for your permit/registration type using this web link: *Used Oil Handlers it is typically the applicant or preparer.* http://www.sos.state.tx.us/tac/index.shtml.
- For further assistance, please see the program area signature requirements and contact information list below. If the CDF is being filled out as a result of an investigation, the CDF can be completed by the investigator but needs to be signed by the Customer. In cases where there is no Customer such as abandoned sites or spills the

CDF should not be signed. *Investigators should contact the program area contact regarding a CDF not being signed.*

- For Water Quality authorizations (CAFO, Wastewater, Storm Water, and Sludge permits and registration), by signing the core data form, the person signing is certifying that they are authorized under **30 Texas Administrative Code §305.44** to sign and submit this document, and can provide documentation in proof of such authorization upon request.

Table 5 – TCEQ Program Area Signature Requirements and Contact Information

TCEQ Program Area	Signatory Regulations (who can sign this form)	Program Area Phone Number
Air New Source (AIRNSR)	Applicant	(512) 239-1250
Air Operating Permits (AirOP)	Applicant	(512) 239-1250
Brownfields (BSA)	Applicant	(512) 239-2200
Concentrated Animal Feeding Operations (CAFO)	30 TAC Ch. 305.44(a)	(512) 239-4671
Dam Safety	Applicant	(512) 239-0400
Districts	30 TAC Ch. 213.4(d)	(512) 239-4691
Dry Cleaners	Owner or Operator	(512) 239-2160
Edwards Aquifer (Austin)	30 TAC Ch. 213.4(d)	(512) 339-2929
Edwards Aquifer (San Antonio)	30 TAC Ch. 213.4(d)	(210) 490-3096
Enclosed Container Registration	335.6 (a,b,c)	(512) 239-6413
Industrial Hazardous Waste Permits (IHW)	Applicant / Preparer	(512) 239-2334
Industrial Hazardous Waste Registrations (IHW)	Applicant / Preparer	(512) 239-6413
Industrial Hazardous Waste -- Corrective Action	Applicant	(512) 239-2343
Innocent Operator Program (IOP)	Applicant	(512) 239-2200
Licensing	Applicant	(512) 239-6300
Medical Waste On-Site Treatment Registration	Owner or Authorized Operator	(512) 239-6413
Medical Waste Transporter Registration	Owner or Authorized Operator	(512) 239-6413

Municipal Solid Waste (MSW)	Applicant / Permittee	(512) 239-2334
On-Site Sewage Facilities	Owner	(512) 239-0914
Petroleum Storage Tank (PST)	Owner or Operator	(512) 239-2160
Public Water Supply (PWS)	30 TAC Ch. 305.44(a)	(512) 239-4691
Sludge – Registration & Permits	30 TAC Ch. 305.44(a)	(512) 239-4671
Sludge - Transporters	30 TAC Ch. 312.142	(512) 239-6413
Storm Water	30 TAC Ch. 305.44(a)	(512) 239-3700
Tires	Applicant	(512) 239-2334
Underground Injection Control	Applicant	(512) 239-2334
Used Oil Registration	Applicant	(512) 239-6413
Waste Water	30 TAC Ch. 305.44(a)	(512) 239-4671
Water Rights (WAM)	Applicant	(512) 239-4691
Water Quality (WQ)	30 TAC Ch. 305.44(a)	(512) 239-4671
Voluntary Cleanup Program (VCP)	Applicant	(512) 239-2200

Form 1 – TCEQ Core Data Form (TCEQ-10400)

TCEQ

TCEQ Use Only

TCEQ Core Data Form

For detailed instructions regarding completion of this form, please read the Core Data Form Instructions or call 512-239-5175.

SECTION I: General Information

1. Reason for Submission *(If other is checked please describe in space provided.)*

☐ New Permit, Registration or Authorization *(Core Data Form should be submitted with the program application.)*

☐ Renewal *(Core Data Form should be submitted with the renewal form)* ☐ Other

2. Customer Reference Number *(if issued)*	Follow this link to search for CN or RN numbers in Central Registry**	**3. Regulated Entity Reference Number** *(if issued)*
CN		RN

SECTION II: Customer Information

4. General Customer Information | **5. Effective Date for Customer Information Updates** (mm/dd/yyyy)

☐ New Customer ☐ Update to Customer Information ☐ Change in Regulated Entity Ownership

☐ Change in Legal Name (Verifiable with the Texas Secretary of State or Texas Comptroller of Public Accounts)

The Customer Name submitted here may be updated automatically based on what is current and active with the Texas Secretary of State (SOS) or Texas Comptroller of Public Accounts (CPA).

6. Customer Legal Name *(If an individual, print last name first: eg: Doe, John)* — *If new Customer, enter previous Customer below:*

7. TX SOS/CPA Filing Number	**8. TX State Tax ID** (11 digits)	**9. Federal Tax ID** (9 digits)	**10. DUNS Number** (if applicable)

11. Type of Customer: ☐ Corporation ☐ Individual Partnership: ☐ General ☐ Limited

Government: ☐ City ☐ County ☐ Federal ☐ State ☐ Other ☐ Sole Proprietorship ☐ Other:

12. Number of Employees
☐ 0-20 ☐ 21-100 ☐ 101-250 ☐ 251-500 ☐ 501 and higher

13. Independently Owned and Operated?
☐ Yes ☐ No

14. Customer Role (Proposed or Actual) – *as it relates to the Regulated Entity listed on this form. Please check one of the following:*

☐ Owner ☐ Operator ☐ Owner & Operator

☐ Occupational Licensee ☐ Responsible Party ☐ Voluntary Cleanup Applicant ☐ Other:

15. Mailing Address:

City		State		ZIP		ZIP + 4	

16. Country Mailing Information *(if outside USA)*	**17. E-Mail Address** *(if applicable)*

18. Telephone Number	**19. Extension or Code**	**20. Fax Number** *(if applicable)*
( ) -		( ) -

SECTION III: Regulated Entity Information

21. General Regulated Entity Information *(If "New Regulated Entity" is selected below this form should be accompanied by a permit application)*

☐ New Regulated Entity ☐ Update to Regulated Entity Name ☐ Update to Regulated Entity Information

The Regulated Entity Name submitted may be updated in order to meet TCEQ Agency Data Standards (removal of organizational endings such as Inc, LP, or LLC.)

22. Regulated Entity Name *(Enter name of the site where the regulated action is taking place.)*

23. Street Address of the Regulated Entity: *(No PO Boxes)*							
	City		State		ZIP		ZIP + 4
24. County							

Enter Physical Location Description if no street address is provided.

25. Description to Physical Location:	

26. Nearest City	State	Nearest ZIP Code

27. Latitude (N) In Decimal:			28. Longitude (W) In Decimal:		
Degrees	Minutes	Seconds	Degrees	Minutes	Seconds

29. Primary SIC Code (4 digits)	30. Secondary SIC Code (4 digits)	31. Primary NAICS Code (5 or 6 digits)	32. Secondary NAICS Code (5 or 6 digits)

33. What is the Primary Business of this entity? *(Do not repeat the SIC or NAICS description.)*

34. Mailing Address:							
	City		State		ZIP		ZIP + 4
35. E-Mail Address:							

36. Telephone Number	37. Extension or Code	38. Fax Number *(if applicable)*
( ) -		( ) -

39. TCEQ Programs and ID Numbers Check all Programs and write in the permits/registration numbers that will be affected by the updates submitted on this form. See the Core Data Form instructions for additional guidance.

☐ Dam Safety	☐ Districts	☐ Edwards Aquifer	☐ Emissions Inventory Air	☐ Industrial Hazardous Waste
☐ Municipal Solid Waste	☐ New Source Review Air	☐ OSSF	☐ Petroleum Storage Tank	☐ PWS
☐ Sludge	☐ Storm Water	☐ Title V Air	☐ Tires	☐ Used Oil
☐ Voluntary Cleanup	☐ Waste Water	☐ Wastewater Agriculture	☐ Water Rights	☐ Other:

SECTION IV: Preparer Information

40. Name:		41. Title:	
42. Telephone Number	43. Ext./Code	44. Fax Number	45. E-Mail Address
( ) -		( ) -	

SECTION V: Authorized Signature

46. By my signature below, I certify, to the best of my knowledge, that the information provided in this form is true and complete, and that I have signature authority to submit this form on behalf of the entity specified in Section II, Field 6 and/or as required for the updates to the ID numbers identified in field 39.

Company:		Job Title:	
Name*(In Print)* :		Phone:	( ) -
Signature:		Date:	

PART VIII: ALL HANDLERS REQUIRED TCEQ REGISTRATION FOR USED OIL HANDLER AND/OR USED OIL FILTER HANDLERS FORM TCEQ-10062

All Used Oil Handlers must file a Core Data Form as above and identify activity that includes used oil transporters, transfer facilities, processors/re-refiners, and burners of off-specification used oil. If you are a used oil handler, you must complete the TCEQ's Registration for Used Oil Handler and/or Used Oil Filter Handlers Form (TCEQ-10062). This form must be submitted with the Core Data Form (TCEQ-10400).

TCEQ instructions for completing this form are provided as a reference document 164. Provided below are instructions based on TCEQ instructions with additional Guide information italicized. However, as always as you fill out the form and you have any questions you should call the TCEQ at the number found at the bottom of the form. You can find a copy of the form at:

https://www.tceq.texas.gov/assets/public/permitting/rrr/forms/10062.pdf

and a copy of the instructions at

https://www.tceq.texas.gov/assets/public/permitting/rrr/forms/10062inst.pdf

See the Used Oil Handlers and Used Oil Containing Materials Management section in this guide for other requirements and activity description in this Guide to clarify your activity applies to one of the filings.

TCEQ – 10062 Form Instructions:

To obtain a Used Oil/Used Oil Filter Handler registration number, an applicant must comply with applicable federal, state and local regulations. Documentation demonstrating compliance must accompany the application, including financial assurance as applicable in 30 TAC Chapter 324 and 40 CFR Part 279 for Used Oil Handlers and 30 TAC

Chapter 328 Subchapter D for Used Oil Filter Handlers.

This form must be submitted with the Core Data Form, TCEQ-10400 attached. To obtain a Used Oil Handler registration number, an applicant must comply with applicable federal, state and local regulations. Documentation demonstrating compliance must accompany the application, including financial assurance and management and safety plans as applicable in 30 TAC Chapter 324 and 40 CFR Part 279. To obtain a Used Oil Filter Handler registration number, an applicant must comply with applicable federal, state and local regulations. Documentation demonstrating compliance must accompany the application, including financial assurance and management and safety plans as applicable in 30 TAC Chapter 328 Subchapter D.

> ***Commentary:*** *As indicated in the note by applying for this permit you agree to insure you meet all the compliance requirements and they must be in place prior to any operation under the permit. When submitting this form you must also attach your financial assurance verification covered in detail later in this guide along with the SPCC plan and Material Analysis Plan requirements. The filing of this form should be the last step in your process.*

A. Use this form for:
   - New (initial) notifications,
   - Subsequent notifications (updates), and
   - Renewals.
   - For updates and renewals, complete only the items that have changed.
2. Facilities with the following activities should register via this form:
   1. Used oil handlers who are a
      - Transporter
      - Transfer facility
      - Re-refiner
      - Burner of off-spec used oil
      - Marketer who first claims used oil meets specs
      - Marketer who directs shipments to burners
      - Processor
      - Used oil filter handlers who are a

- Transporter
- Storage Facility
- Processor

Commentary: *Refer to the Used Oil Handler Guide section to make sure you are identifying the correct activity.*

3. **Items in bold must be filled out for all notifications. Items with asterisks are required for all new registrations. Print clearly or type.** Read the instructions for additional information to determine if this form should be used for your facility and registration type. Please note that incomplete forms will not be processed and may be returned to the sender for completion.
4. **Form Header:** (Bold items required for all notifications. * Items required for all new notifications. Please print clearly or type.)
   1. **Used Oil Handler Registration #:**
      - For new notifications, leave the Used Oil Handler Registration Number blank. Your site will be assigned a Used Oil Handler Registration Number when TCEQ processes the notification.
      - For updates and renewals, enter the last five digits of the 6-character Used Oil Handler Registration Number assigned to your company. Since all Used Oil Handler Registration numbers begin with the letter A, the form has the letter A preprinted. If you are unsure of the Used Oil Handler Registration Number, check the TCEQ Central Registry search, or you can find it on your notice of registration.
      - The Used Oil Handler Registration Number is site specific. If the company moves to a new site location, the company should submit a new notification form to obtain a new Used Oil Handler Registration Number for the new site. Do not use numbers assigned to another site address.
      - When this form is filled out via computer, the Used Oil Handler Registration Number will automatically populate on the other pages of the form.
      - **Effective Date: (Bold)**
      - Indicate the date that new the information provided went into effect. The date cannot be a future date. When this form is filled out via computer, the

effective date will automatically populate on the other pages of the form.

- **Submission Reason: (Bold) (for Parts A-F)**
- When this form is filled out via computer, the submission reason will automatically populate on the other pages of the form.
- New: Mark "New" if this is the first notification for this company/facility or individual at this site address. Items marked with * are required for all new notifications.
- Update/Renewal:
- Mark "Update" if this site already has a Used Oil Handler Registration Number and updated information isbeing provided about the site.
- Mark "Renewal" if the site is a used oil filter handler and is submitting the notification to renew their used oil filter handler registration. Renewals are due by January 25th of each even year. For updates and renewals, indicate the Used Oil Handler Registration Number for the site on all pages submitted.

Note: **Submit Part G for each submission** in conjunction with the parts that pertain to the updates needed. For updates, submitting the entire form is not necessary. For instance, to update the contact information submit Part C and Part G only. Updates can include changing contact, owner/operator, or other information included on this form.

A. **Part A: Site Information**

This is the regulated site requesting the registration. The site is located at the actual physical address where the regulated activity is occurring with the exception of transporters. Transporters should provide the location of their office and/or truck lot in the site information.

1. **EPA ID**
   - Indicate the 12-character Environmental Protection Agency Identification (EPA ID) number for the site, if applicable.
   - For new notifications, leave the EPA ID number blank unless the site has an existing EPA ID. If the site has used oil handler activities, TCEQ will assign the site an EPA ID as needed.

- The EPA ID is site specific. If the company moves to a new site location, the company should submit a new notification form to obtain a new EPA ID for the new site. Do not use numbers assigned to another site address.
- If the company is a transporter located in another state and shipping used oil in Texas, enter the EPA ID Number assigned in the site's home state. The address information in #4-8 should match the address tied to the EPA ID provided.

2. **Regulated Entity Number RN**
   - Enter the 9-digit Regulated Entity Number (RN) assigned by TCEQ if this site (or company, in the case of a transporter) already has one. Leave this item blank if the site does not have a RN.
3. **Site Name (Bold)**
   - Enter the name the site will be known as/by, such as a DBA. Be sure to use a fully descriptive name. (e.g. ABC Company Warehouse 4, not Warehouse 4.) If a site does not have an existing RN, the Registration and Reporting section will use the site name as the regulated entity name.

Note: When this form is filled out via computer, the site name will automatically populate to the other pages of the form.

In accordance with the TCEQ Central Registry Naming Standards, a site name should not contain INC., CO, LP, etc. in the name unless it is an integral part of the site name. This standard also applies to all the following organizational endings:

- Associated, Associates, Association, ASSOC, Company, Corporation, CO, CORP, Incorporated, INC., LC, LLC, Limited Liability Company, Limited Company, LTD, Limited, Partnership, Limited, LP, PA, Professional Limited Liability Company, Professional, PLLC, PC

Notification paperwork received with any of the above organizational endings in the facility name field may be returned for correction unless a justification is provided. To eliminate any delay in processing your notification, ensure you follow these standards. For any questions regarding standards for organizational endings, contact the TCEQ Central Registry at 512-239-5175.

4. **Site Address / Street Address ***
   - Enter the physical street address of the site. Do not provide a P.O. Box. If the site has no street address, complete #5 instead.
5. **Site Description**
   - This field should be blank unless #4 is not available. Enter specific directions for how to reach the site if the physical location of the site has no street address. Be sure to include at least two cross streets for reference. Sample Entry: "E on County Road 101 from US 83, turn left after 1.3 miles, go 0.3 miles".
6. **City ***
   - Enter the city for the physical location of the site. If the site is not within city limits, enter the nearest city in this field.
7. **State ***
   - Enter the state for the physical location of the site. This will usually be Texas unless the site is a registering as a transporter from another state.
8. **Zip+4 ***
   - Enter the zip code for the physical location of the site. If the site is not within a zip code, enter the nearest zip code in this field.
9. **County ***
   - Enter the county in which the site is located. If the site extends over multiple counties, enter the county with the majority of the site.
10. **Primary NAICS ***
    - All new notifications must provide a primary North American Industrial Classification System (NAICS) code. The code should indicate the classification for the ultimate business use of the site. Go to www.census.gov/eos/www/naics/ for a list of NAICS codes.
11. **Site Land Type ***
    - Check the box that best describes the land type of your site. Select only one type: Private, County, District, Federal, Tribal, Municipal, State, or Other. If your site's Land Type could be described as Municipal and an-

other Land Type, such as County, District, or Tribal, do not check Municipal. Instead, choose the other appropriate Land Type.

1. **Financial Assurance**

All handlers must comply with the demonstration of financial responsibility at the time of registration. The agency accepts trust funds, surety bonds, letters of credit or financial test and corporate guarantee. Failure to disclose to the Commission any of the required information may result in loss of state contracts, non-issuance of registration or non-renewal of registration.

- Transporters should include a certificate of insurance (also known as an ACORD form) indicating that the company has at least $500,000 in Single Limit Liability coverage. Additionally, the certificate of insurance should list TCEQ as the certificate holder.
- Used oil transfer facilities, processors, and re-refiners should include proof of financial assurance as outlined in 30 TAC §324.22.
- Used oil filter storage facilities and processors should include proof of financial assurance as outlined in 30 TAC §328.24(e).

Commentary: *As indicated above you must provide your financial responsibility documentation with this filing. This Guide has a section on the financial responsibility requirement but again if you have any questions call the TCEQ if you have any questions.*

2. **Part B: Customer Information**

The customer (owner/operator) is the individual or organization responsible for the site. The information included in this part should refer to the legal entity that is the owner and operator *of the used oil activities* registered on this form.

12. **Customer Number**
    - Enter the 9-digit Customer Number (CN) here if TCEQ has already assigned one to the organization/individual. If the organization/individual does not have a customer number, leave this item blank. A new CN is assigned only if the owner/operator has not yet been issued a CN. A Customer may be involved with many sites and/or licenses but should only have a single CN.

13. **Owner/Operator Name ***
    - This name should be the complete legal name as filed with the Secretary of State's office or Comptroller of Public Accounts, unless the owner/operator is an individual or sole proprietor. Verify the correct legal name for the organization by searching the Secretary of State and/or Texas Comptroller Websites or by calling their help lines. Secretary of State 512-463-5555 http://direct.sos.state.tx.us/acct/acct-login.asp Texas Comptroller 1-800-252-1386 https://ourcpa.cpa.state.tx.us/coa/Index.html
    - The owner/operator listed may be an individual, trust, firm, joint stock company, federal agency, corporation (including a government corporation), partnership, association, state, municipality, commission, political subdivision of a state, or any interstate body responsible for the overall operation and ownership of a site. Note: This is the legal entity which controls the site rather than the plant or site manager. This is usually a company or business name, but may be an individual.
14. **Type of Customer ***

    Check ONE box that best indicates the Customer's relationship to the site. A Limited Liability Company is considered a Corporation. If none of the boxes fit your situation, check "Other" and write in the relationship.
    - **Corporation:** The Customer meets all of the following:
        - Is legally incorporated under the laws of any state or country
        - Is recognized as a corporation by the Texas Secretary of State
        - Has proper operating authority to operate in Texas
        - Examples of Corporations
            - General Corporations
            - Subchapter S Corporations
            - Limited Liability Companies (LLC)
            - Water Supply Corporations
    - **Individual:**
        - An individual is a person regulated by TCEQ but who has not established a business.
    - **Sole Proprietorship – D.B.A.:**

- The Customer in this case is a business owned by only one person and not incorporated. This business may:
  - Be under the person's name
  - Have its own name ("doing business as", or DBA)
  - Have any number of employees
  - On the form, provide the first name, last name and dba name in this format. (e.g. Bryan Smith dba CoolCleaners)

o **Limited Partnership (LP & LLP):**

- A Texas limited partnership is a partnership formed by two or more persons, having one or more general partners and one or more limited partners. The limited partnership operates in accordance with a partnership agreement, written or oral, of the partners as to the affairs of the limited partnership and the conduct of its business. While the partnership agreement is not filed for public record, the limited partnership must file a certificate of limited partnership with the Texas Secretary of State (SOS). The Secretary of State provides a form for the certificate of limited partnership which meets minimum state law requirements.

o **General Partnership:**

- A general partnership is created when two or more persons associate to carry on a business for profit. A partnership generally operates in accordance with a partnership agreement, but here is no requirement that the agreement be in writing and no state-filing requirement. General partnerships are required to file with the county in which they operate. Submit a tax ID if you have a sales tax permit.

o **Government – Federal, State, County or City:**

- The Customer is either an agency of one of these levels of government or the governmental body itself (e.g. Blanco County, City of Houston, Texas Department of Transportation)

- **Government – Other:**
    - The Customer is a utility district, water district, tribal government, college district, council of governments or river authority. (e.g. Lower Colorado River Authority).
- **Other:**
    - Does not fit any of the previous descriptions (e.g. joint venture, trust, or an estate). Enter a short description of the Customer type in the blank provided.

15. **Federal Tax ID (Tax Information)**
    - All businesses, except some small sole proprietors and individuals should have a federal taxpayer identification number (FEIN). Do not enter a Social Security number in this field. Do not use prefixes, dashes or hyphens. For more information on Federal Tax IDs visit https://www.irs.gov/businesses/small-businesses-self-employed.
16. **TX State Tax ID**
    - Leave this field blank if the Customer has not been issued a TX State Tax ID. If one has been issued, enter it here. Check the State Comptroller's website at https://ourcpa.cpa.state.tx.us/coa/Index.html for a listing of the tax ID.
17. **Texas SOS/CPA Filing Number**
    - Leave this field blank if the Customer has not filed with SOS/CPA as a Corporation, Limited Partnership or non-profit organization. If the Customer has filed with SOS/CPA, enter the Filing/Charter Number. Search for the SOS Filing number at http://www.sos.state.tx.us/corp/sosda/index.shtml or call at 512-463-5555.
18. **Number of Employees**
    - Check one box to indicate the number of employees working for the company listed in #13, including subsidiaries. The number should include all locations, not just the location registering.

3. **Part C: On-site owner/operator contact**

Provide the contact information for the individual who represents your site for used oil management issues, often an environmental supervisor or plant manager. Numbers 19-27 are required for all new submissions.

19. **First Name** *
    - Enter the first name of the individual serving as the owner/operator contact for the site.
20. **Last Name** *
    - Enter the last name of the individual serving as the owner/operator contact for the site.
21. **Title** *
    - Enter the title of the individual serving as the owner/operator contact for the site.
22. **Company Name** *
    - Enter the name of the company where the individual serving as the owner/operator contact works. Usually this is the same as #13.
23. **Street Address** *
    - Enter the street address where general correspondence regarding the registration (e.g biennial reporting and renewals), can be sent. Please ensure that the mailing address is deliverable per USPS at https://tools.usps.com/go/ZipLookupAction_input.
24. **City** *
    - Enter the city that correlates to the street address provided in #22.
25. **State** *
    - Enter the state that correlates to the street address provided in #22.
26. **Zip+4** *
    - Enter the zip+4 code that correlates to the street address provided in #22. Please ensure that the zip+4 is correct per USPS at https://tools.usps.com/go/ZipLookupActioninput.
27. **Telephone** *
    - Enter the telephone number to contact the individual serving as the owner/operator contact.
28. **Fax**
    - Enter the fax number to contact the individual serving as the owner/operator contact.

29. Email
    - Enter the email address to contact the individual serving as the owner/operator contact.

I. Part D: Used Oil/Used Oil Filter Activities (for facility registering)

30. **Used Oil Recycling Activity: ***
    - This question pertains to your used oil recycling activities. Check all the categories that apply.
        - o Transporter: The company/individual registering transports used oil.
        - o Transfer Facility: The site stores used oil for more than 24 hours and less than 35 days as part of the normal course of transporting.
        - o Re-refiner: The site re-refines used oil
        - o Burner of off-spec used oil: The site burns off-spec used oil for energy recovery.
        - o Marketer who first claims used oil meets specs: The site first claims that used oil to be burned for energy recovery meets the used oil specifications.
        - o Marketer who directs shipments to burners: The site directs shipments of off-specification used oil to a used oil burner.
        - o Processor- The site is involved in chemical or physical operations designed to (1) produce fuel oils, lubricants, or other products derived from used oil; and/or (2) make used oil more suitable for production of these products; and/or (3) the site stores used oil for more than 35 days. Processors must also describe the processing method in the field below the checkbox for processor.
31. **Used Oil Filter Recycling Activity: ***
    - This question pertains to your used oil filter recycling activities. Check all the categories that apply.
    - o Transporter: The company/individual registering transports used oil filters.

- o Storage Facility: The site stores used oil filters for less than 120 days as part of the normal course of transporting.
- o Processor- The site prepares used oil filters generated by other persons for recycling, steel recovery, energy recovery, and/or proper disposal. Processors must also describe the processing method in the field below the checkbox for processor.

J. Part G: Certification

A certification page with a signature must be included with all submissions. Your signature certifies the following:

- that the information provided in this form is complete and accurate to the best of your knowledge.
- that you understand that the registration listed on this form will be updated with the information provided herein.
- that you have signature authority to submit this form on behalf of the entity listed as the "Site Name".
- that you will abide by all state and federal regulations governing the collection, management, and recycling of used oil and used oil filters.
- Provide information for the person signing form in #32-38.

32. Name *

- Enter the name of the individual who signed the forms.

33. Title

- Enter the title of the individual who signed the forms.

34. Company

- Enter the company name where the individual who signed the forms works.

35. Telephone

- Enter the telephone number to contact the individual who signed the forms.

36. Fax

- Enter the fax number to contact the individual who signed the forms.

37. Email

- Enter the email address to contact the individual who signed the forms.

38. **Signature** *
    - We do not accept documents that have digital or stamped signatures. Forms for new notifications require an original signature, not a duplicate. Forms for updates and renewals may be duplicates of original forms.

K. Mail completed form to:

- Texas Commission on Environmental Quality
  Registration and Reporting Section MC-129
  PO Box 13087
  Austin, TX 78711-3087

Use the following address for any submissions sent via overnight services:

- Texas Commission on Environmental Quality
  Registration and Reporting Section MC-129
  12100 Park 35 Circle Bldg. D
  Austin, TX 78753

L. Updates and renewals only may be faxed to (512) 239-6410. New notifications submitted by fax will be returned to the sender.

M. For any questions about filling out this form not covered above, or for information about the Used Oil Recycling Program, contact us at (512) 239-6413 or wasteval@tceq.texas.gov.

N. Individuals are entitled to request and review their personal information the agency gathers on its forms. They may also have any errors in their information corrected. To review such information, contact the TCEQ Public Information Section at (512) 239-3282.

Commentary: *When the form is complete and all additional documents are prepared mail to the above address WITH your Core Data Form.*

Form 2 – TCEQ Used Oil Handler and/or Used Oil Filter Handlers Form (TCEQ-10062)

For TCEQ use only: MSW UOR/A 8_______ /CO/________________________/INITIAL-UPDATE

TCEQ

Registration for Used Oil Handler and/or Used Oil Filter Handlers

Bold items required for all notifications. * *items required for all new notifications.* Please print clearly or type.

To obtain a Used Oil/Used Oil Filter Handler registration number, an applicant must comply with applicable federal, state and local regulations. Documentation demonstrating compliance must accompany the application, **including financial assurance** as applicable in 30 TAC Chapter 324 and 40 CFR Part 279 for Used Oil Handlers and 30 TAC Chapter 328 Subchapter D for Used Oil Filter Handlers.

Instructions for filling out this form.

Used Oil Handler Registration # (required for all updates):A 8_______ **Effective Date:**__________

Submission Reason: ☐ New ☐ Update ☐ Renewal (Filter Handlers only)

Part A: Site Information (refers to the location being registered)

1. EPA ID:________________________
2. Regulated Entity Number: RN 1________________________
3. **Site Name:** ________________________

Site Address:*

4. Street Address:________________________

   OR (if site does not have a street address)

5. Site description:________________________

   AND

6. City*:________________________
7. State*: TX________ 8. Zip+4:*__________ - ________
9. County*:________________________
10. Primary North American Industry Classification System (NAICS) Code:*__________
11. Site Land Type:*

☐ Private	☐ Federal	☐ State
☐ County	☐ Tribal	☐ Other
☐ District	☐ Municipal	____________

Financial Assurance

All handlers must comply with the demonstration of financial responsibility at the time of registration. The agency accepts trust funds, surety bonds, letters of credit or financial test and corporate guarantee. Failure to disclose to the Commission any of the required information may result in loss of state contracts, non-issuance of registration or non-renewal of registration.

For TCEQ use only: MSW UOR/A 8________ /CO/____________________________/INITIAL-UPDATE

Used Oil Handler Registration # (required for all updates):A 8________ **Effective Date:**__________

Site Name: ____

Submission Reason: ☐ New ☐ Update ☐ Renewal (Filter Handlers only)

**(Items marked with * are required for all new notifications)*

Part B: Customer Information (for the facility registering)

12. Customer number: CN 6____________________

13. Owner/operator name:*____

(Legal Name as filed with the Secretary of State's office)

14. Type of customer:*

☐ Corporation
☐ Individual
☐ Sole Proprietorship- D.B.A.
☐ City Government
☐ County Government
☐ Federal Government
☐ State Government
☐ Other Government
☐ General Partnership
☐ Limited Partnership
☐ Other:____________________

Tax information (Do not enter a Social Security number in this field. For individual ownership type, go to #19)

15. Federal Tax ID:____________________

16. TX State Franchise Tax ID:____________________

17. Texas Secretary of State Filing Number:____________________

18. Number of Employees: ☐ 0-20 ☐ 21-100 ☐ 101-250 ☐ 251-500 ☐ 501 and higher

Part C: On site owner/operator contact*

19. First name:*____________________ 20. Last name:* ____________________

21. Title:*____________________

22. Company Name:* ____

23. Street Address:*____

24. City:*____

25. State:* TX__________ 26. Zip+4:*__________ - ______

27. Telephone:*____________________ 28. Fax: ____________________

29. Email:____

TCEQ Form #10062 (Rev. 06/2017)
For TCEQ Use only: Log #

Page___of____

For TCEQ use only: MSW UOR/A 8_______ /CO/________________________/INITIAL-UPDATE

Used Oil Handler Registration # (required for all updates):A 8_______ **Effective Date:**___________

Site Name: ____

Submission Reason: ☐ New ☐ Update ☐ Renewal (Filter Handlers only)

**(Items marked with * are required for all new notifications)*

Part D: Used Oil/Used Oil Filter Activities (for the facility registering)

30. Used Oil Recycling Activity:* (check all that apply)

☐ Transporter ☐ Transfer Facility ☐ Re-refiner ☐ Burner of off-spec used oil

☐ Marketer who first claims used oil meets specs

☐ Marketer who directs shipments to burners

☐ Processor - describe processing method below:

31. Used Oil Filter Recycling Activity:* (check all that apply)

☐ Transporter ☐ Storage Facility

☐ Processor - describe processing method below:

Part E: Outstanding Fees

Are there any outstanding fees or penalties due to the TCEQ from this owner? ☐ Yes ☐ No
If yes, provide the amount $____________ and the identifying account number____________
The registration form will not be processed until all delinquent fees and/or penalties owed to the TCEQ are paid.

Part F: Facility Certification (required for all transfer/storage facilities, processors, and re refiners)

A simple diagram (blueline, CAD type, or hand drawn diagrams are acceptable) indicating all areas where used oil and used oil filters are transported, stored or processed that corresponds to the below certified square footage for the facility. The amount of financial assurance required is based on the square footage indicated below.

Certification: This document and all attachments were prepared under my direction or supervision according to a system designed to assure that qualified personnel properly gathered and evaluated the information submitted. Based on my inquiry of the person or persons who manage the system, or those persons directly responsible for gathering the information, the information submitted is, to the best of my knowledge and belief, true, accurate, and complete. I am aware that there are significant penalties for submitting false information, including the possibility of fines and imprisonment for knowing violations.

I certify by my signature below that this facility is______ft. by______ ft. = ________square ft.

TCEQ Form #10062 (Rev. 06/2017)
For TCEQ Use only: Log # Page___of____

For TCEQ use only: MSW UOR/A8________/CO/________________________/INITIAL-UPDATE

Part G: Certification (Must be included with all submissions; items marked with * are required.)

By my signature below, I certify, to the best of my knowledge, that the information provided in this form is complete and accurate. I understand that the registration listed on this form will be updated with the information provided herein. I also certify that I have signature authority to submit this form on behalf of the entity listed as the "Site Name", and that I will abide by all state and federal regulations governing the collection, management, and recycling of used oil and used oil filters.

32. Name:* ____________________________________

33. Title: __________________ 34. Company: __________________

35. Telephone: __________________ 36. Fax: __________________

37. Email: ____________________________________

38. Signature:* ____________________________________

Mail completed form to:
Texas Commission on Environmental Quality
Registration and Reporting Section MC-129
PO Box 13087
Austin, TX 78711-3087

Please use the following address for any submissions sent via overnight services:
Texas Commission on Environmental Quality
Registration and Reporting Section MC-129
12100 Park 35 Circle Bldg D
Austin, TX 78753

Fax **updates and renewals only** to (512) 239-6410. New registrations submitted by fax will be returned to the sender.

For any questions about filling out this form not covered in the instructions, or for information about the Used Oil and Used Oil Filter Handler Program, please contact us at (512) 239-6413 or wasteval@tceq.texas.gov.

Individuals are entitled to request and review their personal information the agency gathers on its forms. They may also have any errors in their information corrected. To review such information, please contact the TCEQ Public Information Section at (512) 239-3282.

PART IX: ALL COLLECTION CENTERS TCEQ COLLECTION CENTER REGISTRATION

This section covers the required Texas registration as a Used Oil Collection Center that is a facility that accepts, aggregates, or stores used oil received from generators such as service stations, governments, or businesses (in shipments of not more than 55 gallons of used oil at a time). A center may /licensed/permitted/recognized by a state/co accept used oil from household do-it-yourselfers (DIYs). §279.1 Definitions. Used oil collection center means any site or facility that is registered entity/municipal government to manage used oil and accepts/aggregates and stores used oil collected from used oil generators regulated under subpart C of this part who bring used oil to the collection center in shipments of no more than 55 gallons under the provisions of §279.24. Used oil collection centers may also accept used oil from household do-it yourselfers.

There is an exception to the registration requirement if all you collect is used oil filters as stated by the TCEQ on their web site addressed below where you can also obtain the registration form "A used oil filter collection center (UOFCC) is a facility that accepts, stores, and manages used oil filters collected from do-it-yourselfers. The UOFCCs should remove the free flowing oil from the filters they collect to the greatest extent feasible. Texas Health & Safety Code, Chapter 371 Subchapter E (Sections 371.101-105) UOFCCs do not have to register or report their UOF activities to the TCEQ. TCEQ encourages voluntary UOFCC to help do-it-yourselfers recycle their used oil filters.

> ***Commentary:*** *The requirements for posting as a collection center see the Used Oil Handlers and Used Oil Containing Materials Management section in this guide for other requirements.*

TCEQ instructions for completing this form are provided as a reference document 165. Provided below are instructions based on TCEQ instructions with additional Guide information italicized. However, as always as you fill out the form and you have any questions you should call the TCEQ at the number found at the bottom of the form. You can find a copy of the form at:

https://www.tceq.texas.gov/permitting/registration/used_oil/collection.html

along with other collection center requirements. A copy of the instructions can be found at https://www.tceq.texas.gov/assets/public/permitting/rrr/forms/10062inst.pdf.

Form TCEQ-0533 Instructions: (Rev. 05/2017)

Used Oil Collection Centers as defined in 30 Texas Administrative Code (TAC) §324.7 should register via this form. Use this form for new (initial) notifications as well as subsequent notifications (updates) and renewals. For updates, submit Part F for each submission. Otherwise, submit only the pages with the parts that pertain to the updates needed.

- Read the instructions for additional information to determine if this form should be used for your site. Please note that incomplete forms will not be processed and may be returned to the sender for completion.
- **Items in bold must be filled out for all notifications. Items with * are required for all new notifications. Print clearly or type.**
- You must post and maintain durable and legible signs indicating that you are a public or household used oil collection center. You may print form GI170 for use as signage.
- Signage is found following the form below the instructions.

A. **Form Header:**

1. **Used Oil Collection Registration #:**
    - For new notifications, leave the Used Oil Collection Registration Number blank. Your company will be assigned a Used Oil Collection Registration number when TCEQ processes the notification.
    - For updates, enter the last five digits of the 6-character Used Oil Collection Registration Number assigned to your company. Since all Used Oil Collection Registration numbers begin with the letter C, the form has the letter C preprinted. If you are unsure of the Used Oil Collection Registration Number, check the TCEQ Central Registry search, or you

can find it on your notice of registration.

- The Used Oil Collection Registration Number is site specific. If the company moves to a new site location, the company should submit a new notification form to obtain a new Used Oil Collection Registration Number for the new site. Do not use numbers assigned to another site address.
- When filled out via computer, the Used Oil Collection Registration Number will automatically populate on the other pages of the form.

Commentary: *Ensure that you use the same company name and facility name filed on your Core Data form either prior or the one you are submitting with this form.*

B. **Effective Date: (Bold)**

- Indicate the date that new the information provided went into effect. The date cannot be a future date. When this form is filled out via computer, the effective date will automatically populate on the other pages of the form.

C. **Submission Reason: (Bold)** (for Parts A-E)

- New: Mark "New" if this is the first notification for this company/facility or individual. Items marked with * are required for all new notifications.
- Update: Mark "Update" if this site already has a Used Oil Collection Registration and updated information is being provided about the company.
- Submit Part F for each submission. Otherwise, submit only the pages with the parts that pertain to the updates needed.
- Updates can include change of address or ownership, changing contact information, or other information included on this form.
- When this form is filled out via computer, the submission reason will automatically populate on the other pages of the form.

D. **Part A: Site / Regulatory Entity Information** (refers to the location being registered)

- This is the regulated site for the company requesting the registration. The site is the location where used oil collection occurs.

1. **Regulated Entity Number : RN**
   - Enter the 9-digit Regulated Entity Number (RN) assigned by TCEQ if this company already has one. Leave this item blank if the site does not have a RN.
2. **Facility Name: (Bold)**
   - Enter the name the site will be known as/by, such as a DBA. Be sure to use a fully descriptive name. (e.g. ABC Company Warehouse 4, not Warehouse 4.) If a site/company does not have an existing RN, the Registration and Reporting section will use the company name as the regulated entity name.
   - In accordance with the TCEQ Central Registry Naming Standards, a company/site name should not contain INC., CO, LP, etc. in the name unless it is an integral part of the company/site name. This standard also applies to all the following organizational endings: Associated, Associates, Association, ASSOC, Company, Corporation, CO, CORP, Incorporated, INC., LC, LLC, Limited, Liability Company, Limited Company, LTD, Limited Partnership, Limited, LP, PA, Professional, Limited Liability Company, Professional, PLLC, PC
   - Notification paperwork received with any of the above organizational endings in the facility name field may be returned for correction unless a justification is provided. To eliminate any delay in processing your notification, ensure you follow these standards. For any questions regarding standards for organizational endings, contact the TCEQ Central Registry at 512-239-5175.
   - When this form is filled out via computer, the company name will automatically populate to the other pages of the form.
3. **Site Address / Street Address***
   - Enter the physical street address of the facility/site, which is the loca-

tion where used oil collection occurs. Do not provide a P.O. Box.

4. **City***
   - Enter the city for the physical location of the site.
5. **State***
   - Enter the state for the physical location of the site.
6. **Zip+4***
   - Enter the zip code for the physical location of the site.
7. **County***
   - Enter the county in which the site is located.
8. **Primary NAICS***
   - All new notifications must provide a primary North American Industrial Classification System (NAICS) code. The code should indicate the classification for the ultimate business use of the site. Go to www.census.gov/eos/www/naics/ for a list of NAICS codes.

E. **Part B: Customer Information** (for the company registering)

- The customer is the individual or organization responsible for the company. The information included in this part should refer to the legal entity that is the owner and operator *of the used oil activities* registered on this form.

9. **Customer Number**
   - Enter the 9-digit Customer Number (CN) here if TCEQ has already assigned one to the organization/individual. If the organization/individual does not have a customer number, leave this item blank. A new CN is assigned only if the company has not yet been issued a CN. A Customer may be involved with many sites and/or licenses but should only have a single CN.
10. **Owner/Operator Name***
    - This name should be the complete legal name as filed with the Secretary of State's office or Comptroller of Public Accounts, unless the registrant is an individual or sole proprietor. Verify the correct legal name for the organization/individual by searching the Secretary of State and/or Texas Comptroller Websites or by calling

their help lines.
Secretary of State 512-463-5555
http://direct.sos.state.tx.us/acct/acct-login.asp

Texas Comptroller 1-800-252-1386
https://ourcpa.cpa.state.tx.us/coa/Index.html

- The company listed may be an individual, trust, firm, joint stock company, federal agency, corporation (including a government corporation), and partnership, association, state, municipality, and commission, political subdivision of a state, or any interstate body responsible for the overall operation and ownership of the company.

11. **Type of Customer***
    - Check ONE box that best indicates the Customer's organization type. A Limited Liability Company is considered a Corporation. If none of the boxes fit your situation, check "Other" and write in the type.
    - o **Corporation:** The Customer meets all of the following:
        - ▪ Is legally incorporated under the laws of any state or country
        - ▪ Is recognized as a corporation by the Texas Secretary of State
        - ▪ Has proper operating authority to operate in Texas
        - ▪ Examples of Corporations
            - General Corporations
            - Subchapter S Corporations
            - Limited Liability Companies (LLC)
            - Water Supply Corporations
    - o **Individual:**
        - An individual is a person regulated by TCEQ but who has not established a business.
    - o **Sole Proprietorship – D.B.A.:**
        - The Customer in this case is a business owned by only one

person and not incorporated. This business may:

  - Be under the person's name
  - Have its own name ("doing business as", or DBA)
  - Have any number of employees
  - On the form, provide the first name, last name and dba name in this format. (e.g. Bryan Smith dba CoolCleaners)

- **Limited Partnership (LP & LLP):**
  - A Texas limited partnership is a partnership formed by two or more persons, having one or more general partners and one or more limited partners. The limited partnership operates in accordance with a partnership agreement, written or oral, of the partners as to the affairs of the limited partnership and the conduct of its business. While the partnership agreement is not filed for public record, the limited partnership must file a certificate of limited partnership with the Texas Secretary of State (SOS). The Secretary of State provides a form for the certificate of limited partnership which meets minimum state law requirements.
- **General Partnership:**
- A general partnership is created when two or more persons associate to carry on a business for profit. A partnership generally operates in accordance with a partnership agreement, but here is no requirement that the agreement be in writing and no state-filing requirement. General partnerships are required to file with the county in which they operate. Submit a tax ID if you have a sales tax permit.
- **Government – Federal, State, County or City:**
  - The Customer is either an agency of one of these levels of government or the governmental body itself (e.g. Blanco County, City of Houston, Texas Department of Transportation)
- **Government – Other:**

- The Customer is a utility district, water district, tribal government, college district, council of governments or river authority. (e.g. Lower Colorado River Authority).

o **Other:**

- Does not fit any of the previous descriptions (e.g. joint venture, trust, or an estate). Enter a short description of the Customer type in the blank provided.

F. **Tax Information (Do not enter a Social Security number in this field.** For individual ownership type, go to #16)

12. **Federal Tax ID**
    - All businesses, except some small sole proprietors and individuals should have a federal taxpayer identification number (FEIN). Do not use prefixes, dashes or hyphens. For more information on Federal Tax IDs visit http://www.irs.gov/businesses/small/article/0,,id=97872,00.html.

13. **TX State Tax ID**
    - Leave this field blank if the Customer has not been issued a TX State Tax ID. If one has been issued, enter it here. Check the State Comptroller's website at https://ourcpa.cpa.state.tx.us/coa/Index.html for a listing of the tax ID.

14. **Texas SOS/CPA Filing Number**
    - Leave this field blank if the Customer has not filed with SOS/CPA as a Corporation, Limited Partnership or non-profit organization. If the Customer has filed with SOS/CPA, enter the Filing/Charter Number. Search for the SOS Filing number at http://www.sos.state.tx.us/corp/sosda/index.shtml or call at 512-463-5555.

G. **Part C: On-site owner/operator contact**

- The information provided here should refer to the individual to contact at the site of collection. TCEQ will send all correspondence regarding the registration to the address listed in Part C unless Part D is also filled out. If Part D is filled out, all correspondence will go to the contact listed in Part D. Numbers 15-16 and 18-23 are required for all new submissions.

15. **First Name***
    - Enter the first name of the individual serving as the contact for the registration.
16. **Last Name***
    - Enter the last name of the individual serving as the contact for the registration.
17. **Title***
    - Enter the title of the individual serving as the contact for the registration.
18. **Company Name***
    - Enter the name of the company where the individual serving as the contact for the registration works. Usually this is the same as #10.
19. **Street Address***
    - Enter the street address where general correspondence regarding the registration can be sent. Please ensure that the mailing address is deliverable per USPS at https://tools.usps.com/go/ZipLookupAction_input.
20. **City***
    - Enter the city that correlates to the street address provided in #19.
21. **State***
    - Enter the state that correlates to the street address provided in #19.
22. **Zip+4***
    - Enter the zip+4 code that correlates to the street address provided in #19. Please ensure that the zip+4 is correct per USPS at https://tools.usps.com/go/ZipLookupActioninput.
23. **Telephone***
    - Enter the telephone number to contact the individual serving as the contact for the registration.
24. **Fax**

- Enter the fax number to contact the individual serving as the contact for the registration.

25. **Email**
    - Enter the email address to contact the individual serving as the contact for the registration.

H. Part D: Owner Representative Contact

- If the owner representative contact is different than the on-site owner/operator contact, indicate the contact information for the owner representative contact in #26-36. TCEQ will send all correspondence regarding the registration to the address listed in Part D if Part D is filled out. In order register a separate owner representative contact, provide at least the information in #26-34.

26. **First Name**
    - Enter the first name of the individual serving as the owner representative contact.
27. **Last Name**
    - Enter the last name of the individual serving as the owner representative contact.
28. **Title**
    - Enter the title of the individual serving as the owner representative contact.
29. **Company Name**
    - Enter the name of the company where the individual serving as the owner representative contact works. Usually this is the same as #10.
30. **Street Address**
    - Enter the street address where general correspondence regarding the registration can be sent. Please ensure that the mailing address is deliverable per USPS at https://tools.usps.com/go/ZipLookupAction_input.
31. **City**
    - Enter the city that correlates to the street address provided in #30.

32. **State**
    - Enter the state that correlates to the street address provided in #30.
33. **Zip+4**
    - Enter the zip+4 code that correlates to the street address provided in #30. Please ensure that the zip+4 is correct per USPS at https://tools.usps.com/go/ZipLookupActioninput.
34. **Telephone**
    - Enter the telephone number to contact the individual serving as the owner representative.
35. **Fax**
    - Enter the fax number to contact the individual serving as the owner representative.
36. **Email**
    - Enter the email address to contact the individual serving as the owner representative.

I. **Part E: Waste activities**

37. **Facility Type**
    - Indicate all facility types that apply to the site where used oil collection occurs.
    - Distributors of Oil Products must provide a copy of their permit from the Texas State Comptroller's Office. For additional information about the permits for distributors of oil products (Automotive Oil Sales Fee), visit https://comptroller.texas.gov/taxes/auto-oil/.
    - Note: If you store more than six-55 gallon drums or 1 bulk container of used oil filters, you must register as a Used Oil Filter Storage Facility.

J. **Part F: Certification**

- A certification page with a signature must be included with all submissions. By signing the form, you certify that the information provided on the form is complete and accurate. You also certify that you have signature authority to submit the form on behalf of the entity listed as the "Facility Name", and that you will abide by all state and federal regulations governing the collection, management, and recycling of used oil and used oil filters.

Commentary: *Make sure you understand and comply with this requirement. "By my signature below, I certify, to the best of my knowledge, that the information provided in this form is complete and accurate. I understand that the registration listed on this form will be updated with the information provided herein. I also certify that I have signature authority to submit this form on behalf of the entity listed as the "Site Name", and that I will abide by all state and federal regulations governing the collection, management, and recycling of used oil and used oil filters." As stated under the certification section.*

- **Name:** Enter the name of the individual who signed the forms.
- **Title:** Enter the title of the individual who signed the forms.
- **Telephone:** Enter the telephone number to contact the individual who signed the forms.
- **Company:** Enter the company name where the individual who signed the forms works.
- **Fax:** Enter the fax number to contact the individual who signed the forms.
- **Email:** Enter the email address to contact the individual who signed the forms.
- **Signature:** We do not accept documents that have digital or stamped signatures. Forms for new notifications require an original signature, not a duplicate. Forms for updates may be duplicates of original forms.
- **Mail completed form to:**

Texas Commission on Environmental Quality
Registration and Reporting Section MC-129
PO Box 13087
Austin, TX 78711-3088

- *Use the following address for any submissions sent via overnight services:*

Texas Commission on Environmental Quality
Registration and Reporting Section MC-129
12100 Park 35 Circle Bldg. D
Austin, TX 78753

- Fax **updates and renewals only** to (512) 239-6410. Any new registrations submitted by fax will be returned to the sender.
- For any questions about filling out this form not covered above, or for information about the Used Oil Collection Center Program, contact us at (512) 239-6413 or wasteval@tceq.texas.gov.
- Individuals are entitled to request and review their personal information the agency gathers on its forms. They may also have any errors in their information corrected. To review such information, contact the TCEQ Public Information Section at (512) 239-3282.

Form 3 – TCEQ Registration For Used Oil Collection Centers (TCEQ – 0533)

For TCEQ use only: MSW UOR/C 8________*/CO/*______________________*/INITIAL-UPDATE*

TCEQ

Registration for Used Oil Collection Center

Bold items required for all notifications. * items required for all new notifications.
Please print clearly or type.

You must post and maintain durable and legible signs indicating that you are a public or household used oil collection center. You may print form GI-170 for use as signage.

Instructions for filling out this form.

Used Oil Collection Registration # (required for all updates): C 8__________ **Effective Date:** __________

Submission Reason: ☐ New ☐ Update ☐ Renewal (expires on even years for filters)

Part A: Site/Regulated Entity Information (refers to the location being registered)

1. Regulated Entity Number: RN 1__________
2. **Facility Name:** __________

Site Address:*

3. Street Address:__________
4. City*:__________
5. State*: TX__________ 6. Zip+4:* __________ - __________
7. County*:__________
8. Primary North American Industry Classification System (NAICS) Code:* __________

Part B: Customer Information (for the facility registering)

9. Customer number: CN 6__________
10. Owner/operator name:* __________
    (Legal Name as filed with the Secretary of State's office)
11. Type of customer:*

☐ Corporation	☐ State Government
☐ Individual	☐ Other Government
☐ Sole Proprietorship- D.B.A.	☐ General Partnership
☐ City Government	☐ Limited Partnership
☐ County Government	☐ Other: __________
☐ Federal Government	

Tax information (Do not enter a Social Security number in this field. For individual ownership type, go to #15)

12. Federal Tax ID:__________
13. TX State Franchise Tax ID:__________
14. Texas Secretary of State Filing Number: __________

For TCEQ use only: MSW UOR/C 8________/CO/________________________/INITIAL-UPDATE

Used Oil Collection Registration # (required for all updates): C 8__________ **Effective Date:** __________

Site Name: ____

Submission Reason: ☐ New ☐ Update ☐ Renewal (expires on even years for filters)

*(Items marked with * are required for all new notifications.)

Part C: On site owner/operator contact

15. First name:* ____________________ 16. Last name:* ____________________

17. Title: ____________________ 18. Company Name:* ____________________

19. Street Address:* ____

20. City:* ____

21. State:* TX 22. Zip+4: __________ - ________

23. Telephone:* ____________________ 24. Fax: ____________________

25. Email: ____

Part D: Owner Representative Contact (optional, if different than contact in Part B)

26. First name: ____________________ 27. Last name: ____________________

28. Title: ____________________ 29. Company Name: ____________________

30. Street Address: ____

31. City: ____

32. State: TX 33. Zip+4: __________ - ________

34. Telephone: ____________________ 35. Fax: ____________________

36. Email: ____

Part E: Waste activities (for the facility registering)

37. Facility Type:* (check **all** that apply)

☐ Collection Center for households/do-it-yourselfers (DIY)

☐ Collection Center for Other Generators (Non-households, such as other businesses, governments, service stations, etc.)

☐ Distributor of Oil Products (Must provide a copy of permit from the State Comptroller's Office)

Note: If you store more than six-55 gallon drums or 1 bulk container of used oil filters, you must register as a Used Oil Filter Storage Facility.

For TCEQ use only: MSW UOR/C 8________/*CO/*__________________________/*INITIAL-UPDATE*

Used Oil Collection Registration # (required for all updates): C 8________ **Effective Date:** ____________

Site Name: ____

Part F: Certification (Must be included with all submissions; bold items are required.)

By my signature below, I certify, to the best of my knowledge, that the information provided in this form is complete and accurate. I understand that the registration listed on this form will be updated with the information provided herein. I also certify that I have signature authority to submit this form on behalf of the entity listed as the "Site Name", and that I will abide by all state and federal regulations governing the collection, management, and recycling of used oil and used oil filters.

Name: ____

Title: ______________________ Company: ______________________

Telephone: ______________________ Fax: ______________________

Email: ____

Signature: ____

Mail completed form to:

Texas Commission on Environmental Quality
Registration and Reporting Section MC-129
PO Box 13087
Austin, TX 78711-3087

Please use the following address for any submissions sent via overnight services:

Texas Commission on Environmental Quality
Registration and Reporting Section MC-129
12100 Park 35 Circle Bldg D
Austin, TX 78753

Updates only may be faxed to (512) 239-6410. Initial registrations submitted by fax will be returned to the sender.

For any questions about filling out this form not covered in the instructions, or for information about the Used Oil Collection Center Registration Program, please contact us at (512) 239-6413 or wasteval@tceq.texas.gov.

Individuals are entitled to request and review their personal information the agency gathers on its forms. They may also have any errors in their information corrected. To review such information, please contact the TCEQ Public Information Section at (512) 239-3282.

HOUSEHOLD
USED OIL
COLLECTION
CENTER
Recycle Your Used Oil
& Used Oil Filters Here
IT IS ILLEGAL
to improperly dispose of used oil
and used oil filters in Texas.*
Keep used oil out of our water supplies—
Turn it into fuel and other useful products.
How? Recycle your used oil and used oil filters
here or at a collection center near you.
TCEQ REGISTRATION NUMBER ____________
TEXAS COMMISSION ON
ENVIRONMENTAL QUALITY
*See Section 371.041of the Texas Health and Safety Code.

CENTRO DE
RECOLECCIÓN
DE ACEITE USADO
DEL HOGAR
Recicle su aceite usado y
filtros de aceite usados aquí.
ES CONTRA LA LEY
deshacerse indebidamente de aceite usado
y de filtros de aceite usados en Texas.*
No permita que aceite usado llegue a
nuestras reservas de agua—Conviértalo
en combustible y otros productos útiles.
¿Cómo? Recicle su aceite usado y filtros
de aceite usados aquí o en un centro de
recolección cercano a usted.
NÚMERO DE REGISTRO DE LA TCEQ ________
COMISIÓN DE CALIDAD
AMBIENTAL DE TEXAS
*Vea la Sección 371.041 del Texas Health and Safety
Code (Código de Salud y Seguridad de Texas).

PART X: EPA NUMBER APPLICATION AND UTILIZATION BY TEXAS USED OIL FACILITIES

EPA Form 8700-12 Form Instructions (166)

"Used oil handlers are used oil transporters, transfer facilities, processors/re-refiners, and burners of off-specification used oil. If you are a used oil handler you must register one time through the TCEQ using the EPA's Notification of Regulated Waste Activity Form (EPA 8700-12) and the TCEQ's Registration for Used Oil Handler and/or Used Oil Filter Handlers Form (TCEQ-10062). This form must be submitted with the Core Data Form (TCEQ-10400)."

> ***Commentary:*** *This is an EPA form as indicated but is required to be completed and sent to the TCEQ* ***with all other forms on your first notice as a Used Oil Handler*** *as you must obtain an EPA ID number. The following are the instructions as provided by the EPA but have been significantly edited to identify typical issues addressed by a Used Oil Handler. A complete copy of the instructions, Instructions for Filling Out the RCRA Subtitle C Site Identification Form (Site ID Form)* ***(4)*** *is available from the EPA website at https://www.epa.gov/hwgenerators/instructions-and-form-hazardous-waste-generators-transporters-and-treatment-storage-and.*

EPA's Notification of Regulated Waste Activity Form (EPA 8700-12)

If you download the complete instructions you will also find numerous website references accessible from the instructions that are helpful in understanding RCRA issues as a generator regardless of your class of generator.

Note: *Regarding Instructions for Filling Out the RCRA Subtitle C Site Identification Form (Site ID Form): The information provided below is a redacted version of the actual instructions and the format for those instructions is maintained as much as possible for any referencing you may do.*

In addition to the EPA Form 8700-12 requirements and instructions the General Information is also included for reference. Significant of the additional information is the web site addresses for RCRA Online, State Contacts, Compliance Assistance Centers and the EPA Small Business Ombudsman Office that may also be helpful.

Insure that you address all the sub forms included in the overall Form 8700-12 as several may be not applicable to your current activities but require you to identify that they do not apply.

Form 4 – EPA Form 8700-12

GENERAL INFORMATION

The U.S. Environmental Protection Agency's (EPA's) mission to protect human health and the environment includes the responsibility to effectively manage, with the States, the nation's hazardous waste facilities regulated under the Resource Conservation and Recovery Act of 1976 (RCRA), as amended by the Hazardous and Solid Waste Amendments of 1984 (HSWA). As part of this task, the EPA and the States:

- Collect and maintain information about sites that are conducting RCRA Subtitle C activities via the RCRA Subtitle C Site Identification Form (8700-12);
- Collect and maintain information about the generation, management, and final disposition of the nation's hazardous waste via the Hazardous Waste Report Form (8700-13 A/B); and
- Collect permit information from owners and operators of RCRA facilities where hazardous waste is treated, stored, or disposed via the Hazardous Waste Permit Part A Form (8700-23).

NOTE	Although this document contains information and instructions for completing the forms listed above, it should not be considered a substitute for the regulations in Title 40 of the Code of Federal Regulations (40 CFR). Rather it should be considered a supplement to the regulations and provide additional information not contained in 40 CFR. As a handler of regulated waste, you are responsible for learning and complying with all requirements that apply to you and your regulated waste activities. In addition, this document and the regulations in 40 CFR address only the Federal hazardous waste program. Many States may have notification requirements that differ from the Federal requirements; therefore, those States may use the Federal forms or may choose to use a State form that requires information not requested in the Federal EPA form. Again, it is your responsibility to make sure that you have completed and submitted all forms required under the Federal or your respective State program.

This document is separated into three main sections Notification of RCRA Subtitle C Activities (Site Identification Form), Hazardous Waste Report, and RCRA Hazardous Waste Part A Permit Application. It supersedes all previous documents titled Notification of RCRA Subtitle C Activity, 2015 Hazardous Waste Report, and RCRA Hazardous Waste Part A Permit Application. Please read each section carefully and follow the instructions provided for each applicable set of forms.

WHERE TO GET HELP

FEDERAL REGULATIONS

The Federal regulations can be found at: http://www.gpo.gov/fdsys/.

RCRA LAWS AND REGULATIONS

The RCRA overview, tools, resources, etc. can be found at: https://www.epa.gov/rcra.

RCRA ONLINE

The RCRA Online tool is designed to enable users to locate documents, including publications and other outreach materials that cover a wide range of RCRA issues and topics. The tool can be found at: https://yosemite.epa.gov/osw/rcra.nsf/how+to+use?OpenForm

STATE CONTACTS

We have listed the addresses and phone numbers of the contacts in each State who can answer your questions and help you understand the Federal and State requirements that apply to you. This contact list is located at: https://rcrainfo.epa.gov/rcrainfoweb/documents/contacts.pdf.

Authorized States may have state-specific forms and instructions for reporting and program requirements that are more stringent and broader-in-scope than the federal requirements. Please check with your State contact listed in the above link.

COMPLIANCE ASSISTANCE CENTERS

The EPA's Compliance Assistance Centers help businesses, colleges and universities, local governments, tribes and federal facilities understand and comply with environmental requirements and save money through pollution prevention techniques. Visit the Compliance Assistance Centers at: https://www.epa.gov/compliance/compliance-assistance-centers for a comprehensive source of compliance assistance information and resources.

EPA SMALL BUSINESS OMBUDSMAN OFFICE

1-800-368-5888.

YOUR TRADE ASSOCIATION

If you are a member of an industry-specific trade association, they may have information regarding hazardous wastes that are generated by other members.

CONFIDENTIAL BUSINESS INFORMATION (CBI)

All information you submit via the forms provided in this document can be released to the public, per the Freedom of Information Act, unless it is determined to be confidential by the EPA pursuant to 40 CFR Part 2.

You may not withhold information from the Administrator of the EPA because it is confidential. However, when the Administrator is requested to consider information confidential, it must be treated according to the EPA regulations contained in 40 CFR Part 2, Subpart B. These regulations provide that a business may, if it desires, assert a claim of business confidentiality covering all or part of the information furnished to the EPA. 40 CFR 2.203(b) explains how to assert a claim.

The EPA will treat information covered by such a claim in accordance with the procedures set forth in 40 CFR Part 2, Subpart B. If someone requests release of information covered by a claim of confidentiality, or if the EPA otherwise decides to make a determination as to whether such information is entitled to confidential treatment, the EPA will notify the business. The EPA will not disclose information as to when a claim of confidentiality has been made except to the extent of and in accordance with 40 CFR Part 2, Subpart B. However, if the business does not claim confidentiality when it furnishes the information, the EPA may make the information available to the public without notice to the business.

If your State is authorized to conduct the RCRA Subtitle C program, check with the State if you wish to assert a claim of business confidentiality on your submission. The State may have specific procedures for asserting a claim.

FILLING OUT THE FORMS

SYMBOLS

LIST

The **LIST** symbol denotes references to relevant code lists. Please use only the codes included in the instructions or in the lists of codes provided.

SKIP INSTRUCTIONS

The ➡ symbol denotes directions to skip to the next appropriate section or item to be completed, given certain responses to some questions.

NOTE

The **NOTE** symbol denotes explanatory text of information relevant to filling out the forms.

ALPHANUMERIC FIELDS

Valid characters for alphanumeric fields are limited to:

`~!@#$%^&*()_-+={}[]|\:;'",.?/1234567890ABCDEFGHIJKLMNOPQRSTUVWXYZ

Invalid characters for alphanumeric fields include: <>

If the "<" or ">" symbols are used to indicate less than or greater than, it is recommended that these symbols be replaced with "LT" or "GT."

COMMENTS SECTION OF FORMS

Use the Comments section where applicable to clarify or continue any entry. For the general comment sections, reference the section number and item letter of the entry that is being continued. For example, if your site has more federally regulated hazardous waste codes than can fit in Item 10.B, enter the remaining waste codes in the Comments section and cross-reference Item 10.B. For example, you would write: "Item 10.B, continued: D007."

PAGE NUMBERING OF FORMS

When you have filled out all the appropriate forms on your RCRA Subtitle C submission, number the pages (each piece of paper is a page) consecutively throughout your submission. **Do not** number each set of forms separately, but rather number each page sequentially. The individual page number and the total number of pages in your submission should appear at the bottom of each page (e.g., Page 1 of 7, Page 2 of 7).

If it is necessary to continue information from one form onto another page, make additional copies of the form and number the additional pages with the same page number as the first page, followed by a letter (e.g., page 27, page 27a, page 28, page 28a, 28b). When continuing information on a supplemental page, enter only the information that is being continued.

PAPER FORM REPORTING

Each form is included in this document. If submitting paper copy, photocopy as many forms as you need to complete your submission. Make copies **after** you have written your site name and EPA Identification Number in the top left-hand corner of the form, but **before** you begin filling out the form.

After you have finish filling out the forms, keep a copy for a period of at least three years from the due date of the report as required by 40 CFR 262.40(b).

ELECTRONIC REPORTING

The EPA encourages electronic reporting of RCRA Subtitle C information. Facilities can now enter data via electronic submissions by using the RCRAInfo Industry Application (RIA). The RIA allows for Site Identification Form submissions (myRCRAid) and Hazardous Waste Report submissions (Biennial Report). To see if your State has opted in to the use of the RIA, and obtain instructions on how to file electronically, contact your State or EPA Regional Office.

AUTHORIZATION

Section 3010 of Subtitle C of the Resource Conservation and Recovery Act (RCRA) requires any person who generates, transports, or recycles regulated wastes or who owns or operates a facility for the treatment, storage, or disposal of regulated wastes to notify the U.S. Environmental Protection Agency (EPA) of their activities, including the location and general description of the activities and the regulated wastes handled. Respondents must submit the information required in the Notification of Subtitle C Activity Instructions and Form by completing the RCRA Subtitle C Site Identification Form (EPA Form 8700-12). As required by statute, the EPA promulgated regulations to implement these notification requirements at 40 CFR Parts 260, 261, 262, 263, 264, 265, 266, 270, 273, and 279. The EPA needs this information to determine the universe of persons who generate, handle, and manage these regulated wastes; assign EPA Identification Numbers; and ensure that these regulated wastes are managed in a way that protects human health and the environment as required by RCRA. This is mandatory reporting by the respondents.

The EPA enters notification information submitted by respondents into RCRAInfo, the EPA national database, and assigns EPA Identification Numbers. The EPA uses this information to identify the universe of regulated waste generators, handlers, and managers and their specific regulated waste activities. The EPA also uses the information for tracking and for a variety of enforcement and inspection purposes. Finally, the EPA uses this information to ensure that regulated waste is managed properly, that statutory provisions are upheld, and that regulations are adhered to by facility owners or operators.

Section 3007(b) of RCRA and 40 CFR Part 2, Subpart B, which defines the EPA's general policy on public disclosure of information, both contain provisions for confidentiality. However, the Agency does not anticipate that businesses will assert a claim of confidentiality covering all or part of the Notification of Subtitle C Activity. If such a claim were asserted, the EPA must and will treat the information in accordance with the regulations cited above. The EPA also will assure that the information collection complies with the Privacy Act of 1974 and Office of Management and Budget (OMB) Circular 108.

Estimated Burden: Facilities - The reporting and recordkeeping burden for the Notification of Regulated Waste Activity (initial and subsequent) is estimated to average 1 hour for a facility to complete and submit the form, and keep copies of notification and affiliation agreements on site, as applicable. State Agencies - The recordkeeping burden for the Notification of Regulated Waste Activity (initial and subsequent) is estimated to average 2 hours for State agencies to review and enter notification information into the RCRAInfo database. There is no reporting associated with this requirement, and as such, there is no reporting burden for State agencies.

To comment on the EPA's need for this information, the accuracy of the provided burden estimates, and any suggested methods for minimizing respondent burden, including the use of automated collection techniques, the EPA has established a public docket for the Information Collection Request (ICR) under Docket ID Number EPA-HQ-OLEM–2016– 0182, which is available for online viewing at www.regulations.gov, or in person viewing at the RCRA Docket in the EPA Docket Center (EPA/DC), EPA West, Room B102, 1301 Constitution Avenue, NW, Washington, D.C. The EPA Docket Center Public Reading Room is open from 8:30 a.m. to 4:30 p.m., Monday through Friday, excluding legal holidays. The telephone number for the Reading Room is (202) 566-1744, and the telephone number for the RCRA Docket is (202) 566-0270. An electronic version of the public docket is available at www.regulations.gov.

This site can be used to submit or view public comments, access the index listing of the contents of the public docket, and to access those documents in the public docket that are available electronically. When in the system, select "search," then key in the Docket ID Number identified above. Also, you can send comments to the Office of Information and Regulatory Affairs, Office of Management and Budget, 725 17th Street, NW, Washington, D.C. 20503, Attention: Desk Officer for EPA. Please include the EPA Docket ID Number EPA-HQ-OLEM–2016– 0182 and OMB Control Number 2050-0024 in any correspondence.

INTRODUCTION

These instructions are designed to help you determine if you are subject to requirements under the Resource Conservation and Recovery Act of 1976 (RCRA), as amended by the Hazardous and Solid Waste Amendments of 1984 (HSWA) for notifying the U.S. Environmental Protection Agency (EPA) of your regulated waste activities. Regulated wastes are hazardous wastes as defined by 40 CFR Part 261, universal wastes as defined by 40 CFR Part 273, and used oil as defined by 40 CFR Part 279. If you are managing hazardous secondary material under 40 CFR 260.30, 40 CFR 261.4(a)(23), (24), or (27), you also must notify under 40 CFR 260.42 using the RCRA Subtitle C Site Identification Form and Addendums to the Site Identification Form. In addition, you must notify if:

- you are a facility that either recycles hazardous waste with a RCRA permit (i.e., store the hazardous waste prior to recycling), or recycles hazardous waste without first storing the material you must notify under 40 CFR 261.6 (c)(2)(iv) which references 40 CFR 265.75;
- you are a very small quantity generator (VSQGs), previously called conditionally exempt small quantity generators (CESQGs) and you are taking advantage of the episodic generation provision at 40 CFR 262.232 (a);
- you are large quantity generator (LQG) consolidating waste from VSQGs under the control of the same person under 40 CFR 262.17 (f));
- you are closing either a waste accumulation unit or your facility, both prior to closing and after conducting closure performance operations under 40 CFR 262.17 (a)(8)).

The instructions contained in this document will assist you in obtaining or updating an EPA Identification number by completing and submitting the RCRA Subtitle C Site Identification Form (Site ID Form). RCRA is a Federal law. If you are regulated but do not comply with the RCRA notification requirements, you may be subject to civil penalties.

DETERMINING IF YOU MUST NOTIFY

HOW TO DETERMINE IF YOU MUST NOTIFY EPA OF YOUR SUBTITLE C ACTIVITIES

All persons who generate, transport, recycle, treat, store, or dispose of hazardous waste are required to notify the EPA (or their State agency if the State is authorized to operate its own hazardous waste program) of their hazardous waste activities. Furthermore, if you are managing hazardous secondary material under 40 CFR 260.30, 40 CFR 261.4(a)(23), (24), or (27), you also must notify using the RCRA Subtitle C Site Identification Form and the Addendum to the Site Identification Form: Notification of Hazardous Secondary Material Activity. Lastly, if you are a recognized trader arranging for import or export of hazardous wastes, including hazardous waste managed under the alternate management standards of 40 CFR Part 266, you must notify. These persons must obtain an EPA Identification Number unless their solid waste has been excluded from regulation or their hazardous waste has been exempted as outlined below. These respective notification requirements are found in 40 CFR Parts 260, 261, 262, 263, 264, 265, and 266.

In addition to the discussion below, you will need to refer to 40 CFR Part 261 to help you determine if the waste you handle is both a solid waste and a hazardous waste that is regulated under RCRA. If you need help making this determination after reading these instructions, contact the agency listed for your State. If your facility was granted a solid waste variance under 40 CFR 260.30 prior to July 13, 2015, the management of your hazardous secondary materials under 40 CFR 260.30 is grandfathered under the previous regulations and you are not required to notify.

NOTE	Under the Hazardous Waste Import Regulations, 40 CFR Part 262.84, foreign generators should not apply for an EPA Identification Number. These regulations state that when filling out a U.S. manifest, you must include the name and address of the foreign generator, and the name, address, and EPA Identification Number of the importer. Please contact the U.S. firms involved with your shipments and determine which firm will serve as the U.S. Importer.

To determine if you handle a solid waste that is also a hazardous waste and regulated under RCRA, ask yourself the following questions:

DO I HANDLE A SOLID WASTE?

40 CFR 261.2 defines "solid waste" as any discarded material that is not excluded under 40 CFR 261.4(a) or that is not excluded by variance granted under 40 CFR 260.30 and 260.31. A discarded material is any material which is:

- Abandoned, as explained in 40 CFR 261.2(b); or
- Recycled, as explained in 40 CFR 261.2(c); or
- Considered inherently waste-like as explained in 40 CFR 261.2(d); or
- A military munition identified as a solid waste in 40 CFR 266.202.

The list of general exclusions can be found in 40 CFR 261.4. If the solid waste that you handle has been excluded, either by rule or special variance, then you do not need to notify the EPA for that solid waste unless otherwise stated in the regulations. If your solid waste was not excluded from regulation, you need to determine if it is a hazardous waste that the EPA regulates. The EPA regulates a solid waste as hazardous waste in two ways:

- By specifically listing the solid waste as a hazardous waste and assigning it a unique EPA Hazardous Waste Code Number; or
- By regulating it because it possesses any of four hazardous waste characteristics and assigning it a generic EPA Hazardous Waste Code Number.

IS MY SOLID WASTE SPECIFICALLY LISTED AS A HAZARDOUS WASTE?

40 CFR 261.30 through 261.33 identify certain solid wastes that the EPA has specifically listed as hazardous. Persons who handle listed hazardous waste are subject to regulation and must notify the EPA of their hazardous waste activities unless they are exempted as discussed below. Refer to these regulations to see if your solid waste is included as a "listed hazardous waste." If you are handling a newly regulated hazardous waste and have already notified the EPA prior to that hazardous waste being regulated and already have an EPA Identification Number, you do not need to submit a Subsequent Notification for that newly regulated hazardous waste.

DOES MY SOLID WASTE POSSESS A HAZARDOUS CHARACTERISTIC?

Even if your solid waste is not specifically listed as a hazardous waste, it may still be hazardous because it exhibits certain hazardous characteristics. These characteristics are:

- Ignitability;
- Corrosivity;
- Reactivity; and
- Toxicity.

40 CFR 261.20 through 261.24 explain each of the characteristics and outlines the testing procedures you should use to determine if your solid waste meets these characteristics. Persons who handle characteristic hazardous waste that is regulated must notify the EPA of their activities unless they are exempted, as discussed below. If you are handling a newly regulated hazardous waste and have already notified the EPA prior to that hazardous waste being regulated and already have an EPA Identification Number, you do not need to submit a Subsequent Notification for that newly regulated hazardous waste.

HOW TO DETERMINE IF YOU MUST NOTIFY EPA OF YOUR UNIVERSAL WASTE ACTIVITIES

Under 40 CFR Part 273, Subpart C, Large Quantity Handlers of Universal Waste (LQHUW) who accumulate a total of 5,000 kilograms (kg) or more of universal wastes at any time are required to notify the EPA (or their State agency if the State is authorized to operate its own universal waste program) of their universal waste activities and obtain an EPA Identification Number, unless they have previously notified the EPA of their hazardous waste activities. LQHUWs must notify the EPA of their universal waste activities and obtain an EPA Identification Number before meeting or exceeding the 5,000 kg storage limit. Small Quantity Handlers of Universal Waste are exempt from these notification requirements.

NOTE	Please refer to the regulations in 40 CFR Part 273 to ensure that you are aware of all the requirements that apply to your universal waste handling activities.

HOW TO DETERMINE IF YOU MUST NOTIFY EPA OF YOUR USED OIL MANAGEMENT ACTIVITIES

Under 40 CFR Part 279, Subparts E, F, G, and H, respectively, persons who transport used oil; process or re-refine used oil; burn off-specification used oil for energy recovery; or market used oil fuel, are required to notify the EPA (or their State agency if the State is authorized to operate its own used oil program) and obtain an EPA Identification Number, unless they are exempt as outlined below. Off-specification used oil may be burned for energy recovery in an industrial furnace, boiler, or hazardous waste incinerator subject to regulation under Subpart O of 40 CFR Part 264 or 265.

Used oil transporters; used oil processors/re-refiners; off-specification used oil burners; and used oil fuel marketers who have not previously notified the EPA of their hazardous waste activities or notified under 40 CFR Part 266, Subpart E (replaced by 40 CFR Part 279) must notify the EPA to identify their used oil management activities.

NOTE	Please refer to the regulations in 40 CFR Part 279 to ensure that you are aware of all the requirements that apply to your used oil management activities.

EXEMPTIONS

40 CFR 262.13 and 261.6(a)(3) list certain hazardous wastes that are not subject to RCRA regulation. If the hazardous waste that you handle has been exempted, then you do not need to notify the EPA for that hazardous waste.

USED OIL FOR ENERGY RECOVERY

Used oil that is to be burned for energy recovery and that meets the specification provided under 40 CFR 279.11 is exempt from the regulations. However, the person who first claims that the used oil meets the specification is subject to notification as a used oil fuel marketer and certain other requirements (see 40 CFR Part 279, Subpart H). The burner of fuel that meets the specification in 40 CFR 279.11 is not required to notify.

USED OIL GENERATORS

Used oil generators are not required to notify the EPA.

USED OIL GENERATORS OPERATING USED OIL-FIRED SPACE HEATERS

Persons who burn only used oil that they generate (or used oil received from household do-it-yourself used oil changers) in used oil-fired space heaters are exempt from the notification requirement provided that the device is vented to the outdoors and the device is not designed to have a capacity greater than 0.5 million BTU/hour.

FILING A NOTIFICATION OF RCRA SUBTITLE C ACTIVITIES FORM

OBTAIN OR UPDATE AN EPA ID NUMBER

If you do not currently have an EPA Identification Number and you handle regulated waste or hazardous secondary material directly or as an E-manifest broker, or if you have an EPA Identification Number and need to revise information regarding your site and/or activity, you must submit a Site Identification Form with a reason for submittal of Obtaining or Updating an EPA ID Number for an on-going regulated activity that will continue for a period of time. Circumstances under which you should submit this notification include:

- If you generate, transport, treat, store, or dispose of hazardous waste. or
- If you recycle hazardous wastes. (Recyclable materials are defined as hazardous wastes that are recycled). The recycling process itself is exempt from regulation, but you must notify the EPA and obtain an EPA Identification Number prior to recycling recyclable materials; or
- If you are a large quantity handler of universal waste. (Notification is required for people who have not previously notified the EPA of their hazardous waste activities or who have not already sent a notification to the EPA as required by 40 CFR 273.32); or
- If you transport, process, or re-refine used oil; burn off-specification used oil for energy recovery; or market used oil. (Notification is required for people who have not previously notified the EPA of their hazardous waste activities or have not notified under 40 CFR Part 279 or under 40 CFR Part 266, Subpart E, which was replaced by 40 CFR Part 279.)
- If you are an eligible academic entity opting into 40 CFR Part 262, Subpart K for managing laboratory hazardous wastes AND you have never submitted site identification information, you must submit this form to notify the appropriate State or EPA Regional Office of your activities.

NOTE	You <u>must</u> check with your State to determine if you are eligible to manage laboratory hazardous waste pursuant to 40 CFR Part 262, Subpart K in order for you to notify

- If you will begin managing hazardous secondary material under 40 CFR 260.30, 40 CFR 261.4(a)(23), (24), or (27) you must submit this form, pursuant to 40 CFR 260.42, to notify the appropriate State or EPA Regional Office of your activities.

NOTE	You <u>must</u> check with your State to determine if you are eligible to manage hazardous secondary material under these exclusions in order for you to notify.

- If you are an Electronic Manifest Broker that has a contractual relationship and elects to use the system to obtain, complete and transmit an electronic manifest format supplied by the EPA electronic manifest system for handlers of hazardous wastes.

- If you are either a VSQG or a SQG who, as a result of a planned or unplanned episodic event, generates a quantity of hazardous waste in a calendar month sufficient to cause the facility to move into a more stringent generator category (i.e., VSQG to either an SQG or an LQG; or an SQG to an LQG).

- If your business moves to another location and you are still conducting activities regulated under RCRA Subtitle C.
- If the contact for your site changes.
- If the ownership of your site changes.
- If an additional owner has been added or replaced since you submitted your last notification.
- If the type of RCRA Subtitle C activity you conduct changes.
- If you have previously submitted site identification information and are an eligible academic entity opting into or withdrawing from 40 CFR Part 262, Subpart K for managing laboratory hazardous wastes, you must use this form.

NOTE	You must check with your State to determine if you are eligible to manage laboratory hazardous waste pursuant to 40 CFR Part 262, Subpart K in order for you to notify.

- If you are managing or will stop managing hazardous secondary material under 40 260.30, 40 CFR 261.4(a)(23), (24), or (27), you are required to re-notify by March 1 of each even-numbered year pursuant to 40 CFR 260.42.

NOTE	If your facility was granted a solid waste variance under 40 CFR 260.30 prior to July 13, 2015, the management of your hazardous secondary materials under 40 CFR 260.30 is grandfathered under the previous regulations and you are not required to notify.

HOW MANY FORMS SHOULD I FILE?

If you fall under any of the regulations above to notify EPA, you must submit one Site Identification Form along with the applicable Addendums. For example, if you manage hazardous secondary material under 40 CFR 260.30, 40 CFR 261.4(a)(23), (24), or (27), along with the Site Identification Form you should submit the Notification of Hazardous Secondary Material Activity addendum. If you are an SQG or VSQG generating hazardous waste from a planned or unplanned episodic event pursuant to 40 CFR262 Subpart L, along with Site Identification Form you should submit the Episodic Generator addendum. And if you are receiving hazardous waste from VSQGs under the control of the same person, along with the Site Identification Form you should submit the LQG Consolidation of VSQG Hazardous Waste addendum.

WHERE SHOULD I SEND MY COMPLETED FORM?

Click *here* to find access to a contact list containing the address for your State or EPA Regional Office where you should send your completed Site ID Form. The list contains contact names, addresses, phone numbers, and e-mail addresses that you can use to obtain additional information.

Many States use the forms included in this document; some also require additional information. Other States require that you complete and submit a State-specific form. Information about which form to use is included with the contact list located at the web page noted above. Even if you use the included form, you should check with your State to determine if you need to submit additional information. Also, contact your State if you have any questions about your submission.

After your completed Site ID Form for Obtaining an EPA Identification number or for Obtaining an Electronic Manifest Broker is received and processed, you will be sent a written acknowledgement that will include your EPA Identification Number. You must use this number on all communications with the EPA regarding your regulated waste activities for this site.

INSTRUCTIONS FOR FILLING OUT THE RCRA SUBTITLE C SITE IDENTIFICATION FORM

Type or print, in black ink, an "X" in all items that apply (if "Yes", type or print an "X" in the "Y" box, if "No", type or print an "X" in the "N" box) and then type or print an "X" in all other boxes that apply. In Item 19, provide the required ink signatures. Signatures must be original. Stamped or photocopied signatures are not acceptable. Enter your site's EPA Identification Number in the top left-hand corner on all pages of the form; for an initial notification for this site, leave the EPA identification Number blank. Use Item 18 – Comments to clarify or provide additional information for any entry. When entering information in the comments section, enter the item number and box letter to which the comment refers. If you must use additional sheets for comments, enter your site's EPA Identification Number in the top left-hand corner of each sheet.

ITEM 1 – REASON FOR SUBMITTAL

Place an "X" in the appropriate box to indicate whether you are submitting this form to obtain or update an EPA ID Number for an on-going regulated activity; as a component of the Hazardous Waste Report; to notify that regulated activity is no longer occurring at your site; to obtain or update an EPA ID Number for conducting electronic manifest broker activities; or as a component of a First or a Revised Hazardous Waste Part A Permit Application.

OBTAINING OR UPDATING AN EPA ID NUMBER FOR AN ON-GOING REGULATED ACTIVITY THAT WILL CONTINUE FOR A PERIOD OF TIME. (INCLUDES HSM ACTIVITY)

- If your waste activity is regulated under Subtitle C of the Resource Conservation and Recovery Act (RCRA) and the rules promulgated pursuant to the Act (specifically 40 CFR Parts 260-299), you <u>must</u> submit this form to notify the appropriate State or EPA Regional Office of your regulated waste activities and **obtain an EPA Identification Number**.

- If you are an eligible academic entity opting into 40 CFR Part 262, Subpart K for managing laboratory hazardous wastes **AND** you have never submitted site identification information, you must submit this form to notify the appropriate State or EPA Regional Office of your activities.

NOTE	You must check with your State to determine if you are eligible to manage laboratory hazardous waste pursuant to 40 CFR Part 262, Subpart K in order for you to notify.

- If you are a recognized trader arranging for export or import of hazardous waste, including those managed under the alternate standards of 40 CFR Parts 266 and 273; or an exporter or importer of spent lead acid batteries (SLABs), you must submit this form to notify the appropriate State or EPA Regional Office of your activities.

- You must use this form to **submit a subsequent notification** if your site already has an EPA Identification Number and you wish to change information (e.g., generator status, new site contact person, new owner, new mailing address, new regulated waste activity, etc.).

- If you have previously submitted site identification information and are notifying (or re-notifying) that you will begin managing, are managing, or have stopped managing hazardous secondary material under 40 CFR 260.30, 40 CFR 261.4(a)(23), (24), or (27), you must submit this form, pursuant to 40 CFR 260.42, to notify the appropriate State or Regional Office of your activities.

NOTE	You must check with your State to determine if you are eligible to manage hazardous secondary material under these exclusions in order for you to notify.

- If you are a very small quantity generator (VSQG), previously called conditionally exempt small quantity generator (CESQG), taking advantage of the episodic generation provision at 40 CFR 262.232 (a), you must submit this form to report your episodic event.

- You must use this form to if you are a small quantity generator (SQG) re-notifying, beginning in 2021 and every four (4) years thereafter, unless a state program has more frequent reporting/notification requirements (See 40 CFR 262.18).

- You must use this form if you are a large quantity generator (LQG):
  - Consolidating wastes from VSQGs under the control of the same person. (See 40 CFR 262.17 (f)). Such LQGs must complete the Addendum to the Site Identification Form: LQG Consolidation of VSQG Hazardous Waste, and identify the RCRA Identification Number of the VSQG (if applicable), name, address, emergency contact phone number and contact name of every VSQG that they are receiving hazardous wastes from.
  - Closing either a waste accumulation unit (optional) or their facility, both prior to closing and after conducting closure performance operations (See 40 CFR 262.17 (a)(8)).

SUBMITTING AS A COMPONENT OF THE HAZARDOUS WASTE REPORT

If you are required to submit a Hazardous Waste Report indicating the amount of hazardous waste you generate, treat, recycle, dispose, on-site or ship off-site for subsequent treatment, recycling and disposal, or receive from off-site, you must fill out this form. A Site ID Form submitted with a Hazardous Waste Report is equivalent to a subsequent notification.

- *Site was a TSD facility and/or generator of ≥ 1,000 kg of hazardous waste, > 1 kg of acute hazardous waste, or > 100 kg of acute hazardous waste spill cleanup in one or more months of the report year (or State Equivalent LQG regulations)*

  The purpose of this check box is to distinguish between sites that meet the criteria and are required to file a report versus those who file voluntarily or by State-only requirement but were not a treatment, storage, and disposal facility (TSDF) or a Large Quantity Generator (LQG) during the report year. Sites required to file the report should place an "X" in this box, while non-LQG/TSD sites should not. For more information about who must file a report, refer to the *Who Must File a Hazardous Waste Report* section.

NOTIFIYING THAT REGULATED ACTIVITY IS NO LONGER OCCURING AT YOUR SITE

If you are no longer conducting ANY RCRA Subtitle C federal or state regulated hazardous waste activities, listed on the Site Identification Form, then you can use this Reason for Submittal to deactivate your EPA ID number.

OBTAINING OR UPDATING AN EPA ID NUMBER FOR CONDUCTING ELECTRONIC MANIFEST BROKER ACTIVITIES

If you are requesting an EPA ID in order to create and broker manifest transactions for handlers of hazardous waste, then select this Reason for Submittal. An Electronic Manifest Broker is considered a user of the electronic manifest system that has a contractual relationship and elects to use the system to obtain, complete and transmit an electronic manifest form supplied by the EPA electronic manifest system for handlers of hazardous wastes. This designation is for users of the electronic manifest system, defined in 40 CFR 260.10 as a person that elects to use the system to obtain, complete and transmit an electronic manifest format supplied by the EPA electronic manifest system.

SUBMITTING A NEW OR REVISED PART A (PERMIT) FORM

If your site is planning to treat, store, or dispose of hazardous waste on-site in a unit that is not exempt from obtaining a hazardous waste permit, you must submit this form as a component of the Part A Permit Application. Also, if the activity at this site (treatment, storage, or disposal) became newly regulated under RCRA Subtitle C and the rules promulgated pursuant to the Act (specifically 40 CFR Parts 260-299), you must submit this form as part of the Part A Permit Application. Also, you must submit a Revised Part A Permit Application to reflect changes that have occurred at your site, you must submit this form as part of your Revised Part A Permit Application.

ITEM 2 – SITE EPA ID NUMBER

Provide your EPA Identification Number in Item 2 **for this site**. The first two characters of the EPA Identification Number must be a valid State postal code. Be sure to include your EPA Identification Number at the top of all pages of the form (as well as on any attachments to the Site ID Form).

NOTE	If this is your initial notification for this site, leave the EPA Identification Number blank and proceed to Item 3.

ITEM 3 AND 4 – SITE NAME AND SITE LOCATION ADDRESS

Provide the legal name of your site and a complete location address. Please note that the address you give for Item 4, Site Location Address, must be a physical address, not a post office box or route number. Only foreign hazardous waste transporters, with their headquarters located outside the U.S., may provide a Site Location Country outside of the U.S.

NOTE	A new EPA Identification Number is **required** if you change the location of your site

ITEM 5 – SITE MAILING ADDRESS

Provide the Site Mailing Address. If the Mailing Address and the Site Location Address (Item 4) are the same, you can check the "Same as Location Address" checkbox.

ITEM 6 – SITE LAND TYPE

Place an "X" in the box that **best describes** the land type of your site. Select only one type: Private, County, District, Federal, Tribal (see below), Municipal, State, or Other. If your site's Land Type could be described as Municipal **and** another Land Type, such as County, District, or Tribal, do not place an "X" in Municipal. Instead, choose the other appropriate Land Type. (For example, if your site's Land Type is both Municipal and County, you would place an "X" in the box for County.) You may explain this in Item 18 – Comments.

Tribal–The land which your site is on belongs to one of the tribes/entities on the list of Federally recognized American Indian tribes and Alaskan Native entities located at: http://www.epa.gov/tribal/whereyoulive/tribes-a-z.htm.

ITEM 7 – NORTH AMERICAN INDUSTRY CLASSIFICATION SYSTEM (NAICS) CODE(S)

Box A (Primary) must be completed. Completing Boxes B-D is recommended, if applicable.

BOX A

Provide the North American Industry Classification System (NAICS) code that best describes your site's **primary** business production process for your products or services. Referencing the latest version of

NAICS codes, use the 6-digit code (most specific description) if available for your business; if not, use the 5-digit code; do not enter any four (4) or less digit codes.

Check with your accounting or business staff to determine your NAICS code(s); the NAICS code is used in tax reporting and other business reports. You can obtain additional information about NAICS codes at http://www.census.gov/eos/www/naics.

BOXES B – D

List other NAICS codes that describe the other business production processes for your site. Referencing the latest version of NAICS codes, use the 6-digit code (most specific description) if available for your business; if not, use the 5-digit code; do not enter any four (4) or less digit codes.

NOTE	The Census Bureau has published NAICS Code effective January 1, 2017. Please verify that your NAICS codes are still applicable. You can obtain additional information about the 2017 NAICS codes at: http://www.census.gov/eos/www/naics

ITEM 8 – SITE CONTACT INFORMATION

Enter the name, title, business address, e-mail address, telephone number, extension, and fax number of the individual who should be contacted regarding the information submitted in the Site ID Form. A subsequent notification is recommended when the Site Contact Person changes. **Do not** enter other contact persons here; if there are other persons, who may be contacted about this submission, list them and their contact information in Item 18 – Comments. If the person completing this form is not the primary site RCRA hazardous waste contact, enter the primary site RCRA hazardous waste contact here and add the contact information for the person completing the form in Item 18 – Comments.

NOTE	This is NOT the Facility Permit Contact's information. The Facility Permit Contact information should be entered on the RCRA Hazardous Waste Part A Permit Application.

ITEM 9 – LEGAL OWNER AND OPERATOR OF THE SITE

This section should be used to indicate all owners and operators of this site. If your Reason for Submittal is for an Electronic Manifest Broker whose site of business is an office only, and you do not otherwise physically generate, treat, store, recycle or dispose of hazardous waste on site, you do not have to fill out this item.

A. NAME OF SITE'S LEGAL OWNER

Provide the name of your site's legal owner(s). This includes owner(s) of the building(s) and land. Please review these definitions:

Owner – The person who owns a RCRA site or part of a RCRA site. **Note:** This includes the owner(s) of the building(s) and/or land. This may be an individual, company, or business name. See **Person**.

Person – An individual, trust, firm, joint stock company, Federal Agency, corporation (including a government corporation), partnership, association, State, municipality, commission, political subdivision of a State, or any interstate body.

DATE BECAME AN OWNER

Indicate the date on which the above entity became the owner of your site. Enter dates as in this example: For April 22, 2015, enter 04/22/2015. This is optional information.

OWNER TYPE

Place an "X' in the box that **best describes** the owner type for your site. Select only one type: Private, County, District, Federal, Tribal (see below), Municipal, State, or Other. If your site's Owner Type could be described as Municipal **and** another Owner Type, such as County, District, or Tribal, do not place an "X" in Municipal. Instead, choose the other appropriate Owner Type. (For example, if your site's Owner Type is both Municipal and County, you would place an "X" in the box for County.) You may explain this in Item 18 – Comments.

Tribal - A member of one of the tribes/entities on the list of Federally recognized American Indian tribes and Alaskan Native entities located at: http://www.epa.gov/tribal/whereyoulive/tribes-a-z.htm.

LEGAL OWNER ADDRESS

Enter the address of the legal owner. If the address and the Location of Site (Item 4) are the same, you can check the "Same as Location Address" checkbox.

ADDITIONAL OWNER INFORMATION

Enter the e-mail, telephone number, extension, and fax number of the legal owner.

Use the Comments section to list any additional owners, their names, the dates they became owners, owner type, mailing address, and which owner(s), if any, are no longer owners since your last submission of this form. If necessary, attach a separate sheet of paper. Remember to enter your site's EPA Identification Number in the top left-hand corner of each sheet.

B. NAME OF SITE'S LEGAL OPERATOR

Provide the name of your site's operator. Please review these definitions:

Operator – The person responsible for the overall operation of a RCRA site. **Note:** This is the legal entity which controls the RCRA site operation rather than the plant or site manager. This is usually a company or business name, but may be an individual. See **Person**.

Person – An individual, trust, firm, joint stock company, Federal Agency, corporation (including a government corporation), partnership, association, State, municipality, commission, political subdivision of a State, or any interstate body.

DATE BECAME AN OPERATOR

Indicate the date on which the above entity became the operator of your site. Enter dates as in this example: For April 22, 2015, enter 04/22/2015. This is optional information.

OPERATOR TYPE

Place an "X" in the box that **best describes** the operator type for your site. Select only one type: Private, County, District, Federal, Tribal (see below), Municipal, State, or Other. If your site's Operator Type could be described as Municipal **and** another Operator Type, such as County, District, or Tribal, do not place an "X" in Municipal. Instead, choose the other appropriate Operator Type. (For example, if your site's Operator Type is both Municipal and County, you would place an "X" in the box for County.) You may explain this in Item 18 – Comments.

Tribal - A member of one of the tribes/entities on the list of Federally recognized American Indian tribes and Alaskan Native entities located at: http://www.epa.gov/tribal/whereyoulive/tribes-a-z.htm.

LEGAL OPERATOR ADDRESS

Enter the address of the legal operator. If the address and the Location of Site (Item 4) are the same, you can check the "Same as Location Address" checkbox.

ADDITIONAL OPERATOR INFORMATION

Enter the e-mail, telephone number, extension, and fax number of the operator.

Use the Comments section to list any additional operators, their names, the dates they became operators, operator type, mailing address, and which operator(s), if any, are no longer operators since your last submission of this form. If necessary, attach a separate sheet of paper. Remember to enter your site's EPA Identification Number in the top left-hand corner of each sheet.

NOTE	A subsequent notification is recommended when the owner or operator of a site changes. Because an EPA Identification Number is site-specific, the new owner will keep the existing EPA Identification Number for that location. If your business moves to another location, the owner or operator must notify the State or EPA Regional Office of this change. Since your business has changed locations, a new EPA Identification Number will be assigned.

ITEM 10 – TYPE OF REGULATED WASTE ACTIVITY (AT YOUR SITE)

Place an "X" in box "Y" or box "N" as appropriate for all **current** activities at this site **(as of the date submitting the form)**; complete any additional boxes as instructed. **Current** activities mean activities that are in effect when the form is submitted or those that the site plans to begin after EPA Identification Number assignment. The information you provide in Item 10 will be considered current as of the date you certify the form. If the site is no longer a generator as of the date you certify the form, you should mark the "N" (not a generator) box for Generator of Hazardous Waste.

10.A HAZARDOUS WASTE ACTIVITIES

NOTE	Listed below are the Federal generator status definitions. If, however, the State where your site is located has definitions different from the Federal definitions, you must use the State definitions.

10.A.1. GENERATOR OF HAZARDOUS WASTE

If you generate a hazardous waste that is listed in 40 CFR 261.31 through 261.33 or identified by one or more hazardous waste characteristic(s) contained in 40 CFR 261.21 through 261.24, place an "X" in the appropriate box for the quantity of hazardous waste that is generated per calendar month. The regulations for hazardous waste generators are found in 40 CFR Part 262—specifically 40 CFR 262.14 for very small quantity generators (VSQGs), 40 CFR 262.16 for small quantity generators (SQGs), and 40 CFR 262.17 for large quantity generators (LQGs). Consult these regulations and your State for details about how the regulations apply to your situation. Below is a brief description of the three types of hazardous waste generators.

If "Yes", place an "X" in only one of the following – a, b, or c. Otherwise, **place an "X" in the "N" box.**

a. **LQG: Large Quantity Generator**
For purposes of providing information in this report, the site is a Large Quantity Generator (LQG) if the site generates **any** of the following amounts in a calendar month:
- (i) Generates, in any calendar month, (including quantities imported by importer site) 1,000 kilograms (2,200 pounds) or more of non-acute RCRA hazardous waste; **or**
- (ii) Generates, in a calendar month, more than 1 kilogram (2.2 pounds) of any RCRA acute hazardous waste listed in sections 261.31 or 261.33(e); **or**
- (iii) Generates, in any calendar month, more than 100 kilograms (220 pounds) of residue or contaminated soil, waste, or other debris resulting from the cleanup of a spill, into or on any land or water, of any RCRA acute hazardous waste listed in sections 261.31 or 261.33(e).

NOTE	If, in addition to being a LQG, you recycle hazardous wastes at your site, mark both this box and Item 10.A.6. Hazardous secondary material managed under 40 CFR 260.30, 40 CFR 261.4(a)(23), (24), or (27) DOES NOT count towards your generator status. However, you must check with your State to determine if you are eligible to manage hazardous secondary material under these exclusions.

b. **SQG: Small Quantity Generator**
This site is a SQG if the site meets **all** of the following criteria:
- (i) Generates, in any calendar month, greater than 100 kilograms (220 pounds) but less than 1,000 kilograms (2,200 pounds) of non-acute hazardous waste; **and**
- (ii) Generates, in any calendar month, less than or equal to 1 kilogram (2.2 pounds) of acute hazardous waste listed in sections 261.31 or 261.33(e); **and**

(iii) Generates, in any calendar month, less than or equal to 100 kilograms (220 pounds) of any residue or contaminated soil, water, or other debris resulting from the cleanup of a spill, into or on any land or water, of any acute hazardous waste listed in sections 261.31 or 261.33(e).

c. **VSQG: Very Small Quantity Generator**:
This site is a VSQG if the site meets **all** of the following criteria:

(i) Generates in any calendar month, less than or equal to 100 kilograms (220 pounds) of hazardous waste; **and**
(ii) Generates in any calendar month, less than or equal to 1 kilogram (2.2 pounds) of acute hazardous wastes listed in sections 261.31, or 261.33(e); **and**
(iii) Generates in any calendar month, less than or equal to 100 kilograms (220 pounds) of any residue or contaminated soil, waste, or other debris resulting from the cleanup of a spill, into or on any land or water, of any acute hazardous wastes listed in sections 261.31, or 261.33(e).

10.A.2. SHORT-TERM GENERATORS

Place an "X" in the "Y" box if the site is normally not a generator of hazardous waste, but is currently generating hazardous waste only as the result of a one-time, non-recurring, temporary event that is not related to normal production processes. In other words, short-term generators produce hazardous waste from a particular activity for a limited time and then cease conducting that activity. Short-term generators are not considered episodic generators because episodic generators (i.e., VSQGs and SQGs) regularly generate hazardous waste as part of their operations, but elevate to a higher generator category as a result of a planned or unplanned event. Examples of short-term generators include: (1) one-time highway bridge waste generation; (2) underground storage tank removals; (3) generation of off-specification or out-of-date chemicals at a site that normally doesn't generate hazardous waste; (4) remediation or spill clean-up at sites with no previous RCRA EPA Identification Number, and (5) site or production process decommissions by a new operator. If you mark "Y", you must provide an explanation of your short-term generation event in Item 18 –Comments. Otherwise, **place an "X" in the "N" box**.

10.A.3. MIXED WASTE GENERATOR

Place an "X" in the "Y" box if you are a generator of mixed waste (waste that is both hazardous and radioactive). Otherwise, **place an "X" in the "N" box**. RCRA defines "mixed waste" as waste that contains both hazardous waste and source, special nuclear, or by-product material subject to the Atomic Energy Act (AEA), RCRA Section 1004(41), 42 U.S.C. 6903 (63 FR 17414; April 9, 1998).

10.A.4. TREATER, STORER, OR DISPOSER OF HAZARDOUS WASTE

If you treat, store, or dispose of hazardous waste, **place an "X" in the "Y" box**. A RCRA Hazardous Waste Part B Permit is **required** for this activity. Contact the appropriate office for your State for more information. The Federal regulations for owners and operators of permitted treatment, storage, and disposal facilities (TSDFs) are found in 40 CFR Parts 264, 265, 266, and 270.

Place an "X" in the "N" box if any of the following conditions are true for your facility:

- This facility does not receive hazardous waste from other generators and ships all waste off-site for management within the regulatory timeframe.
- This facility is only involved with on-going post-closure activities, corrective actions under the Hazardous and Solid Waste Amendments of 1984 (HSWA), or a consent order under a non-traditional permit or without a RCRA permit being required.
- Receives waste from off-site but does not store greater than 10 days before re-shipping (i.e., transfer facility).

NOTE	If your site is a destination facility for universal wastes in addition to being a TSDF for other RCRA hazardous wastes, place an "X" in the "Y" box for both this box **and** Item 11.B.2.

10.A.5. RECEIVES HAZARDOUS WASTE FROM OFF-SITE

If you received hazardous waste from another site, whether this waste was received as a commercial transaction or waste received from a restricted group of off-site generators, **place an "X" in the "Y" box.** Otherwise, **place an "X" in the "N" box**. Item 10.A.5 "Receives Hazardous Waste from Off-site" is not for transfer facilities. If you are a transfer facility receiving hazardous waste from off-site, you should mark item 11.A.1.b (Transfer Facility).

10.A.6. RECYCLER OF HAZARDOUS WASTE

Place an "X" in the "Y" box if you recycle regulated hazardous wastes (recyclable materials) at your site. Otherwise, **place an "X" in the "N" box**. If you mark "Y", then mark the subsequent box that identifies whether you recycle regulated hazardous wastes, with or without storage prior to recycling. The Federal regulations for owners and operators of sites that recycle hazardous waste are found in 40 CFR 261.6. You also may be subject to other Federal and State regulations; in some cases, a permit is required.

NOTE	The 2016 Hazardous Waste Generator Improvements Final Rule requires that both facilities that do store prior to recycling and facilities that do not store prior to recycling submit a Hazardous Waste Report. If your site, in addition to being a recycling site for hazardous waste, treats, stores, or disposes of hazardous waste, place an "X" in the "Y" box for both this box **and** Item 10.A.4. If your site is a destination facility for universal wastes in addition to being a recycling site for other RCRA hazardous wastes, place an "X" in the "Y" box for both this box **and** Item 11.B.2.

10.A.7. EXEMPT BOILER AND/OR INDUSTRIAL FURNACE

If "Yes", place an "X" in all that apply. Otherwise, **place an "X" in the "N" box**.

a. **Small Quantity On-Site Burner Exemption**
You burn small quantities of hazardous waste in an on-site boiler or industrial furnace in accordance with the conditions in 40 CFR 266.108, place an "X" in the box to indicate that you qualify for the Small Quantity On-Site Burner Exemption.

b. **Smelting, Melting, and Refining Furnace Exemption**
You process hazardous wastes in a smelting, melting, or refining furnace solely for metals recovery, as described in 40 CFR 266.100(d), or to recover economically significant amounts of precious metals, as described in 40 CFR 266.100(g), or if you process hazardous wastes in a lead recovery furnace to recover lead, as described in 40 CFR 266.100(h), place an "X' in the box to indicate that you qualify for the Smelting, Melting, and Refining Furnace Exemption.

10.B. WASTE CODES FOR FEDERALLY REGULATED HAZARDOUS WASTES

Please list the waste codes of the Federal hazardous wastes (described in 40 CFR Part 261) handled at your site. List them in the order they are presented in the regulations using the appropriate 4-digit code(s) (e.g., D001, D003, F007, U112).

NOTE	If you handle more hazardous wastes than will fit under Item 10.B, please continue under Item 18 – Comments or on an extra sheet. Remember to include your EPA Identification Number on the top of each page. If you handle a large number of codes, you may copy the list in this document and mark the ones that you handle. Attach any additional sheets to the Site ID Form. Remember to include your EPA Identification Number on the top of each page.

LIST	Click *here* for a list of the nationally-defined Hazardous Waste Codes.

10.C. WASTE CODES FOR STATE-REGULATED (NON-FEDERAL) HAZARDOUS WASTES

If you manage State-regulated hazardous wastes that have a State waste code, enter the appropriate code(s) in the box(es) provided. Please list the waste codes of the State-regulated hazardous wastes handled at your site in the order they are presented in the regulations.

NOTE	If you handle more hazardous wastes than will fit under Item 10.C, please continue under Item 18 – Comments or on an extra sheet. Remember to include your EPA Identification Number on the top of each page.

ITEM 11– ADDITIONAL REGULATED WASTE ACTIVITIES

11.A OTHER WASTE ACTIVITIES

Place an "X" in the "Y" or "N" box as appropriate for all additional current regulated waste activities at this site **(as of the date submitting the form)**; complete any additional boxes as instructed. **Current** activities mean activities that are in effect when the form is submitted or those that the site plans to begin after EPA Identification Number assignment. The information you provide in Item 11 will be considered current as of the date you certify the form

11.A.1. TRANSPORTER OF HAZARDOUS WASTE

If "Y", place an "X" in all that apply. Otherwise, **place an "X" in the "N" box**.

a. **Transporter**
You transport hazardous waste within the U.S. The Federal regulations for hazardous waste transporters are found in 40 CFR Part 263.

b. **Transfer Facility**
You are a hazardous waste transfer facility, at your site, if you hold manifested hazardous waste(s) at your site for a period of ten (10) days or less while the waste is in transit. The Federal regulations for hazardous waste transfer facilities are found in 40 CFR 263.12.

11.A.2. UNDERGROUND INJECTION CONTROL

If you generate, treat, store, or dispose of hazardous waste and place the waste or its residuals into an underground injection well (e.g., a Class I well) located at your site, **place an "X" in the "Y" box**. Otherwise, **place an "X" in the "N" box**. The Federal regulations for owners and operators of underground injection wells are found in 40 CFR Part 148.

11.A.3. UNITED STATES IMPORTER OF HAZARDOUS WASTE

Place an "X" in the "Y" box if you import hazardous waste from a site located in a foreign country into the U.S. Refer to 40 CFR 262.10(e) and 40 CFR 262.84 for additional information. Otherwise, **place an "X" in the "N" box.**

11.A.4. RECOGNIZED TRADER

Place an "X" in the "Y" box if you are a recognized trader, defined in 40 CFR 260.10 as a person domiciled in the United States, who acts to arrange and facilitate transboundary movements of wastes destined for recovery or disposal operations, either by purchasing from and subsequently selling to United States and foreign facilities, or by acting under arrangements with a United States waste facility to arrange for the export or import of the waste. Otherwise, **place an "X" in the "N" box**. Mark all that apply.

a. **Importer**
b. **Exporter**

11.A.5. IMPORTER/EXPORTER OF SPENT LEAD-ACID BATTERIES (SLABS) UNDER 40 CFR PART 266 SUBPART G

Place an "X" in the "Y" box if you are an importer or exporter of spent lead-acid batteries (SLABs) being managed domestically under 40 CFR 266 Subpart G to obtain an EPA Identification number (see 40 CFR 266.80(a)(6), (8) - (10)). Otherwise, **place an "X" in the "N" box**. Mark all that apply.

a. **Importer**
b. **Exporter**

11.B UNIVERSAL WASTE ACTIVITIES

Refer to your State-specific requirements and definitions for universal waste. Also, refer to 40 CFR 261.9 and 40 CFR Part 273 for the Federal regulations covering universal waste. **Complete parts 1 and 2.**

11.B.1 LARGE QUANTITY HANDLER OF UNIVERSAL WASTE (LQHUW)

You are a Large Quantity Handler of Universal Waste (LQHUW) if you accumulate a total of 5,000 kg or more total of universal wastes (batteries, pesticides, mercury-containing equipment, or lamps – calculated collectively) at any time. This designation is retained through the end of the calendar year in which the 5,000 kg limit is met or exceeded. **Place an "X" in the "Y" box,** then **place an "X"** in the appropriate box(es) to indicate the type(s) of universal wastes managed at your site. Otherwise, **place an "X" in the "N" box**. If your State has other additional universal wastes, indicate what they are by **placing an "X"** in the corresponding box(es) (11.B.1.e - g).

11.B.2 DESTINATION FACILITY FOR UNIVERSAL WASTE

Place an "X" in the "Y" box if you treat, dispose, or recycle universal wastes on-site. Otherwise, **place an "X" in the "N" box**. A hazardous waste permit is required if you treat or dispose of universal wastes; a permit may be required if you recycle universal wastes.

NOTE	If your site, in addition to being a destination facility for universal wastes, is also a TSDF for RCRA hazardous wastes, **place an "X" in the "Y" box** for both this **and** Item 10.A.4. In addition, if your site recycles RCRA hazardous wastes, **Place an "X" in the "Y" box** for both this **and** Item 10.A.6.

11.C. USED OIL ACTIVITIES

Place an "X" in the appropriate box(es) to indicate which used oil management activities are taking place at this site. Otherwise, **place an "X" in the "N" box**. The Federal regulations for used oil management are found in 40 CFR Part 279. Also, the facility should check with its State to find out if there are additional State-specific reporting requirements for used oil activities. **Complete all parts 1 through 4.**

11.C.1. USED OIL TRANSPORTER

Place an "X" in the "Y" box, then **place an "X" in all that apply.** Otherwise, **place an "X" in the "N" box.**

a. **Transporter**
You transport used oil within the U.S. The Federal regulations for used oil transporters are found in 40 CFR 279.40-47.

b. **Transfer Facility (at your site)**
You own or operate a used oil transfer facility. The Federal regulations for used oil transfer facilities are found in 40 CFR 279.40-47.

11.C.2. USED OIL PROCESSOR AND/OR RE-REFINER

Place an "X" in the "Y" box, then **place an "X" in all that apply.** Otherwise, **place an "X" in the "N" box.**

a. **Processor**
You process used oil. The Federal regulations for processors of used oil are found in 40 CFR 279.50-59.

b. **Re-refiner**
You refine used oil. The Federal regulations for re-refiners of used oil are found in 40 CFR 279.50-59.

11.C.3. OFF-SPECIFICATION USED OIL BURNER

Place an "X" in the "Y" box, to indicate that you are conducting this used oil management activity. Otherwise, **place an "X" in the "N" box.**

11.C.4. USED OIL FUEL MARKETER

Place an "X" in the "Y" box, then **place an "X" in all that apply.** Otherwise, **place an "X" in the "N" box.**

a. **Marketer Who Directs Shipment of Off-Specification Used Oil to Off-Specification Used Oil Burners**
You are a marketer who directs shipment of off-specification used oil to off-specification used oil burners. The Federal regulations for used oil fuel marketers are found in 40 CFR 279.70-75.

b. **Marketer Who First Claims the Used Oil Meets the Specification**
You are the first to claim that used oil meets the used oil specifications established in 40 CFR 279.11.

NOTE	If either of these boxes is marked, you must also notify (or have previously notified) as a used oil transporter (11.C.1), used oil processor/re-refiner (11.C.2), or off-specification used oil fuel burner (11.C.3), unless you are a used oil generator. (Used oil generators are not required to notify.)

ITEM 12 – ELIGIBLE ACADEMIC ENTITIES WITH LABORATORIES

NOTE	40 CFR 262, Subpart K must be in effect in your State in order to report as an eligible academic entity with laboratories. See EPA's website for more information about these regulations at http://www2.epa.gov/hwgenerators/regulations-hazardous-waste-generated-academic-laboratories

Subpart K is an optional alternative set of requirements for eligible academic entities with laboratories. Certain generators (i.e., eligible academic entities as defined in 40 CFR 262.200) are eligible to operate under Subpart K for management of their hazardous wastes in laboratories in lieu of 40 CFR 262.15 (or 40 CFR 262.14 for VSQGs). Eligible academic entities with laboratories that generate hazardous waste that elect to opt into Subpart K, are currently operating under Subpart K, or subsequently withdraw from Subpart K must complete this section to meet the notification requirements of this Subpart. Refer to 40 CFR 262.203 and 40 CFR 262.204.

NOTE	Eligible academic entities with laboratories must complete a separate Site ID Form for each site (i.e., EPA Identification Number) that is managing hazardous waste under Subpart K. All laboratories with the same EPA Identification Number will be regulated under this Subpart. If eligible academic entities with laboratories withdraw from Subpart K, all laboratories with the same EPA Identification Number associated with the withdrawal from Subpart K will be regulated under 40 CFR 262.15 (or 40 CFR 262.14 for VSQGs).

12.A OPTING INTO OR CURRENTLY OPERATING UNDER 40 CFR PART 262, SUBPART K FOR THE MANAGEMENT OF HAZARDOUS WASTES IN LABORATORIES

Place an "X" in the "Y" box, if you are an eligible academic entity and you elect to opt into or are currently operating under 40 CFR 262, Subpart K for the hazardous wastes generated in your laboratories. Otherwise, **place an "X" in the "N" box**. If you mark "Y" for this box, you must place an "X" in at least one of the following to indicate your type of eligible academic entity. **Place an "X" in all that apply**:

a. **College or University**
   You are an eligible college or university if you are a private or public, post-secondary, degree-granting, academic institution, that is accredited by an accrediting agency listed annually by the U.S. Department of Education.

b. **Teaching Hospital that is owned by or has a formal written affiliation agreement with a college or university**
   You are an eligible teaching hospital if you are a hospital that trains students to become physicians, nurses, or other health personnel and is either: (1) owned by a college or university, or (2) has a master affiliation agreement and program letter of agreement, as defined by the Accreditation Council for Graduate Medical Education, with an accredited medical program or medical school.

c. **Non-profit Institute that is owned by or has a formal written affiliation agreement with a college or university**
You are an eligible non-profit institute if you are an organization that conducts research as its primary function and files as a non-profit organization under the tax code of 26 U.S.C. 501(c)(3) and is either: (1) owned by a college or university, or (2) has a formal written affiliation agreement with a college or university that establishes a relationship between institutions for the purposes of research and/or education and is signed by authorized representatives from each institution. A relationship on a project-by-project or grant-by-grant basis is not considered a formal written affiliation agreement.

12.B WITHDRAWING FROM 40 CFR 262, SUBPART K FOR THE MANAGEMENT OF HAZARDOUS WASTES IN LABORATORIES

Place an "X" in the "Y" box, if you have previously elected to opt into 40 CFR Part 262, Subpart K and are now withdrawing from participation in this optional set of alternative requirements for hazardous waste generation in laboratories. Withdrawing generators will automatically revert to regulation under 40 CFR 262.15 requirements (or 40 CFR 262.14 for VSQGs). If marking "Y" for this box, please include comments in Item 18 – Comments that explain your reasons for withdrawing from Subpart K. Otherwise, **place an "X" in the "N" box.**

ITEM 13 – EPISODIC GENERATION

Place an "X" in the "Y" box, if you are a VSQG or SQG notifying that you are taking advantage of the episodic generator event provision in 40 CFR 262.232. This provision allows a VSQG or an SQG to generate additional quantities of hazardous waste—temporarily exceeding its normal generator category limits—and still maintain its existing generator category, provided it complies with the specified conditions identified in 40 CFR 262.232 (a) and (b). Otherwise, **place an "X" in the "N" box.**

NOTE	If you **mark "Y,"** you must fill out the Addendum to the Site ID Form: Episodic Generator.

The generator may use this provision once per calendar year with the ability to petition for a second event. However, if the first event is planned, the petition must be for a second event that is unplanned, or vice-versa. It is recommended you review the regulation at 40 CFR 262.233 to understand what is required of a generator should you choose to take advantage of this petition process.

Although not inclusive, examples of planned episodic events include tank cleanouts, short-term construction projects, short-term site remediation, equipment maintenance during plant shutdowns, removal of excess chemical inventories, and site and production process decommissions by a new operator. Unplanned episodic events, which EPA expects would be less frequent, include production process upsets, product recalls, accidental spills, or "acts of nature," such as a tornado, hurricane, or flood. If you are taking advantage of this provision, you must complete the Addendum to the Site Identification Form for Episodic Generation. Information to be completed includes:

- the type of episodic event (i.e., planned or unplanned),
- the name and telephone number of an emergency contact at the site,

- the beginning or start date of the episodic event, and expected completion date (no later than 60 days from beginning date),
- the reason for the episodic event (event description),
- identification of the applicable federal waste codes (and state waste codes, if applicable), and
- the estimated total quantity of hazardous wastes that will be generated as a result of the episodic event.

ITEM 14 – LQG CONSOLIDATION OF VSQG HAZARDOUS WASTE

The 2016 Hazardous Waste Generator Improvements Final Rule allows LQGs to receive and consolidate hazardous wastes from VSQGs if the VSQGs are under the control of the same "person" as defined in 40 CFR 260.10.

NOTE	"Control," for the purposes of this section, means the power to direct the policies of the generator, whether by the ownership of stock, voting rights, or otherwise. Contractors who operate generator facilities on behalf of a different person as defined in 40 CFR 260.10 are not be deemed to "control" such generators. If you **mark "Y,"** you must fill out the Addendum to the Site ID Form: LQG Consolidation of VSQG Hazardous Waste.

Place an "X" in the "Y" box, if you are an LQG taking advantage of the provision found at 40 CFR 262.17 (f), you must notify (or re-notify) EPA or your authorized State. Otherwise, **place an "X" in the "N" box**. Information to be completed for each VSQG you are receiving hazardous waste from including:

- EPA Identification number (if applicable),
- the site name,
- address,
- contact name, and
- telephone number.

ITEM 15 – NOTIFICATION OF LQG SITE CLOSURE FOR A CENTRAL ACCUMULATION AREA (CAA) OR ENTIRE FACILITY

The 2016 Hazardous Waste Generator Improvements Final Rule requires LQGs to notify EPA no later than 30 days prior to closing their facility. They must also notify EPA within 90 days after closing the facility and having complied with the closure performance standards of 40 CFR 262.17 (a)(8) (iii) or 40 CFR 262.17 (a)(8)(iv), or notify EPA that they cannot meet the closure performance standards. Optionally, an LQG may notify that they are closing a central accumulation area.

Place an "X" in the "Y" box, then **complete the appropriate boxes: 15.A – D.** Otherwise, **place an "X" in the "N" box**.

15.1 CENTRAL ACCUMULATION AREA (CAA) OR ENTIRE FACILITY

Indicate if you are closing a central accumulation area or are closing the entire facility.

15.2 EXPECTED CLOSURE DATE

Provide the date (mm/dd/yyyy) that you expect to close the CAA or the entire facility.

15.3 REQUESTING NEW CLOSURE DATE

If you cannot complete the closure of your facility (or central accumulation area) within 90 days of starting the closure process, indicate the new closure date (mm/dd/yyyy). Explain in Item 18 - Comment why you are requesting the additional time.

15.4 DATE CLOSED

Provide the date (mm/dd/yyyy) that the facility or CAA closed.

a. In compliance with the closure performance standards 40 CFR 262.17(a)(8):
Indicate if the facility or CAA closed in compliance with the closure performance standards in 40 CFR 262.17(a)(8).

b. Not in compliance with the closure performance standards 40 CFR 262.17(a)(8):
Indicate if the facility or CAA closed but failed to meet closure performance standards in 40 CFR 262.17(a)(8).

ITEM 16 – NOTIFICATION OF HAZARDOUS SECONDARY MATERIAL (HSM) ACTIVITY

NOTE	40 CFR 260.42 must be in effect in your State in order to manage hazardous secondary material under these regulations.

16.A NOTIFICATION OF MANAGEMENT OF HAZARDOUS SECONDARY MATERIAL

Place an "X" in the "Y" box if you are notifying under 40 CFR 260.42 that you will begin managing, are still managing, or will stop managing hazardous secondary material under 40 CFR 260.30, 40 CFR 261.4(a)(23), (24), or (27). Otherwise, **place an "X" in the "N" box.**

NOTE	If you mark "Y," you must fill out the Addendum to the Site ID Form: Notification of Hazardous Secondary Material Activity.

16.B NOTIFICATION OF RECYCLING OF HAZARDOUS SECONDARY MATERIAL

Place an "X" in the "Y" box if you are submitting a legitimate recycling notification under 40 CFR 260.43(a)(4)(iii). Recycling of hazardous secondary materials for the purpose of the exclusions or exemptions from hazardous waste regulations must be legitimate. You must notify EPA if the product of your recycling process has levels of hazardous constituents that are not comparable to a legitimate product or intermediate, but the recycling is still legitimate. The recycling can be shown to be legitimate based on lack of exposure from toxics in the product, lack of the bioavailability of the toxics in the product, or other relevant considerations which show that the recycled product does not contain levels of hazardous constituents that pose a significant human health or environmental risk. Documentation of legitimacy must be maintained on-site for three years. Provide additional information regarding your recycling in Item 18 - Comments. Otherwise, **place an "X" in the "N" box**. Notification is not required if the product of the recycling process is comparable to a legitimate product or intermediate per 40 CFR 260.43(a)(4)(i) or (ii).

ITEM 17 – ELECTRONIC MANIFEST BROKER

Place an "X" in the "Y" box if you are a person as defined in 40 CFR 260.10 that elects to use the electronic manifest system to obtain, complete and transmit an electronic manifest format supplied by the EPA electronic manifest system under a contractual relationship with a hazardous waste generator. Otherwise, **place an "X" in the "N" box**.

ITEM 18 – COMMENTS

Use this section as needed to provide additional information for Items 1 through 17. Include the item number and box letter (if any) for each comment you make. You may attach additional sheets if needed. Remember to include your EPA Identification Number on the top of each page.

ITEM 19 – CERTIFICATION

This certification must be signed and dated by the generator(s), owner(s), operator(s), or authorized representative(s) of the site. See 40 CFR 270.11 for more information on signatories in general. An "authorized representative" is a person responsible for the overall operation of the site or an operational unit (i.e., a plant manager or superintendent, or a person of equivalent responsibility). To qualify as an "authorized representative," generator, owner, operator, or responsible official must submit a written authorization to the Director in an authorized state or the EPA Regional Director in non-authorized states.

NOTE	All Site ID Form submissions must include this certification to be complete.

ADDENDUM TO THE SITE IDENTIFICATION FORM:
NOTIFICATION OF HAZARDOUS SECONDARY MATERIAL (HSM) ACTIVITY

YOU MUST FILL OUT THIS ADDENDUM IF:

- You are located in a State that allows you to manage excluded hazardous secondary material under 40 CFR 260.30, 40 CFR 261.4(a)(23), (24), or (27), (or state equivalent) **AND**

- You are or will be managing excluded HSM in compliance with 40 CFR 260.30, 261.4(a)(23), (24), or (27) (or state equivalent) or have stopped managing excluded HSM in compliance with the exclusion(s) and do not expect to manage any amount of excluded HSM under the exclusion(s) for at least one year. Do not include any information regarding your hazardous waste activities in this section. Note: If your facility was granted a solid waste variance under 40 CFR 260.30 prior to July 13, 2015, your management of HSM under 40 CFR 260.30 is grandfathered under the previous regulations and you are not required to notify for the HSM management activity excluded under 40 CFR 260.30. See EPA's website for more information about these regulations at https://www.epa.gov/hw.

Complete all Items 1 and 2.

NOTE	You must be managing excluded hazardous secondary material in compliance with 40 CFR 260.30, 40 CFR 261.4(a)(23), (24), or (27) (or State equivalent). Do not include any information regarding your hazardous wastes in this section. See 80 FR 1694 or https://www.epa.gov/hw for more information on these exclusions. You must submit a completed Site ID Form, including this Addendum, prior to operating under the exclusion(s) and by March 1 of each even-numbered year thereafter to your regulatory authority using the Site ID Form as pursuant to 40 CFR 260.42. Persons who must satisfy this notification requirement can submit this information at the same time as their Hazardous Waste Report (which is also due by March 1 of each even-numbered year). If you stop managing hazardous secondary material in accordance with the exclusion(s) and do not expect to manage any amount of hazardous secondary material under the exclusion(s) for at least one year, you must also submit a completed Site ID Form, including this Addendum, within thirty (30) days pursuant to 40 CFR 260.42.

ITEM 1 – REASON FOR NOTIFICATION

Place an "X" in the box for the reason that applies to you:

FACILITY WILL BEGIN MANAGING EXCLUDED HAZARDOUS SECONDARY MATERIAL AS OF (MM/DD/YYYY)

Place an "X" in this box if you are notifying that you will begin managing hazardous secondary material under the exclusion(s).

- Facilities must notify prior to operating under the exclusion(s).
- If placing an "X" in this box, list the date (mm/dd/yyyy) when you will begin managing hazardous secondary material under 40 CFR 260.30, 40 CFR 261.4(a)(23), (24), or (27).

NOTE	If the facility had previously notified that it will stop managing hazardous secondary material in the past but will now begin anew, list the next planned start date.

FACILITY IS STILL MANAGING EXCLUDED HAZARDOUS SECONDARY MATERIAL/RE-NOTIFYING AS REQUIRED BY MARCH 1 OF EACH EVEN-NUMBERED YEAR

Place an "X" in this box if you are re-notifying that you are still managing hazardous secondary material under the exclusion(s).

- Facilities must re-notify by March 1st of each even-numbered year.
- If placing an "X" in this box, you do not have to list a date.

NOTE	You must have previously notified that you began managing hazardous secondary material in order to check this box.

FACILITY HAS STOPPED MANAGING EXCLUDED HAZARDOUS SECONDARY MATERIAL AS OF (MM/DD/YYYY) AND IS NOTIFYING AS REQUIRED

Place an "X" in this box, if you are notifying that you have stopped managing hazardous secondary material under the exclusion(s) and do not expect to manage any amount of hazardous secondary material for at least one year (pursuant to 40 CFR 260.42(b)). List the date when you stopped managing hazardous secondary material. Enter the date in "mm/dd/yyyy" format.

- Facilities must notify within 30 days of when they stopped managing hazardous secondary material. You are considered to have stopped managing hazardous secondary material if: (1) you stop managing hazardous secondary material completely (e.g., you cease operations); (2) you choose to manage the hazardous secondary material as hazardous waste; (3) you undergo closure and request release from financial assurance per 40 CFR 261.143(h) or 40 CFR 264.143; or (4) you temporarily suspend management of hazardous secondary material for at least one year.

- Only place an "X" in this box if you have stopped managing all hazardous secondary material under the exclusion(s). For example, if your facility only stopped managing one hazardous secondary material, but continued to manage another hazardous secondary material, you would leave this box blank since your facility continues to manage some amount of hazardous secondary material.

- If you submit a notification that you have stopped managing hazardous secondary material, you do not need to re-notify (unless you choose to manage hazardous secondary material again, in which case you would have to submit a notification prior to managing). After submitting a stop notification, you can leave the Addendum blank for subsequent submissions, including any subsequent Hazardous Waste Report submissions.

ITEM 2 – DESCRIPTION OF EXCLUDED HAZARDOUS SECONDARY MATERIAL (HSM) ACTIVITY

In the table provided on this Addendum to the Site Identification Form, list your appropriate facility code, each waste code for the hazardous secondary material you manage, the estimated and actual quantities, in short tons, for each hazardous secondary material, and the appropriate land-based code for how you manage the hazardous secondary material. Do not include any information regarding your hazardous wastes in this section. See examples below on how to answer this question.

a. **Facility Code**
Using the nationally-defined Facility Codes enter the appropriate 2-digit code(s) that correctly describes your facility. If more than one code applies to your facility, enter each 2-digit code on a separate row. Each hazardous secondary material should be reported by facility code.

LIST	Click *here* for a list of the nationally-defined Facility Codes.

b. **Waste Code(s) for HSM**
Use the box provided to enter the appropriate 4-digit hazardous waste code(s) that would apply to your hazardous secondary material if you managed it as hazardous waste (i.e., the waste code(s) that would apply if you did not manage your material in accordance with 40 CFR 260.30, 40 CFR 261.4(a)(23), (24), or (27).

NOTE	If you list more codes or manage more hazardous secondary material than will fit in the table under Item 2, please continue the Site Identification Form under Item 18 – Comments, or on an extra sheet. Remember to include your EPA Identification Number on the top of each page.

LIST	Click *here* for a list of the nationally-defined Hazardous Waste Codes.

c. **Estimate Short Tons of Excluded HSM to be Managed Annually**
In the box provided, enter your estimated quantity (using short tons) of hazardous secondary material you expect to manage annually. Convert all physical quantities (e.g., gallons, cubic yards, kilograms,

metric tons, etc.) to short tons (1 short ton = 2,000 pounds) and round to the nearest ton (no decimals).

NOTE	Your estimated quantity should be for the entire amount of hazardous secondary material to be reclaimed NOT just the quantity of constituent or product reclaimed.

d. **Actual Short Tons of Excluded HSM Managed During the Most Recent Odd-Numbered Year**
Report the quantity (using short tons) of each hazardous secondary material you actually managed during the most recent odd-numbered year. For example, if you are submitting this notification on February 20, 2018, enter the amount you actually managed during 2017 (i.e., the quantity you managed from January 1, 2017 to December 31, 2017). Convert all physical quantities (e.g., gallons, cubic yards, kilograms, metric tons, etc.) to short tons (1 short ton = 2,000 pounds) and round to the nearest ton (no decimals). If this is your initial notification, enter "0."

NOTE	Your actual quantity should be for the entire amount of hazardous secondary material that was sent for reclamation NOT just the quantity of constituent or product reclaimed.

e. **Land-based Unit Code**
Using the nationally-defined Land-based Unit Codes, enter in the 2-digit code that best describes the land-based unit you use or will use to manage the hazardous secondary material. If you do not use any land-based units, enter "NA." If you use the code "OT" (Other), please describe your land-based unit in Item 18 – Comments. If more than one land-based unit code applies to a hazardous secondary material, list it separately using another row.

LIST	Click *here* for a list of the nationally-defined Land-based Unit Codes.

EXAMPLES FOR REPORTING HAZARDOUS SECONDARY MATERIAL ACTIVITY

EXAMPLE 1

A pharmaceutical manufacturer generates spent solvents that are characteristic for ignitability (D001). The manufacturer plans to manage spent solvents under 40 CFR 261.4(a)(23) and 261.4(a)(24) by sending some amount to a reclaimer within its own company and the rest off-site to a recycling facility within the U.S. The manufacturer will not manage any spent solvents in a land-based unit. Following the regulations, the manufacturer submits an initial notification prior to managing its spent solvents under the exclusions. The facility would report its hazardous secondary material activity as follows:

a. Facility code	b. Waste Code(s) for hazardous secondary material (HSM)	c. Estimated short tons of HSM to be managed annually	d. Actual short tons of HSM managed during the most recent odd-numbered year	e. Land-based unit code
02	D001	15	0	NA
06	D001	40	0	NA

EXAMPLE 2

A steel manufacturer generates electric arc furnace dust and spent pickle liquor from one of its steel operations. The manufacturer sends electric arc furnace dust (K061) off-site to a recycling facility within the U.S. and reclaims spent pickle liquor (K062) on-site. Neither hazardous secondary material is managed in a land-based unit. The steel manufacturer has managed both hazardous secondary material under 40 CFR 261.4(a)(23) and 261.4(a)(24) for a number of years and it is now time to re-notify. The facility would report its hazardous secondary material activity as follows:

a. Facility code	b. Waste Code(s) for hazardous secondary material (HSM)	c. Estimated short tons of HSM to be managed annually	d. Actual short tons of HSM managed during the most recent odd-numbered year	e. Land-based unit code
01	K062	60	52	NA
06	K061	20,000	22,468	NA

EXAMPLE 3

A RCRA-permitted recycling facility has been receiving and reclaiming spent solvents under 40 CFR 261.4(a)(23) and 261.4(a)(24) for a number of years. The facility receives and reclaims spent solvents from multiple hazardous secondary material generators, some of which are within the same company. No spent solvents are managed in a land-based unit. It is now time to re-notify. The facility would report its hazardous secondary material activity as follows:

a. Facility code	b. Waste Code(s) for hazardous secondary material (HSM)	c. Estimated short tons of HSM to be managed annually	d. Actual short tons of HSM managed during the most recent odd-numbered year	e. Land-based unit code
03	D001; F002; F003; F005	6,000	7,533	NA
03	D001; D035; F002; F003	1,500	918	NA
07	D001; F002; F003; F005	3,000	3,509	NA
07	D001; D038; F002; F003	1,000	523	NA

EXAMPLE 4

A smelting operation generates furnace bricks that are characteristic for chromium (D007) and sends them off-site for recycling. Before shipping the bricks off-site, the facility manages some of the bricks in a containment building and the rest in a pile on the land. The facility has been managing the bricks under 40 CFR 261.4(a)(24) for a number of years and must now re-notify. The facility would report its activity as follows:

a. Facility code	b. Waste Code(s) for hazardous secondary material (HSM)	c. Estimated short tons of HSM to be managed annually	d. Actual short tons of HSM managed during the most recent odd-numbered year	e. Land-based unit code
06	D007	200	235	NA
06	D007	115	126	PL

EXAMPLE 5

A RCRA-permitted intermediate facility has been managing wastewater treatment sludges from electroplating operations (F006) for the past seven years but, due to company consolidation, it will soon shut down. In accordance with 40 CFR 260.42, the facility notifies that it will stop managing hazardous secondary material. The facility would report its activity as follows:

a. Facility code	b. Waste Code(s) for hazardous secondary material (HSM)	c. Estimated short tons of HSM to be managed annually	d. Actual short tons of HSM managed during the most recent odd-numbered year	e. Land-based unit code
08	F005	0	5,034	NA

ADDENDUM TO THE SITE IDENTIFICATION FORM: LQG CONSOLIDATION OF VSQGS HAZARDOUS WASTES

YOU MUST FILL OUT THIS ADDENDUM IF:

You must fill out this Addendum if you are an LQG taking advantage of the new provision at 40 CFR 262.17(f) allowing you to consolidate hazardous wastes from VSQGs under the control of the same person (as defined in 40 CFR 260.10). For each VSQG that the LQG is receiving hazardous waste from, the LQG must identify the EPA Identification Number, if applicable, the name of the site, the site address, contact name, phone number, and email address.

ITEM 1 – EPA ID NUMBER

Provide the EPA Identification Number for the VSQG whose waste you are consolidating, if applicable. A VSQG may have an EPA ID Number either because it's State requires it, or because it may have been an SQG or LQG at one time, or for another reason. If the VSQG does not have an EPA ID Number, leave this blank.

ITEM 2 – SITE NAME

Provide the legal name of the VSQG.

ITEMS 3-6 – SITE LOCATION

Provide the complete location address for the VSQG. Please note that the address must be a physical address, not a post office box or route number.

ITEM 7-9 –CONTACT INFORMATION

Enter the telephone number, name of the individual who should be contacted for information about the VSQG, and their email address.

ADDENDUM TO THE SITE IDENTIFICATION FORM: EPISODIC GENERATOR

YOU MUST FILL OUT THIS SECTION IF:

You must fill out this Addendum if you are a VSQG or SQG taking advantage of the provision in 40 CFR 262 Subpart L allowing you to generate additional quantities of hazardous waste—temporarily exceeding your normal generator category limits — and still maintain your existing generator category, if complying with the specified conditions identified at 40 CFR 262.232(a) and (b). These conditions include management standards, completing the event, and removing all hazardous waste from the site within sixty (60) days.

NOTE	You cannot use this Addendum when submitting your Hazardous Waste Report.

Examples of planned episodic events include tank cleanouts, short-term construction projects, short-term site remediation, equipment maintenance during plant shutdowns, removal of excess chemical inventories, and site and production process decommissions by a new operator. Unplanned episodic events, which EPA expects would be less frequent, include production process upsets, product recalls, accidental spills, or "acts of nature," such as a tornado, hurricane, or flood. Generators conducting episodic events must notify with the type of event (planned or unplanned), an emergency contact and phone number, the start and end date of the episodic event (must be sixty (60) days or fewer apart), a description of the event, federal and state waste codes of wastes being generated, and the expected amount of waste to be generated.

ITEM 1-2 – PLANNED/UNPLANNED EVENT

Indicate whether the event being conducted is planned or unplanned. Furthermore, indicate the reason for the planned or unplanned event. If none of the reasons listed apply, mark "Other" and describe the event in Item 18 - Comments.

ITEM 3-4 – EMERGENCY CONTACT INFORMATION

Provide an emergency contact phone number and contact name for the individual who should be contacted regarding the information relating to this episodic event.

ITEM 5-6 – BEGINNING AND END DATES

Provide the estimated start date and end date of the event. The event must be complete within sixty (60) days of the start date.

WASTE 1 – 3

For each waste stream produced as a result of the episodic event, provide a description of the waste generated, the estimated quantity generated, and the applicable federal and/or state hazardous waste codes. If necessary, attach a separate sheet of paper. Remember to enter your site's EPA Identification Number in the top left-hand corner of each sheet.

7. WASTE DESCRIPTION

Provide a short narrative description of the hazardous waste, such as:

- General type;
- Source;
- Type of hazard; and
- Generic chemical name or primary hazardous constituents.

8. ESTIMATED QUANTITY

Provide an estimated quantity, **in pounds**, of hazardous waste to be generated as a result of the episodic event.

9. FEDERAL AND/OR STATE HAZARDOUS WASTE CODES

Provide the federal hazardous waste codes and/or the state hazardous waste codes for the hazardous wastes you expect to generate during the episodic event.

LIST	Click *here* for a list of the nationally-defined Hazardous Waste Unit Codes.

OMB# 2050-0024; Expires 05/31/2020

United States Environmental Protection Agency RCRA SUBTITLE C SITE IDENTIFICATION FORM	

1. Reason for Submittal (Select only one.)

☐	Obtaining or updating an EPA ID number for an on-going regulated activity that will continue for a period of time. (Includes HSM activity)
☐	Submitting as a component of the Hazardous Waste Report for _________ (Reporting Year)
	☐ Site was a TSD facility and/or generator of > 1,000 kg of hazardous waste, > 1 kg of acute hazardous waste, or > 100 kg of acute hazardous waste spill cleanup in **one or more months of the reporting year** (or State equivalent LQG regulations)
☐	Notifying that regulated activity is no longer occurring at this Site
☐	Obtaining or updating an EPA ID number for conducting Electronic Manifest Broker activities
☐	Submitting a new or revised Part A Form

2. Site EPA ID Number

3. Site Name

4. Site Location Address

Street Address		
City, Town, or Village		County
State	Country	Zip Code

5. Site Mailing Address ☐ Same as Location Address

Street Address		
City, Town, or Village		
State	Country	Zip Code

6. Site Land Type

☐ Private ☐ County ☐ District ☐ Federal ☐ Tribal ☐ Municipal ☐ State ☐ Other

7. North American Industry Classification System (NAICS) Code(s) for the Site (at least 5-digit codes)

A. (Primary)	C.
B.	D.

EPA Form 8700-12, 8700-13 A/B, 8700-23

Page __ of __

EPA ID Number | | | | | | | | | | | | | OMB# 2050-0024; Expires 05/31/2020

8. Site Contact Information ☐ Same as Location Address

First Name	MI	Last Name
Title		
Street Address		
City, Town, or Village		
State	Country	Zip Code
Email		
Phone	Ext	Fax

9. Legal Owner and Operator of the Site

A. Name of Site's Legal Owner ☐ Same as Location Address

Full Name	Date Became Owner (mm/dd/yyyy)
Owner Type ☐ Private ☐ County ☐ District ☐ Federal ☐ Tribal ☐ Municipal ☐ State ☐ Other	
Street Address	
City, Town, or Village	

State	Country	Zip Code
Email		
Phone	Ext	Fax
Comments		

B. Name of Site's Legal Operator ☐ Same as Location Address

Full Name	Date Became Operator (mm/dd/yyyy)
Operator Type ☐ Private ☐ County ☐ District ☐ Federal ☐ Tribal ☐ Municipal ☐ State ☐ Other	
Street Address	
City, Town, or Village	

State	Country	Zip Code
Email		
Phone	Ext	Fax
Comments		

EPA ID Number | | | | | | | | | | | | | OMB# 2050-0024; Expires 05/31/2020

10. Type of Regulated Waste Activity (at your site)
Mark "Yes" or "No" for all current activities (as of the date submitting the form); complete any additional boxes as instructed.

A. Hazardous Waste Activities

☐ Y ☐ N	1. Generator of Hazardous Waste—If "Yes", mark only one of the following—a, b, c		
	☐	a. LQG	-Generates, in any calendar month (includes quantities imported by importer site) 1,000 kg/mo (2,200 lb/mo) or more of non-acute hazardous waste; or - Generates, in any calendar month, or accumulates at any time, more than 1 kg/mo (2.2 lb/mo) of acute hazardous waste; or - Generates, in any calendar month or accumulates at any time, more than 100 kg/mo (220 lb/mo) of acute hazardous spill cleanup material.
	☐	b. SQG	100 to 1,000 kg/mo (220-2,200 lb/mo) of non-acute hazardous waste and no more than 1 kg (2.2 lb) of acute hazardous waste and no more than 100 kg (220 lb) of any acute hazardous spill cleanup material.
	☐	c. VSQG	Less than or equal to 100 kg/mo (220 lb/mo) of non-acute hazardous waste.
If "Yes" above, indicate other generator activities in 2 and 3, as applicable.			
☐ Y ☐ N	2. Short-Term Generator (generates from a short-term or one-time event and not from on-going processes). If "Yes", provide an explanation in the Comments section.		
☐ Y ☐ N	3. Mixed Waste (hazardous and radioactive) Generator		
☐ Y ☐ N	4. Treater, Storer or Disposer of Hazardous Waste—Note: A hazardous waste Part B permit is required for these activities.		
☐ Y ☐ N	5. Receives Hazardous Waste from Off-site		
☐ Y ☐ N	6. Recycler of Hazardous Waste		
	☐	a. Recycler who stores prior to recycling	
	☐	b. Recycler who does not store prior to recycling	
☐ Y ☐ N	7. Exempt Boiler and/or Industrial Furnace—If "Yes", mark all that apply.		
	☐	a. Small Quantity On-site Burner Exemption	
	☐	b. Smelting, Melting, and Refining Furnace Exemption	

B. Waste Codes for Federally Regulated Hazardous Wastes. Please list the waste codes of the Federal hazardous wastes handled at your site. List them in the order they are presented in the regulations (e.g. D001, D003, F007, U112). Use an additional page if more spaces are needed.

C. Waste Codes for State Regulated (non-Federal) Hazardous Wastes. Please list the waste codes of the State hazardous wastes handled at your site. List them in the order they are presented in the regulations. Use an additional page if more spaces are needed.

EPA ID Number ☐☐☐☐☐☐☐☐☐☐☐☐ OMB# 2050-0024; Expires 05/31/2020

11. Additional Regulated Waste Activities (NOTE: Refer to your State regulations to determine if a separate permit is required.)

A. Other Waste Activities

☐ Y ☐ N		1. Transporter of Hazardous Waste—If "Yes", mark all that apply.
	☐	a. Transporter
	☐	b. Transfer Facility (at your site)
☐ Y ☐ N		2. Underground Injection Control
☐ Y ☐ N		3. United States Importer of Hazardous Waste
☐ Y ☐ N		4. Recognized Trader—If "Yes", mark all that apply.
	☐	a. Importer
	☐	b. Exporter
☐ Y ☐ N		5. Importer/Exporter of Spent Lead-Acid Batteries (SLABs) under 40 CFR 266 Subpart G—If "Yes", mark all that apply.
	☐	a. Importer
	☐	b. Exporter

B. Universal Waste Activities

☐ Y ☐ N		1. Large Quantity Handler of Universal Waste (you accumulate 5,000 kg or more) - If "Yes" mark all that apply. Note: Refer to your State regulations to determine what is regulated.
	☐	a. Batteries
	☐	b. Pesticides
	☐	c. Mercury containing equipment
	☐	d. Lamps
	☐	e. Other (specify) ____________
	☐	f. Other (specify) ____________
	☐	g. Other (specify) ____________
☐ Y ☐ N		2. Destination Facility for Universal Waste Note: A hazardous waste permit may be required for this activity.

C. Used Oil Activities

☐ Y ☐ N		1. Used Oil Transporter—If "Yes", mark all that apply.
	☐	a. Transporter
	☐	b. Transfer Facility (at your site)
☐ Y ☐ N		2. Used Oil Processor and/or Re-refiner—If "Yes", mark all that apply.
	☐	a. Processor
	☐	b. Re-refiner
☐ Y ☐ N		3. Off-Specification Used Oil Burner
☐ Y ☐ N		4. Used Oil Fuel Marketer—If "Yes", mark all that apply.
	☐	a. Marketer Who Directs Shipment of Off-Specification Used Oil to Off-Specification Used Oil Burner
	☐	b. Marketer Who First Claims the Used Oil Meets the Specifications

EPA ID Number | | | | | | | | | | | | | OMB# 2050-0024; Expires 05/31/2020

12. Eligible Academic Entities with Laboratories—Notification for opting into or withdrawing from managing laboratory hazardous wastes pursuant to 40 CFR 262 Subpart K.

☐ Y ☐ N	A. Opting into or currently operating under 40 CFR 262 Subpart K for the management of hazardous wastes in laboratories—If "Yes", mark all that apply. Note: See the item-by-item instructions for definitions of types of eligible academic entities.	
	☐	1. College or University
	☐	2. Teaching Hospital that is owned by or has a formal written affiliation with a college or university
	☐	3. Non-profit Institute that is owned by or has a formal written affiliation with a college or univer-
☐ Y ☐ N	B. Withdrawing from 40 CFR 262 Subpart K for the management of hazardous wastes in laboratories.	

13. Episodic Generation

☐ Y ☐ N	Are you an SQG or VSQG generating hazardous waste from a planned or unplanned episodic event, lasting no more than 60 days, that moves you to a higher generator category. If "Yes", you must fill out the Addendum for Episodic Generator.

14. LQG Consolidation of VSQG Hazardous Waste

☐ Y ☐ N	Are you an LQG notifying of consolidating VSQG Hazardous Waste Under the Control of the Same Person pursuant to 40 CFR 262.17(f)? If "Yes", you must fill out the Addendum for LQG Consolidation of VSQGs hazardous waste.

15. Notification of LQG Site Closure for a Central Accumulation Area (CAA) (optional) OR Entire Facility (required)

☐ Y ☐ N	LQG Site Closure of a Central Accumulation Area (CAA) or Entire Facility.
	A. ☐ Central Accumulation Area (CAA) ☐ Entire Facility
	B. Expected closure date: ________ mm/dd/yyyy
	C. Requesting new closure date: ________ mm/dd/yyyy
	D. Date closed : mm/dd/yyyy ☐ 1. In compliance with the closure performance standards 40 CFR 262.17(a)(8) ☐ 2. Not in compliance with the closure performance standards 40 CFR 262.17(a)(8)

16. Notification of Hazardous Secondary Material (HSM) Activity

☐ Y ☐ N	A. Are you notifying under 40 CFR 260.42 that you will begin managing, are managing, or will stop managing hazardous secondary material under 40 CFR 260.30, 40 CFR 261.4(a)(23), (24), or (27)? If "Yes", you must fill out the Addendum to the Site Identification Form for Managing Hazardous Secondary Material.
☐ Y ☐ N	B. Are you notifying under 40 CFR 260.43(a)(4)(iii) that the product of your recycling process has levels of hazardous constituents that are not comparable to or unable to be compared to a legitimate product or intermediate but that the recycling is still legitimate? If "Yes", you may provide explanation in Comments section. You must also document that your recycling is still legitimate and maintain that documentation on site.

17. Electronic Manifest Broker

☐ Y ☐ N	Are you notifying as a person, as defined in 40 CFR 260.10, electing to use the EPA electronic manifest system to obtain, complete, and transmit an electronic manifest under a contractual relationship with a hazardous waste generator?

EPA Form 8700-12, 8700-13 A/B, 8700-23 Page __ of __

EPA ID Number | | | | | | | | | | | | OMB# 2050-0024; Expires 05/31/2020

18. Comments (include item number for each comment)

19. Certification I certify under penalty of law that this document and all attachments were prepared under my direction or supervision in accordance with a system designed to assure that qualified personnel properly gather and evaluate the information submitted. Based on my inquiry of the person or persons who manage the system, or those persons directly responsible for gathering the information, the information submitted is, to the best of my knowledge and belief, true, accurate, and complete. I am aware that there are significant penalties for submitting false information, including the possibility of fines and imprisonment for knowing violations. **Note: For the RCRA Hazardous Waste Part A permit Application, all owners and operators must sign (see 40 CFR 270.10(b) and 270.11).**

Signature of legal owner, operator or authorized representative	Date (mm/dd/yyyy)
Printed Name (First, Middle Initial Last)	Title
Email	

Signature of legal owner, operator or authorized representative	Date (mm/dd/yyyy)
Printed Name (First, Middle Initial Last)	Title
Email	

EPA ID Number | | | | | | | | | | | | | OMB# 2050-0024; Expires 05/31/2020

ADDENDUM TO THE SITE IDENTIFICATION FORM: NOTIFICATION OF HAZARDOUS SECONDARY MATERIAL ACTIVITY

ONLY fill out this form if:

- You are located in a State that allows you to manage excluded hazardous secondary material (HSM) under 40 CFR 261.2(30), 261.4(a)(23), (24), or (27) (or state equivalent; See https://www.epa.gov/epawaste/hazard/dsw/statespf.htm for a list of eligible states; AND
- You are or will be managing excluded HSM in compliance with 40 CFR 260.30, 261.4(a)(23), (24), or (27) (or state equivalent) or have stopped managing excluded HSM in compliance with the exclusion(s) and do not expect to manage any amount of excluded HSM under the exclusion(s) for at least one year. Do not include any information regarding your hazardous waste activities in this section. Note: If your facility was granted a solid waste variance under 40 CFR 260.30 prior to July 13, 2015, your management of HSM under 40 CFR 260.30 is grandfathered under the previous regulations and you are not required to notify for the HSM management activity excluded under 40 CFR 260.30.

1. Reason for Notification (Include dates where requested)

☐ Facility will begin managing excluded HSM as of ______________ (mm/dd/yyyy).

☐ Facility is still managing excluded HSM/re-notifying as required by March 1 of each even-numbered year.

☐ Facility has stopped managing excluded HSM as of ______________ (mm/dd/yyyy) and is notifying as required.

2. Description of Excluded HSM Activity. Please list the appropriate codes (see Code List section of the instructions) and quantities, in short tons, to describe your excluded HSM activity ONLY (do not include any information regarding your hazardous wastes). Use additional pages if more space is needed.

A. Facility Code	B. Waste Code(s) for HSM	C. Estimate Short Tons of excluded HSM to be managed annually	D. Actual Short Tons of excluded HSM that was managed during the most recent odd-numbered year	E. Land-based Unit Code

EPA ID Number | | | | | | | | | | | | | OMB# 2050-0024; Expires 05/31/2020

ADDENDUM TO THE SITE IDENTIFICATION FORM: EPISODIC GENERATOR

ONLY fill out this form if:

- You are an SQG or VSQG generating hazardous waste from a planned or unplanned episodic event, lasting no more then 60 days, that moves the generator to a higher generator category pursuant to 40 CFR 262 Subpart L. Note: Only one planned and one unplanned episodic event are allowed within one year; otherwise, you must follow the requirements of the higher generator category. Use additional pages if more space is needed.

Episodic Event	
1. Planned ☐ Excess chemical inventory removal ☐ Tank cleanouts ☐ Short-term construction or demolition ☐ Equipment maintenance during plant shutdowns ☐ Other ________________	2. Unplanned ☐ Accidental spills ☐ Production process upsets ☐ Product recalls ☐ "Acts of nature" (Tornado, hurricane, flood, etc.) ☐ Other ________________
3. Emergency Contact Phone	4. Emergency Contact Name
5. Beginning Date ________ (mm/dd/yyyy)	6. End Date ________ (mm/dd/yyyy)

Waste 1

7. Waste Description					8. Estimated Quantity (in pounds)
9. Federal and/or State Hazardous Waste Codes					

Waste 2

7. Waste Description					8. Estimated Quantity (in pounds)
9. Federal and/or State Hazardous Waste Codes					

Waste 3

7. Waste Description					8. Estimated Quantity (in pounds)
9. Federal and/or State Hazardous Waste Codes					

ADDENDUM TO THE SITE IDENTIFICATION FORM:
LQG CONSOLIDATION OF VSQG HAZARDOUS WASTE

ONLY fill out this form if:

- You are an LQG receiving hazardous waste from VSQGs under the control of the same person. Use additional pages if more space is needed.

VSQG 1		
1. EPA ID Number (if assigned)	2. Name	
3. Street Address		
4. City, Town, or Village	5. State	6. Zip Code
7. Contact Phone Number	8. Contact Name	
9. Email		

VSQG 2		
1. EPA ID Number (if assigned)	2. Name	
3. Street Address		
4. City, Town, or Village	5. State	6. Zip Code
7. Contact Phone Number	8. Contact Name	
9. Email		

VSQG 3		
1. EPA ID Number (if assigned)	2. Name	
3. Street Address		
4. City, Town, or Village	5. State	6. Zip Code
7. Contact Phone Number	8. Contact Name	
9. Email		

PART XI: USED OIL HANDLER REPORT REQUIREMENTS

TCEQ Report requirements for Used Oil Handlers are required by Used Oil Processors, Re-Refiners, Used Oil Filter Handlers, Collection Centers, and Do-it-yourselfer Used Oil Collection Centers as required by the Texas Health & Safety Code 371.026 **(167)** and 40 Code of Federal Regulations Part 279, Section 279.57(b).**(168)** The reporting requirements are a biennial report for processors, re-refiners and filter handlers using the Biennial Reporting Form For Used Oil Processors, Re-Refiners and Filter Handers TCEQ-0648 **(167)** form. Collection centers along with Do-it-yourself collection centers have an annual requirement using the 20XX Annual Reporting Form for Used Oil Collection Centers TCEQ-0567**(156)** form.

Used Oil Handler activities are identified below for your reference with qualifying statements to insure your applicability.

Following the activities are instructions for completing the required forms and blank copies (Forms 1 and 2) for reference. Neither forms have TCEQ detailed instructions, however below limited instructions are provided with comments ***italicized and bolded.***

Both forms are available on line as PDF auto entry forms with the TCEQ-0648 form at:

https://www.tceq.texas.gov/assets/public/permitting/rrr/forms/0648.pdf and the TCEQ -0567

form at: https://www.tceq.texas.gov/assets/public/permitting/rrr/forms/0567.pdf.

USED OIL HANDLERS:

1. **Used Oil Processors and Re-Refiners:** Processing means any chemical or physical operation designed to produce or facilitate the production of fuel oils, lubricants, or other used oil-derived products from used oil. These operations include

but are not limited to blending used oil with virgin petroleum products, blending used oils to meet fuel specification, filtration, simple distillation, chemical or physical separation, and re-refining. Processing, however, does not include generators processing used oil generated onsite for use on-site. If transporters or transfer facilities filter oil from oil-bearing transformers before returning it to its original use, they would not be subject to processing standards.

a. **Re-Refining Versus Processing:** Re-refining is not the same as processing even though they may be subject to many of the same requirements. The major difference is that re-refined used oil has sufficient quality to be used again as a lubricating oil. By contrast, processing produces a used oil of lower quality, which can be used as a fuel for producing electricity or other purposes. Processed used oil can also be blended for marine fuel or other use. It can even be mixed with asphalts.
b. **Report Requirement: (YES)** RULE §324.12 **(115)** Processors and Re-refiners Biennial report. The biennial report required by 40 CFR §279.57(b) covering each odd numbered year must be provided to the commission by December 1 of the odd numbered year if all used oil operations have been completed for that year. If not, you must submit the report by January 25 of the following even numbered year. The information must be entered on a commission form. Mail the report to the Texas Commission on Environmental Quality, Used Oil Recycling Program.

2. **Used Oil Filter Handler: TAC §328.22. Applicability:** Applies to a used oil filter only if the filter has been determined to be nonhazardous or exempt from being hazardous waste due to draining of used oil per Title 40 Code of Federal Regulations, §261.4(b) **(13)** and the filter has not been: (1) separated into its component parts; or (2) burned for: (A) steel recovery; or (B) energy recovery. (b) **This subchapter does not apply to**: (1) an industrial generator that is: (A) registered with the agency as an industrial or hazardous waste facility; or (B) under the waste management authority of a state agency other than the commission. (2) a do-it-yourselfer. Adopted December 1, 1999 Effective December 23, 1999.
   a. **TAC §328.21(116) Used Oil Filter Handler activities: TAC §328.21** Processor – A person that processes used oil filters generated by another per-

son. The term does not include a generator that consolidates, drains, or crushes used oil filters for off-site recycling or disposal. **TAC §328.21 (116) Storage facility** - A location that stores used oil filters before transportation, processing, recycling, or disposal of the filters. The term does not include the location of a generator. **TAC §328.21 Transporter** – A person that transports used oil filters to a location for storage, processing, recycling, or disposal.

b. **Reporting Requirements: (YES) TAC §328.24. Reporting:** (a) A transporter, storage facility, or processor must report by January 25th of every even-numbered year and reporting information must be mailed to the TCEQ.

3. **Used Oil Collection Centers:** A collection center is a facility that accepts, aggregates, or stores used oil received from generators such as service stations, governments, or businesses (in shipments of not more than 55 gallons of used oil at a time). A center may /licensed/permitted/recognized by a state/co accept used oil from household do-it-yourselfers (DIYs). §279.1 Definitions. Used oil collection center means any site or facility that is registered entity/municipal government to manage used oil and accepts/aggregates and stores used oil collected from used oil generators regulated under subpart C of this part who bring used oil to the collection center in shipments of no more than 55 gallons under the provisions of §279.24. Used oil collection centers may also accept used oil from household do-it yourselfers.
    a. **Reporting Requirements: (YES)** Collection Centers TAC §324.7(3)(E) **(115)** report annually the amount of household and non-household used oil collected by January 25 of each year on a commission form. Mixtures of household used oil and non-household used oil shall be considered non-household used oil. Mail the report to the Texas Commission on Environmental Quality, Used Oil Recycling Program.

4. **Do-it-yourselfer Used Oil Collection Centers:** Do-it-yourselfer (DIY) is a used oil collection center is any site or facility that accepts/aggregates and stores used oil collected only from household do-it-yourselfers and must comply with all same re-

quirements as a Used Oil Collection Center.

   a. **Reporting Requirements: (YES)** Collection Centers TAC §324.7(1) (E) **(115)** annually report the amount of household used oil collected by January 25 of each year on a commission form. Mail the report to the Texas Commission on Environmental Quality, Used Oil Recycling Program.

5. **Used Oil Transporters and Transfer Facilities:** Used oil transporters may consolidate or aggregate used oil for the purposes of transporting and may conduct incidental processing operations that occur during the normal course of transportation (e.g., separating water from used oil that has settled during the aggregation or transportation process). Transporters who otherwise produce used oil-derived products or make these products more amenable for production are considered used oil processors. Like generators, used oil transporters who engage in other used oil management practices must follow the applicable standards in Part 279. For example, a transporter who processes used oil must also comply with the requirements for used oil processors and re-refiners in Part 279, Subpart F, including transporters who blend used oil with virgin oils to meet specification. Transporters, however, may conduct incidental processing operations and consolidation for the purposes of transportation in compliance with §279.41 without being considered a processor or re-refiner.
   a. Reporting Requirements: No reporting requirements.

6. **Used Oil Transfer Facilities Exception:** Used oil transfer facilities are transportation related facilities including loading docks, parking areas, storage areas, and other areas where shipments of used oil are held for more than 24 hours during the normal course of transportation and not longer than 35 days. So, unless you meet a transporter exception and the transfer exceptions you are still subject to the 9. (A) section below on the record keeping requirements as a processor at a minimum.
   a. Reporting Requirements: No reporting requirements.

7. **Used Oil Fuel Marketers:** Per 40 CFR 279.11, a used oil fuel marketer as any person who either (1) directs a shipment of off-specification used oil from that facility to a used oil burner, or (2) first claims that the used oil going to be burned for

energy recovery meets the specification requirements.

   a. Used oil marketers can be divided into two categories:
      1. Those that market off-specification used oil and;
      2. Those that market oil that meets specification.
   b. Reporting Requirements: No reporting requirements.

8. **Off-Specification Used Oil Marketers:** Used oil marketers handling off-specification oil are subject to the prohibitions found in §279.71, as well as analysis of used oil fuel and notification to EPA of location and description of used oil management activities.
   a. Reporting Requirements: No reporting requirements.

9. **Used Oil Aggregation Points:** 40 CFR §279.32 A used oil aggregation point is any site or facility that accepts, aggregates, and/or stores used oil collected only from other used oil generation sites owned or operated by the owner or operator of the aggregation point, from which used oil is transported to the aggregation point in shipments of no more than 55 gallons under the provisions of §279.24(b). Used oil aggregation points may also accept used oil from household do-it-yourselfers.
   a. Reporting Requirements: No reporting requirements.

10. **Used Oil Generators:** 40 CFR §279.1 identifies a used oil generator as any person, by site, whose act or process produces used oil or whose act first causes used oil to become subject to regulation. Used oil generators ranging from multinational corporations to local service stations, used oil generators are subject to limited standards under Part 279, Subpart C. Reporting Requirements: No reporting requirements.

PART XII: BIENNIAL REPORTING FORM FOR USED OIL PROCESSORS, RE-REFINERS AND FILTER HANDLERS TCEQ-0648 REQUIREMENTS (167)

As always, if you have any questions regarding the form or need to verify your information follow the instruction indicated on the form as follows: If you have questions on how to fill out this form or about the Used Oil Recycling Program, please contact us at 512/239-6832, option 2. Individuals are entitled to request and review their personal information the agency gathers on its forms. They may also have any errors in their information corrected. Any administrative changes must be made on the TCEQ Core Data Form (TCEQ 10400). To review such information, contact TCEQ at **512/239-3282.**

Information Applicable to all Used Oil Handlers (TCEQ-0648)

Please return completed form to the TCEQ by JANUARY 25, 20XX: ***Fill in the year the form is being filed for.***

Any administrative changes must be made on the TCEQ Core Data Form (TCEQ 10400).

> ***Commentary:*** *Please ensure all changes related to your Core Data form previously filed to receive your permit(s) are current; if corrections were made following the last report, verify your TCEQ changes are in place prior to completing this report in order to avoid any confusion at the TCEQ as to your appropriate compliance.*

1. TCEQ Registration #:

Commentary: *This number must match your permit number; should verify this prior to entry.*

2. EPA ID #: ***Verify this is the number received following your EPA Form 8700-12 filed with your initial permit filing.***
3. Name of Facility: ***This must match your Core Data***
4. Facility Contact Person/Title: ***This person should be familiar with all data contained in this report but does not have to be the person who completed the report.***
5. Facility Mailing Address: ***Must match your Core Data Form.***
6. Facility Location Address: ***Must match your Core Data Form.***
7. Phone: ***Must match your Core Data Form.***
8. Please review the above information. Any necessary corrections and changes are to be made on the TCEQ Core Data form: ***Again, a reminder on the form to insure a Core Data form is filed to make any site corrections.***
9. Please complete the information as applicable: Fill out The required section for your used oil handler activity leave the rest blank.

A. Used Oil Re-refiners Report

1. Total Gallons Re-refined:

Commentary: *This number must be backed by documentation typically is tank gauging documents, and bill of ladings etc.; and must cover the reporting period.*

2. Describe Re-refined Method:

Commentary: *Keep this simple in generic terms but enough information that a person who understands re-refining would understand what you are doing.*

Used Oil Processors Report

Indicate each processing method on a separate line.

1. Total Gallons Processed: *Total is ALL used oil processed using all processing methods identified below.*
2. Describe Processing Method: *As indicated each processing method needs a separate line and again keep it simple.*

Directions to Complete Biennial (Annual Identified On Form) Report for Filters

Commentary: *Note that the form implies "Please return completed form to the TCEQ by January 25, 20XX" seems to imply an annual requirement; however, it is just not a clear statement for verification see the following two regulations Health and Safety Code (HSC) §371.104 (3) below and TAC §328.24 (4) below that identify a biennial reporting requirement not an annual requirement as the report form indicates. HSC §371.104(c)(2) "report to the commission biennially the number of used oil filters the person transported, stored, or processed in the two years preceding the date of the report.". TAC §328.24(c)" Transporters, storage facilities, and processors must renew their registrations by January 25th of every even-numbered year, and must report by January 25th of every even-numbered year, in accordance with paragraphs (1)-(4) of this subsection." However, the report form that may be sent to permit holders by the TCEQ required to submit biennial reports does not have this unclear statement and an example is shown in this section as Example Reports but does not have a TCEQ form number even though it does meet all the reporting requirements identified in the regulations.*

1. *Please complete the information as applicable:* TCEQ NOTES: ***(a) can provide some challenges as many facilities collect filters in drums and totes and even roll-offs but you must be able to documents your accounting process.***
   a. Please report total used oil filters in pounds or 55-gallon drum volumetric equivalents.
   b. Provide TCEQ Registration # for recipients.
   c. Please attach additional pages as necessary to provide complete information.

d. Used Oil Filter Transporters Report Total Used Oil Filters Transported (55-gallon drum volumetric equivalents): ***Enter the Number of 55 gallon drums or volume equivalents OR the pounds of filters you have transported. After entering number of drums or equivalent pounds enter.*** SUM TOTAL TRANSPORTED
e. Transported to (Recipient Name and TCEQ Registration Number): ***Enter the name of the location you transported filters to and their TCEQ Registration adding another page as needed if you have more than 2.***
f. Used Oil Filter Storage Facility Report
   1. Total Used Oil Filters Stored (55-gallon drum volumetric equivalents): ***Enter the Number of 55 gallon drums or volume equivalents OR the pounds of filters you have stored at your storage facility then enter total ponds or drums*** SUM TOTAL SHIPPED.
   2. Shipped to (Recipient Name and TCEQ Registration Number): ***Enter the name of the facility where you shipped your filters to and their TCEQ registration number and add another page if you have more than 2.***
g. Used Oil Filter Processors Report (indicate each processing method and number of filters on a separate line): ***Enter the total number of filters you processed using all methods you used at your facility as 55 gallon drum equivalents or the pounds for each process. Enter SUM*** Total Processed: (55-gallon drum volumetric equivalents)
   1. Describe Each Processing Method:

Commentary: *Be brief but ensure the report reviewer can understand what each process is. A process method typically starts when you transfer bulk filters from receiving containers then do something to them such as crushing, shredding, heating or cubing or a combination of processes and ends when you have extracted the oil then transfer filters to containers for shipment off-site. Again, add another page if more than two processing methods are used.*

h. Used Oil Handled (Generated) From Filters. ***Enter Total Gallons Used Oil From Filters: Enter the total gallons you send to each used oil pro-***

cessor, re-refiner, or storage facility then enter the Total Gallons sent to all processors and identify them by EPA ID number and their TCEQ Registration Number.

Failure to disclose to the Commission any of the required information may result in loss of state contracts, non-issuance of registration or non-renewal of registration.

Commentary: ***Reinforcement of what should hopefully be obvious that your data must be valid.***

1. CERTIFICATION STATEMENT:

I certify that the above information is true and correct to the best of my knowledge and that I will abide by TCEQ rules governing the recycling of used oil, the collection, management and recycling of used oil filters.

Commentary: *Signing this certification must be taken very seriously and in signing it you must be able to attest to the information as real liabilities exist so if you are unsure about anything entered on this form you must get it cleared up prior to submitting. This guide does not provide legal advice but from a compliance enforcement standpoint you subject your company and yourself to significant issues.*

2. Signature:
3. Date:
4. Name (Type or Print):

   i. Any changes will have to be reported on the TCEQ Core Data Form, TCEQ-10400 and/or form TCEQ-10062 and attached to this report:

Commentary: *Even though the TCEQ allows for changes in your information provided related to your previous Core Data form or your registration form TCEQ-10062 to be identified when filing your report it is not recommended.*

j. Mailing Address:

Texas Commission on Environmental Quality
Used Oil Recycling Program
P.O. Box 13087, (MC-129)
Austin, Texas 78711-3087:

Mail your completed report to this address.

Form 5 – TCEQ Biennial Reporting Form For Used Oil Processors, Re-Refiners and Filter Handers (TCEQ-0648)

TCEQ

BIENNIAL REPORTING FORM FOR USED OIL PROCESSORS, RE-REFINERS AND FILTER HANDLERS

Please return completed form to the T C E Q by JANUARY 25, 20 ☐

**Any administrative changes must be made on the T C E Q Core Data Form (TCEQ 10400).*

T C E Q Registration #: ________ E P A ID #: ________

Name of Facility: ________

Facility Contact Person/Title: ________

Facility Mailing Address: ________

Facility Location Address: ________

Phone: ________

**Please review the above information. Any necessary corrections and changes are to be made on the TCEQ Core Data form*

Please complete the information as applicable:

Used Oil Re-refiners Report

Total Gallons Re-refined: ________

Describe Re-refined Method:

____

Used Oil Processors Report (Indicate each processing method on a separate line.)

Total Gallons Processed: ________

Describe Processing Method:

____

Directions to Complete Annual Report for Filters

Please complete the information as applicable:

NOTES:

- Please report total used oil filters in pounds or 55-gallon drum volumetric equivalents.
- Provide TCEQ Registration # for recipients.
- Please attach additional pages as necessary to provide complete information.

If you have questions on how to fill out this form or about the Used Oil Recycling Program, please contact us at 512/239-6832, option 2. Individuals are entitled to request and review their personal information the agency gathers on its forms. They may also have any errors in their information corrected. Any administrative changes must be made on the TCEQ Core Data Form (TCEQ 10400). To review such information, contact us at 512/239-3282.

Used Oil Filter Transporters Report

Total Used Oil Filters Transported

[ ] (55-gallon drum volumetric equivalents)
[ ] lbs
[ ] lbs
[ ] lbs SUM TOTAL TRANSPORTED

Transported to (Recipient Name and TCEQ Registration Number):

NAME: [ ] T C E Q# [ ] :
NAME: [ ] T C E Q# [ ] :

Used Oil Filter Storage Facility Report

Total Used Oil Filters Stored

[ ] (55-gallon drum volumetric equivalents)
[ ] lbs
[ ] lbs
[ ] lbs SUM TOTAL SHIPPED

Shipped to (Recipient Name and TCEQ Registration Number):

NAME: [ ] T C E Q# [ ] :
NAME: [ ] T C E Q# [ ] :

Used Oil Filter Processors Report (indicate each processing method and number of filters on a separate line)

Total Used Oil Filters Processed

[ ] (55-gallon drum volumetric equivalents)
[ ] lbs
[ ] lbs
[ ] lbs SUM TOTAL PROCESSED

Describe Each Processing Method:

Processing Method: [ ]
Processing Method: [ ]

Used Oil Handled (Generated) From Filters

Total Gallons Used Oil From Filters	E P A ID Number of Recipient	T C E Q Registration Number of Recipient
[ ] Total Gallons	[ ]	[ ]
[ ] Total Gallons	[ ]	[ ]
[ ] Total Gallons	[ ]	[ ]
GRAND TOTAL GALLONS		
[ ] Grand Total Gallons		

Failure to disclose to the Commission any of the required information may result in loss of state contracts, non-issuance of registration or non-renewal of registration.

CERTIFICATION STATEMENT

I certify that the above information is true and correct to the best of my knowledge and that I will abide by TCEQ rules governing the recycling of used oil, the collection, management and recycling of used oil filters.

Signature: ______________________ Date: [ ]

Signature Line 1

Name (Type or Print): [ ]

Print Form
Reset Form

Any changes will have to be reported on the TCEQ Core Data Form, TCEQ-10400 and/or form TCEQ-10062 and attached to this report.

We appreciate your cooperation in completing this report which is required by the Texas Health & Safety Code 371.026 and 40 Code of Federal Regulations Part 279, Section 279.57(b).

Texas Commission on Environmental Quality
Used Oil Recycling Program
P.O. Box 13087, (MC-129)
Austin, Texas 78711-3087

PART XIII: 20XX ANNUAL REPORTING FORM FOR USED OIL COLLECTION CENTERS

As always, if you have any questions regarding the form or need to verify your information follow the instruction indicated on the form as follows: If you have questions on how to fill out this form or about the Used Oil Recycling Program, please contact us at 512/239-6832, option 2. Individuals are entitled to request and review their personal information the agency gathers on its forms. They may also have any errors in their information corrected. Any administrative changes must be made on the TCEQ Core Data Form (TCEQ 10400). To review such information, contact TCEQ at 512/239- 3282.

TCEQ-0567 Form Instructions (156)

Please return completed form to the TCEQ by JANUARY 25, 20XX (***XX=Year of Report).***

Information applicable to all Collection Centers: ***See II (1) above.***

a. TCEQ registration #:
b. Name of Collection Center:
c. Mailing Address:
d. Physical Address of Collection Center:
e. Contact Representative/Title:
f. Phone:

Please review the above information and make any necessary corrections and changes on the TCEQ Core Data Form (TCEQ-10400).

Used Oil Collection: *Fill in applicable gallons for the categories identified. Remember you are required to account for oil received by the categories listed* **between January**

1, 20XX and December 31, (report in gallons)

a. From Household Do-It-Yourselfers - Gallons
b. From Other Generators - Gallons
c. From Your Facility (optional) – Gallons

2. **Who picks up your used oil? (If there are more than two transporters, please continue list on the back of this page.):**

> ***Commentary:*** *This information must be accurate and should be on any bill of lading transportation document you received and signed when your oil was picked up.*

Transporter Name:

a. Address:
b. Phone:
c. Transporter Name:
d. Address:
e. Phone:

3. Failure to disclose to the Commission any of the required information may result in loss of state contracts, non-issuance of registration or non-renewal of registration.

> ***Commentary:*** *Reinforcement of what should hopefully be obvious that your data must be valid.*

7. CERTIFICATION STATEMENT:

> ***Commentary:*** *Signing this certification must be taken very seriously and in signing it you must be able to attest to the information as real liabilities exist so if you are unsure about anything entered on this form you must get it cleared up prior to submitting. This guide does not provide legal advice but from a compliance enforcement standpoint you subject your company and*

yourself to significant issues.

a. I certify that the above information is true and correct to the best of my knowledge and that I will abide by TCEQ rules governing the collection, management and recycling of used automotive oil and used oil filters.
b. Signature:
c. Printed Name:
d. Date:
e. Any changes will have to be reported on the TCEQ Core Data Form TCEQ-10400 and/form TCEQ-0533 and attached to this report.

Commentary: *Even though the TCEQ allows for changes in your information provided related to your previous Core Data form or your registration form TCEQ-10062 to be identified when filing your report it is not recommended.*

Mailing Address: **Texas Commission on Environmental Quality Used Oil Recycling Program/Registration and Reporting Section (MC-129), P.O. Box 13087, Austin, Texas 78711-3087**

Additional Transporters: As indicated if you have more than 2 enter the others on this page.

a. Transporter Name:
b. Address:
c. Phone:

Form 6 – TCEQ 20XX Annual Reporting Form for Used Oil Collection Centers (and DIY Collection Centers (TCEQ-0567)

Print Form
Reset Form

TCEQ

20☐ ANNUAL REPORTING FORM FOR USED OIL COLLECTION CENTERS

If you have questions on how to fill out this form or about the Used Oil Recycling Program, please contact us at 512/239-6832, option 2. Individuals are entitled to request and review their personal information the agency gathers on its forms. They may also have any errors in their information corrected. Any administrative changes must be made on the TCEQ Core Data Form (TCEQ 10400). To review such information, contact us at 512/239-3282.

*****Please return completed form to the TCEQ by JANUARY 25, 20☐. *****

TCEQ Registration #: ☐

Name of Collection Center: ☐

Mailing Address: ☐

Physical Address of Collection Center: ☐

Contact Representative/Title: ☐ Phone: ☐

*****Please review the above information and make any necessary corrections and changes on the TCEQ Core Data Form (TCEQ-10400) *****

USED OIL COLLECTED
between January 1, 20☐ and December 31, 20☐ (report in gallons)

From Household Do-It-Yourselfers	Gallons	From Other Generators	gallons	From Your Facility (optional)	gallons

Who picks up your used oil? (If there are more than two transporters, please continue list on the back of this page.)

Transporter Name: ☐
Address: ☐ Phone: ☐

Transporter Name: ☐
Address: ☐ Phone: ☐

Failure to disclose to the Commission any of the required information may result in loss of state contracts, non-issuance of registration or non-renewal of registration.

CERTIFICATION STATEMENT

I certify that the above information is true and correct to the best of my knowledge and that I will abide by TCEQ rules governing the collection, management and recycling of used automotive oil and used oil filters.

Signature: ______________________ Date: ☐
Signature Line 1

Print Name: ☐

Any changes will have to be reported on the TCEQ Core Data Form TCEQ-10400 and/form TCEQ-0533 and attached to this report.

We appreciate your cooperation in completing this report which is required by the Texas Administrative Code Sections 324.7(1)(E) and 324.7(2)(E).

Texas Commission on Environmental Quality
Used Oil Recycling Program/Registration and Reporting Section (MC-129)
P.O. Box 13087, Austin, Texas 78711-3087

Print Form | Reset Form

Additional Transporters

Transporter Name:

Address: Phone:

Transporter Name:

Address: Phone:

Transporter Name:

Address: Phone:

Transporter Name:

Address: Phone:

Transporter Name:

Address: Phone:

Transporter Name:

Address: Phone:

Transporter Name:

Address: Phone:

Transporter Name:

Address: Phone:

PART XIV: FINANCIAL ASSURANCE FOR USED OIL RECYCLERS

Part of your permit requirements for any Used Oil Permit holder is that you have some form of financial assurance in place in the event that your actions create an environmental liability associated with any release to soil. Your Financial Assurance document must be filed at the same time you file for your permit. The following information is drawn from the TCEQ website available from:

http://tceq.state.tx.us/adminservices/financial-assurance/revenue/used_oil_recycling.html/#overview (170)

The guidance includes the following statement: "Used Oil Handlers are allowed to choose from several types of financial assurance mechanisms as set out in Subchapter L of Chapter 37. The description of how various financial assurance mechanisms work and their specific criteria are set out in Subchapter C of Chapter 37. The required wording of the mechanisms can be reviewed and downloaded from Subchapter D of Chapter 37. ***"What also needs to be identified is that as you review the various chapters identified above and others you will find an example of the submission format required identified as "Attached Graphic."***

The following remark from the TCEQ website emphasizes the importance of reviewing the submission format; TCEQ will only accept the "exact" language identified:

"Wording of Mechanisms: For Used Oil Handlers, the financial assurance mechanism you submit to us must be worded exactly as set out in our Rules except that the instructions in parenthesis are to be replaced with the relevant information and the parenthesis deleted. One way to ensure that your mechanism is worded correctly is to download the document from our website and inform your financial institution not to add or delete language. Please be advised that our agency cannot accept the financial assurance mechanism if it does not incorporate the exact language as set out in our Rules."

The information available at the site identified above is identified with the following outline found at the site. The information provided in this section is compiled from various financial assurance cites based on information available from the TCEQ site.

TCEQ Website Financial Assurance for Used Oil Recycling

- Financial assurance requirements and contact information pertaining to used oil recycling.
- General Overview
- Questions? We Can Help
- Cost Estimates
- Types of Financial Assurance Mechanisms
- Wording of Mechanisms

General Overview

Please note that there are different financial assurance requirements depending upon whether you are registered as a "transporter" of used oil or whether you are registered as a "used oil handler". A used oil handler includes owners or operators of a used oil transfer, processing re-refining, or off-specification used oil burning facility. Please refer below to Types of Financial Assurance Mechanisms for more details.

Both used oil handlers and used oil transporters should refer to Subchapter L of Chapter 37 which establishes requirements and mechanisms for demonstrating financial assurance for soil remediation and insurance. The amount of financial assurance required is determined by the Registration and Reporting (R&R) Section and is discussed in further detail under Cost Estimates below.

Referenced sections of SUBCHAPTER L: FINANCIAL ASSURANCE FOR USED OIL RECYCLING §§37.2001, 37.2003, 37.2011, 37.2013, 37.2015, 37.2021 Effective February 12, 2003 **(170)**

1. **§37.2001. Applicability**. This subchapter applies to used oil transporters required to provide evidence of financial assurance under §324.22 of this title (relating to

Soil Remediation for Used Oil Handlers). This subchapter also applies to an owner or operator of a used oil transfer, processing, re-refining, and off specification used oil burning facilities, hereinafter referred to as "used oil handlers," which are required to provide evidence of financial assurance under §324.22 of this title. This subchapter does not apply to state or federal governmental entities whose debts and liabilities are the debts and liabilities of a state or the United States. This subchapter establishes requirements and mechanisms for demonstrating financial assurance for soil remediation and automobile insurance.

a. The following §324.22 is the most detailed and complicated as referenced above in determining your specific application and cost estimation guidance so again read the requirements carefully and call the TCEQ or seek other professional advice if you have any questions. When you submit your documentation the TCEQ will also review for applicability but any follow-up will create delays in your obtaining your permit.

§324.22. Soil Remediation Requirements for Used Oil Handlers.

(1) This section applies to transporters of used oil who are seeking registration under this chapter. It also applies to owners and operators of used oil transfer, processing, re-refining, and off-specification used oil burning facilities referred to as "used oil handlers." It does not apply to a used oil handler which is owned or otherwise effectively controlled by the owners or operators where the used oil is generated.

(2) Used oil handlers, subject to the soil remediation requirements of subsection (c) or (d) of this section, and transporters of used oil, must meet the requirements of Chapter 37, Subchapter L of this title (relating to Financial Assurance for Used Oil Recycling).

(3) Used oil handlers meeting the requirements of this subsection must provide financial assurance for soil remediation in the amounts specified.

- A used oil handler must, within 30 days after an increase in the active area of the facility which results in a higher financial assurance requirement, provide for increased financial assurance.
- Additionally, a used oil handler must, at a minimum, up-

date its financial assurance annually to cover any increased cost due to inflation and to account for any other appropriate adjustments, including a lower financial assurance amount.

- The active area of the facility is the earthen area at the facility over which any transportation, storage, or processing of used oil occurs. Records demonstrating the size of the active area of the facility and related financial assurance are to be maintained in the facility's operating record; however, the original financial assurance mechanism must be submitted to the commission per §37.2015 of this title (relating to Submission of Documents).

Commentary: *This section is extremely important to understand and means all areas like containment, parking lots, lay down areas, access roads, and driveways etc.;.The word "active" above from personal experience is interpreted very broadly.*

The amount required for financial assurance is:

Commentary: *Please pay special attention to this section and you must be able to document your calculation for the specific application.*

(1) for a facility with an active area of over 1,000 square feet up to 10,000 square feet, $410 for each 1,000-square-foot increment;
(2) for a facility with an active area of over 10,000 square feet up to 100,000 square feet, $4,100 for each 10,000-square-foot increment;
(3) for a facility with an active area of over 100,000 square feet up to 1 million square feet, $41,000 for each 100,000 square-foot increment and $4,100 for each 10,000 square-foot increment;
(4) for a facility with an active area of over 1 million square feet, $410,000 for each 1-million-square foot increment, $41,000 for each 100,000 square-foot increment, and $4,100 for each 10,000 square-foot increment; or

(d) Used oil handlers may meet the following alternate requirements.

Commentary: *This option provides a method of avoiding the financial assurance costs but is a significant challenge in any facility as identified in the example facility because it is extremely difficult to avoid any area where soil is present from being identified as (3) below as just one of the difficulties.*

(1) Used oil handlers must:
- (A) annually provide a certification statement to the executive director that the used oil handler is in compliance with the applicable requirements of this chapter; and
- (B) obtain certification from a Registered Professional Engineer or other qualified independent professional that the used oil facility units have been designed and constructed in accordance with appropriate design standards, and that the units exhibit mechanical integrity. Such a certification must be obtained for each unit added to the facility, and for each unit that has undergone repair to restore mechanical integrity, within 90 days of the addition or completion of repair;

(2) Used oil handlers must ensure that spills in quantities of 25 gallons or greater are reported to the agency in accordance with the spill reporting requirements of Chapter 327 of this title (relating to Spill Prevention and Control);

(3) Used oil handler facilities must be provided with secondary containment for all areas where used oil is stored, transferred, or otherwise handled, including, but not limited to, loading docks, parking areas, storage areas, and any other areas where shipments of used oil are held for more than 24 hours; and the facility's used oil tanks, containers, and secondary containment must be constructed, operated, and maintained to conform to the requirements of Title 40 Code of Federal Regulations §§264.174, 264.193(c) - (f), and 264.195(b), as if the used oil were hazardous waste, or to conform to the following:
- (A) the secondary containment must be:
  - (i) stationary;
  - (ii) constructed of non-earthen material (e.g., concrete);

(iii) maintained free of cracks, gaps, or holes; and

(iv) overlain or underlain with a synthetic liner at least 40 mils thick;

(B) the secondary containment must:

(i) be large enough to contain a catastrophic spill of 100% of the capacity of the largest used oil storage, transfer, or other handling equipment or device; and

(ii) have at least 12 inches of freeboard or sufficient freeboard to hold the precipitation which would be collected, including any run-on or infiltration of precipitation, as a result of a 25-year, 24-hour rainfall event;

(C) the secondary containment system must prevent the release of used oil or other accumulated liquid from the secondary containment system to the soil, ground water, or surface water until removed;

(D) used oil or other accumulated liquid must be removed from the secondary containment system within 24 hours from discovery, or in as timely manner as possible;

(4) Used oil handlers must provide spill response capability to adequately respond to a catastrophic spill of 100% of the capacity of the largest used oil storage, transfer, or other handling equipment or device, plus 10% of the remaining storage capacity; and

(5) Used oil handlers must meet the requirements of subsection (c) of this section, but the amount of financial assurance provided is 10% of the amount that would otherwise be required under subsection (c). (e) As required, processors or re-refiners who store or process used oil in aboveground tanks must, at closure of a tank system, demonstrate financial assurance in the amount of the cost to comply with the closure requirements of 40 CFR §279.54(h). If the used oil handler cannot demonstrate that all contaminated soils are removed or decontaminated as required in 40 CFR §279.54(h), the used oil handler must further demonstrate financial assurance in the amount required to cover the soil and perform post closure in accordance with the closure and post closure care requirements that apply to hazardous waste landfills under 40 CFR §§265.310, 265.117-265.120, and 265.145.

2. **§37.2003. Definitions.** Definitions for terms that appear throughout this sub-

chapter may be found in Subchapter A of this chapter (relating to General Financial Assurance Requirements), as well as Chapter 324 of this title (relating to Used Oil), except the term "closure" for purposes of this subchapter includes the term "soil remediation" as used in Chapter 324 of this title.

3. §37.2011. Financial Assurance Requirements for Used Oil Handlers. In addition to the requirements of this subchapter, used oil handlers who must demonstrate financial assurance for soil remediation must do so in an amount as specified in §324.22(c) or (d) of this title (relating to Financial Responsibility Technical Requirements) and must comply with Subchapters A - D of this chapter (relating to General Financial Assurance Requirements; Financial Assurance Requirements for Closure, Post Closure, and Corrective Action; Financial Assurance Mechanisms for Closure, Post Closure, and Corrective Action; and Wording of the Mechanisms for Closure, Post Closure, and Corrective Action).
4. §37.2013. Financial Assurance Mechanisms for Used Oil Handlers. A used oil handler subject to this subchapter may use any of the financial assurance mechanisms as specified in Subchapter C of this chapter (relating to Financial Assurance Mechanisms Texas Commission on Environmental Quality Page 2 Chapter 37 - Financial Assurance for Closure, Post Closure, and Corrective Action) to demonstrate financial assurance for soil remediation except a pay-in trust fund may not be used. ***For your reference a redacted listing of applicable mechanisms and management of those mechanisms are provided below:***

§37.201. Trust Fund. (172)

(a) An owner or operator may satisfy the requirements of financial assurance by establishing either a fully funded trust or a pay-in trust which conforms to the requirements of this section, in addition to the requirements specified in Subchapters A and B of this chapter (relating to General Financial Assurance Requirements and Financial Assurance Requirements for Closure, Post Closure, and Corrective Action), and submitting an originally signed duplicate of the executed trust agreement to the executive director.

(b) The trustee must be an entity which has the authority to act as a trustee and whose trust operations are regulated and examined by a federal or state agency.

(c) The wording of the trust agreement must be identical to the wording specified in

§37.301(a) of this title (relating to Trust Agreement) including a formal certification of acknowledgment as specified in §37.301(b) of this title.

(d) Schedule A of the trust agreement as specified in §37.301(a) of this title must be updated within 60 days after an approved change in the amount of the current cost estimate or annual inflation adjustments.

(e) A fully funded trust requires that the initial payment into the trust fund be at least equal to the current cost estimate, or when a combination of mechanisms are used in accordance with '37.41 of this title (relating to Use of Multiple Financial Assurance Mechanisms), the initial payment plus the amount of the combined mechanism(s) must be at least equal to the current cost estimate. A receipt from the trustee for the initial payment must be submitted by the owner or operator to the executive director with the originally signed duplicate of the trust agreement.

(f) In the case of a pay-in trust for closure or post closure, payments into the trust fund must be made annually by the owner or operator over the term of the initial permit or over the remaining life of the facility, whichever is shorter. In the case of a pay-in trust for corrective action for known releases, Texas Commission on Environmental Quality Page 2 Chapter 37 - Financial Assurance the payments into the trust fund must be made annually by the owner or operator over one-half of the estimated length of the corrective action program. The periods referred to in this subsection are the pay-in periods. The payments into the trust fund must be made in accordance with this subsection. During the period of post closure, a pay-in trust for post closure may not be used.

(1) For a pay-in trust used to demonstrate financial assurance for closure and post closure, the first payment into the fund must be at least equal to the current cost estimate for closure or post closure, less the amount of the combined mechanisms, divided by the number of years in the pay in period. Subsequent payments must be made no later than 30 days after each anniversary date of the first payment. The amount of subsequent payments must be determined by the following formula: Next Payment = CE – CV / Y where CE is the current closure and post closure cost estimates, CV is the current value of the trust fund, and Y is the number of years remaining in the pay-in period.

(2) For a pay-in trust used to demonstrate financial assurance for corrective action, the first payment into the trust fund must be at least equal to one-half of the

current cost estimate for corrective action, less the amount of the combined mechanisms, divided by the number of years in the corrective action pay-in period. The amount of subsequent payments must be determined by the following formula: Next Payment = RB - CV / Y where RB is the most recent estimate of the required trust fund balance for corrective action, CV is the current value of the trust fund, and Y is the number of years remaining in the pay-in period.

(3) The owner or operator may accelerate payments into the trust fund or may deposit the full amount of the current cost estimate at the time the fund is established. However, the owner or operator must maintain the value of the fund at no less than the value that the fund would have if annual payments were made as specified in paragraphs (1) or (2) of this subsection.

(4) If the owner or operator establishes a trust fund after having used another financial assurance mechanism, the first payment must be at least equal to the amount that the fund would contain if the trust fund was established when the permit was initially issued, and subsequent payments must be made as specified in paragraphs (1) or (2) of this subsection.

(g) After the initial payment for a fully funded trust or after the pay-in period is completed for a pay-in trust, whenever the current cost estimate changes, the owner or operator must compare the new Texas Commission on Environmental Quality Page 3 Chapter 37 - Financial Assurance estimate with the trustee=s most recent annual valuation of the trust fund. If the value of the fund is less than the amount of the new estimate, the owner or operator, within 30 days after the change in the current cost estimate, must either deposit an amount into the fund so that its value after this deposit at least equals the amount of the current cost estimate, or obtain an additional financial assurance mechanism as specified in this subchapter to cover the difference.

(h) If the value of the trust fund is greater than the total amount of the current cost estimate, the owner or operator may submit a written request to the executive director for release of the amount in excess of the current cost estimate.

(i) Within 60 days after receiving a request from the owner or operator for release of funds as specified in subsection (h) of this section, the executive director shall instruct the trustee to release to the owner or operator such funds as the executive director specifies in writing.

(j) An owner or operator or any other person authorized by the executive director to perform closure, post closure, or corrective action may request reimbursement for closure, post closure, or corrective action expenditures by submitting itemized bills to the executive director. The request shall include an explanation of the expenses and all applicable itemized bills. The owner or operator may request reimbursements for partial closure only if sufficient funds are remaining in the trust fund to cover the maximum costs of closing the facility over its remaining operating life. After receiving bills for closure, post closure, or corrective action activities, the executive director shall instruct the trustee to make reimbursement in such amounts as the executive director specifies in writing, if the executive director determines that the partial or final closure, post closure, or corrective action expenditures are in accordance with the approved closure plan, post closure plan, or corrective action activities, or are otherwise justified. If the executive director has reason to believe that the cost of closure, post closure, or corrective action over the remaining life of the facility will be greater than the value of the trust fund, the executive director may withhold reimbursement of such amounts as deemed prudent until it is determined, in accordance with Subchapters A and B of this chapter (relating to General Financial Assurance Requirements and Financial Assurance Requirements for Closure, Post Closure, and Corrective Action) that the owner or operator is no longer required to maintain financial assurance for final closure, post closure, or corrective action at the facility.

(k) If an owner or operator substitutes other financial assurance as specified in this section for all or part of the trust fund, the owner or operator may submit a written request to the executive director for release of the amount in excess of the current cost estimate covered by the trust fund.

§37.211. Surety Bond Guaranteeing Payment. (172)

(a) An owner or operator may satisfy the requirements of financial assurance by obtaining a surety bond which conforms to the requirements of this section, in addition to the requirements specified in Subchapters A and B of this chapter (relating to General Financial Assurance Requirements Texas Commission on Environmental Quality Page 4 Chapter 37 - Financial Assurance and Financial Assurance Requirements for Closure, Post Closure, and Corrective Action), and submitting an originally signed surety bond to the executive director.

(b) The surety company issuing the bond must, at a minimum, be among those listed as acceptable sureties on Federal bonds in Circular 570 of the United States Department of the Treasury.

(c) The wording of the surety bond must be identical to the wording specified in §37.311 of this title (relating to Payment Bond).

(d) The bond must guarantee that the owner or operator shall:

(1) fund the standby trust fund as required in §37.161 of this title (relating to Establishment of a Standby Trust) in an amount equal to the penal sum of the bond before the beginning of final closure of, or corrective action at, the facility;

(2) fund the standby trust fund as required in §37.161 of this title in an amount equal to the penal sum within 15 days after an administrative order to begin final closure or corrective action issued by the executive director becomes final, or within 15 days after an order to begin final closure or corrective action is issued by the United States district court or other court of competent jurisdiction; or

(3) provide alternate financial assurance as specified in this subchapter, and obtain the executive director=s written approval of the assurance provided, within 90 days after receipt by both the owner or operator and the executive director of a notice of cancellation of the bond from the surety.

(e) Under the terms of the bond, the surety shall become liable on the bond obligation when the owner or operator fails to perform as guaranteed by the bond.

(f) The penal sum of the bond must be in an amount at least equal to the current cost estimate, except as provided in §37.41 of this title (relating to Use of Multiple Financial Assurance Mechanisms), §37.51 of this title (relating to Use of a Financial Assurance Mechanism for Multiple Facilities), or §37.52 of this title (relating to Use of a Universal Financial Assurance Mechanism for Multiple Facilities and Program Areas).

(g) Under the terms of the bond, the surety may cancel the bond by sending notice of cancellation by certified mail to the owner or operator and to the executive director. Cancellation may not occur, however, during the 120 days beginning on the date of receipt of the notice of cancellation by both the owner or operator and the executive director, as evidenced by the return receipts. Adopted February 24, 2000 Effective March 21, 2000

§37.221. Surety Bond Guaranteeing Performance. (172)

(a) An owner or operator may satisfy the requirements of financial assurance by obtaining a surety bond which conforms to the requirements of this section, in addition to the requirements Texas Commission on Environmental Quality Page 5 Chapter 37 - Financial Assurance specified in Subchapters A and B of this chapter (relating to General Financial Assurance Requirements and Financial Assurance Requirements for Closure, Post Closure, and Corrective Action), and submitting an originally signed surety bond to the executive director.

(b) The bond must, at a minimum, be among those listed as acceptable sureties on federal bonds in Circular 570 of the United States Department of Treasury.

(c) The wording of the surety bond must be identical to the wording specified in §37.321 of this title (relating to Performance Bond).

(d) A surety bond guaranteeing performance of closure, post closure, or corrective action must guarantee that the owner or operator shall:

 (1) perform closure or post closure in accordance with the closure plan, post closure plan, and other applicable requirements of the permit, or perform corrective action in accordance with the permit or other applicable requirements; and

 (2) provide alternate financial assurance as specified in this subchapter, and obtain the executive director=s written approval of the assurance provided within 90 days after receipt by both the owner or operator and the executive director of a notice of cancellation of the bond from the surety.

(e) Under the terms of the bond, the surety shall become liable on the bond obligation when the owner or operator fails to perform as guaranteed by the bond. Following a determination by the executive director that the owner or operator has failed to perform closure or post closure in accordance with the closure plan, post closure plan, or other applicable requirements of the permit, or has failed to perform corrective action in accordance with the permit or other applicable requirements, under terms of the bond, the surety shall either perform closure, post closure, or corrective action as guaranteed by the bond or deposit the amount of the penal sum of the bond into a standby trust, in accordance with §37.161 of this title (relating to Establishment of a Standby

Trust).

(f) Under the terms of the bond, the surety may cancel the bond by sending notice of cancellation by certified mail to the owner or operator and to the executive director. Cancellation of the bond may not occur, however, during the 120 days beginning on the date of the receipt of the notice of cancellation by both the owner or operator and the executive director, as evidenced by the return receipts. If the owner or operator fails to provide an alternate financial assurance mechanism as specified in this subchapter within 90 days of the receipt of notice of cancellation from the surety to the executive director and to the owner or operator, and obtain written approval of the alternate assurance from the executive director, the surety shall be required to perform under the terms of the bond.

(g) The penal sum of the bond must be in an amount at least equal to the current cost estimate, except as provided in §37.41 of this title (relating to Use of Multiple Financial Assurance Mechanisms) or §37.52 of this title (relating to Use of a Universal Financial Assurance Mechanism for Multiple Facilities and Program Areas). Texas Commission on Environmental Quality Page 6 Chapter 37 - Financial Assurance

(h) The surety shall not be liable for deficiencies in the performance of closure, post closure, or corrective action by the owner or operator after the executive director releases the owner or operator from the requirements of this section, in accordance with Subchapter A of this chapter.

§37.231. Irrevocable Standby Letter of Credit. (172)

(a) An owner or operator may satisfy the requirements of financial assurance by obtaining an irrevocable standby letter of credit which conforms to the requirements of this section, in addition to the requirements specified in Subchapters A and B of this chapter (relating to General Financial Assurance Requirements and Financial Assurance Requirements for Closure, Post Closure, and Corrective Action), and submit an originally signed irrevocable standby letter of credit to the executive director.

(b) The financial institution issuing the irrevocable standby letter of credit shall be an entity that has the authority to issue irrevocable standby letters of credit and whose operations are regulated and examined by a federal or state agency.

(c) The wording of the irrevocable standby letter of credit must be identical to the wording specified in §37.331 of this title (relating to Irrevocable Standby Letter of Credit).

(d) The originally signed irrevocable standby letter of credit must be accompanied by a letter from the owner or operator referring to the irrevocable standby letter of credit by number, issuing institution, and date, and providing the following information for each facility: the permit number, name and physical and mailing addresses of the facility, and the amount of funds assured for closure, post closure, or corrective action by the irrevocable standby letter of credit.

(e) The letter of credit must be irrevocable and issued for a period of at least one year. The irrevocable standby letter of credit must provide that the expiration date shall be automatically extended for a period of at least one year unless, at least 120 days before the current expiration date, the issuing institution notifies both the owner or operator and the executive director by certified mail of a decision not to extend the expiration date. Under the terms of the irrevocable standby letter of credit, the 120 days shall begin on the date when both the owner or operator and the executive director have received the notice, as evidenced by the return receipts.

(f) The irrevocable standby letter of credit must be issued in an amount at least equal to the current cost estimate, except as provided in §37.41 of this title (relating to Use of Multiple Financial Assurance Mechanisms), §37.51 of this title (relating to Use of a Financial Assurance Mechanism for Multiple Facilities), or §37.52 of this title (relating to Use of a Universal Financial Assurance Mechanism for Multiple Facilities and Program Areas).

(g) Following a determination that the owner or operator has failed to perform closure or post closure in accordance with the closure plan, post closure plan, and other applicable requirements of the Texas Commission on Environmental Quality Page 7 Chapter 37 - Financial Assurance permit, or has failed to perform corrective action in accordance with the permit or other applicable requirements, the executive director may draw on the irrevocable standby letter of credit.

(h) If the owner or operator does not establish alternate financial assurance as specified in this subchapter and obtain written approval of such alternate assurance from the executive director within 90 days after receipt by both the owner or operator and the executive director of a notice from the issuing institution that it has decided not to extend the irrevocable standby letter of credit beyond the current expiration date, the executive director shall draw on the irrevocable standby letter of credit. The executive director may delay the drawing if the issuing institution grants an extension of the term of the letter of credit. During the last 30 days of any such extension, the executive director shall draw on the irrevocable standby letter of credit if the owner or operator has failed to provide alternate financial assurance as specified in this subchapter and obtain written approval of such assurance from the executive director.

(i) Upon termination, in accordance with §37.61 of this title (relating to Termination of Mechanisms), the executive director shall return the irrevocable standby letter of credit to the issuing institution.

§37.241. Insurance. (172)

(a) An owner or operator may satisfy the requirements of financial assurance by obtaining insurance which conforms to the requirements of this section, in addition to the requirements specified in Subchapters A and B of this chapter (relating to General Financial Assurance Requirements and Financial Assurance Requirements for Closure, Post Closure, and Corrective Action), and submitting an originally signed certificate to the executive director.

(b) At a minimum, the insurer must be licensed to transact the business of insurance, or eligible to provide insurance as an excess or surplus lines insurer, in one or more states.

(c) The wording of the certificate of insurance must be identical to the wording specified in §37.341 of this title (relating to Certificate of Insurance).

(d) The insurance policy must be issued for a face amount at least equal to the current cost estimate for closure, post closure, or corrective action, except when a combination of mechanisms are used in accordance with §37.41 of this title (relating to Use of Multiple Financial Assurance Mechanisms) or §37.52 of this title (relating to Use of a Universal Financial Assurance Mechanism for Multiple Facilities and Program Areas). Actual payments by the insurer shall not change the face amount, although the insurer's future liability shall be lowered by the amount of the payments.

(e) The insurance policy must guarantee that funds shall be available to provide for closure, post closure, or corrective action of the facility. The policy shall also guarantee that once closure, post closure, or corrective action begins, the issuer shall be responsible for paying out funds, up to an Texas Commission on Environmental Quality Page 8 Chapter 37 - Financial Assurance amount equal to the face amount of the policy, upon the direction of the executive director, to such party or parties as the executive director specifies.

(f) An owner or operator or any other person authorized to perform closure, post closure, or corrective action may request reimbursement for closure, post closure, or corrective action expenditures by submitting itemized bills to the executive director. The request shall include an explanation of the expenses and all applicable itemized bills. The owner or operator may request reimbursement for partial closure only if the remaining value of the policy is sufficient to cover the maximum costs of closing the facility over its remaining operating life. Within 60 days after receiving bills for closure, post closure, or corrective action activities, the executive director shall determine whether the closure, post closure, or corrective action expenditures are in accordance with the approved closure, post closure, or corrective action activities or are otherwise justified, and if so, shall instruct the insurer to make reimbursement in such amounts as the executive director specifies in writing. If the executive director has reason to believe that the maximum cost of closure, post closure, or corrective action over the remaining life of the facility will be greater than the face amount of the policy, the executive director may withhold reimbursement of such amounts as deemed prudent until the executive director determines, in accordance with Subchapters A and B of this chapter, that the owner or operator is no longer required to maintain financial assurance requirements for closure, post closure, or corrective action of the facility. If the executive director does not instruct the insurer to make such reimbursements, the executive director shall provide the owner or operator with a detailed written statement of reasons.

(g) The owner or operator must maintain the policy in full force and effect until the executive director consents to termination of the policy. Failure to pay the premium, without substitution of alternate financial assurance as specified in this subchapter, shall constitute a violation of these regulations, warranting such remedy as the executive director deems necessary. Such violation shall be deemed to begin upon receipt by the executive director of a notice of future cancellation, termination, or failure to renew due to nonpayment of the premium, rather than upon the date of expiration of the policy.

(h) The policy must provide that the insurer may not cancel, terminate, or fail to renew the policy except for failure to pay the premium. The automatic renewal of the policy shall, at a minimum, provide the insured with the option of renewal at the face amount of the expiring policy. If there is a failure to pay the premium, the insurer may elect to cancel, terminate, or fail to renew the policy by sending notice by certified mail to the owner or operator and the executive director. Cancellation, termination, or failure to renew may not occur, however, during 120 days beginning with the date of receipt of the notice by both the executive director and the owner or operator, as evidenced by the return receipts.

(i) Cancellation, termination, or failure to renew may not occur and the policy shall remain in full force and effect in the event that on or before the date of expiration:

(1) the executive director deems the facility abandoned; or Texas Commission on Environmental Quality Page 9 Chapter 37 - Financial Assurance

(2) the permit expires, is terminated, is revoked, or a new or renewal permit is denied; or

(3) closure is ordered by the executive director of the commission or by a United States district court or other court of competent jurisdiction; or

(4) the owner or operator is named as debtor in a voluntary or involuntary proceeding under Title 11 (Bankruptcy), United States Code; or

(5) the premium due is paid.

(j) Each policy must contain a provision allowing assignment of the policy to a successor owner or operator. Such assignment may be conditional upon consent of the insurer, provided such consent is not unreasonably refused.

(k) For insurance policies providing coverage for post closure, commencing on the date that liability to make payments pursuant to the policy accrues, the insurer will thereafter annually increase the face amount of the policy. Such increase must be equivalent to the face amount of the policy, less any payments made, multiplied by an amount equivalent to 85% of the most recent investment rate or of the equivalent coupon issue yield announced by the United States Treasury for 26-week Treasury securities.

§37.251. Financial Test. (172)

(a) An owner or operator may satisfy the requirements of financial assurance by establishing a financial test which conforms to the requirements of this section, in addition to the requirements specified in Subchapters A and B of this chapter (relating to General Financial Assurance Requirements and Financial Assurance Requirements for Closure, Post Closure, and Corrective Action).

(b) To pass this test, the owner or operator must meet the criteria of either paragraph (1) or (2) of this subsection:

- (a) the owner or operator must have:
  - (A) two of the following three ratios: a ratio of total liabilities to net worth less than 2.0; a ratio of the sum of net income plus depreciation, depletion, and amortization to total liabilities greater than 0.1; and a ratio of current assets to current liabilities greater than 1.5; and
  - (B) net working capital and tangible net worth each at least six times the sum of the current cost estimates, liability coverage requirements, and any other financial assurance obligations under the Texas Commission on Environmental Quality (TCEQ) or other federal or state environmental regulations assured by a financial test; and Texas Commission on Environmental Quality Page 10 Chapter 37 - Financial Assurance
  - (C) tangible net worth of at least $10 million; and
  - (D) assets located in the United States amounting to at least 90% of the owner's or operator's total assets or at least six times the sum of the current cost estimates, liability coverage requirements, and any other financial assurance obligations under the TCEQ or other federal or state environmental regulations assured by a financial test;
- (b) the owner or operator must have:
  - (A) a current rating for his most recent bond issuance of AAA, AA, A, or BBB as issued by Standard and Poor's or Aaa, Aa, A, or Baa as issued by Moody's; and
  - (B) tangible net worth at least six times the sum of the current cost estimates, liability coverage requirements, and any other financial assurance obligations under the TCEQ or other federal or state environmental regulations assured by a financial test; and
  - (C) tangible net worth of at least $10 million; and

(D) assets located in the United States amounting to at least 90% of the owner's or operator's total assets or at least six times the sum of the current cost estimates, liability coverage requirements, and any other financial assurance obligations under the TCEQ or other federal or state environmental regulations assured by a financial test.

(c) To demonstrate that the requirements of the test are being met, the owner or operator shall submit the following items to the executive director:

(1) a letter signed by the owner's or operator's chief financial officer worded identically to the wording specified in §37.351 of this title (relating to Financial Test). If an owner or operator is using the financial test to demonstrate assurance for closure, post closure, or corrective action as specified in Subchapter B of this chapter and liability coverage as specified in Subchapter E of this chapter (relating to Financial Assurance Requirements for Liability Coverage), the owner or operator must submit the letter specified in the Financial Test for Liability, Part B in §37.651 of this title (relating to Financial Test for Liability) to cover both forms of financial responsibility. A separate letter as specified in §37.351 of this title is not required; and

(2) a copy of the owner's or operator's independently audited year-end financial statements for the latest fiscal year including the "unqualified opinion" of the auditor; and

(3) a special report from the owner's or operator's independent certified public accountant to the owner or operator stating that: Texas Commission on Environmental Quality Page 11 Chapter 37 - Financial Assurance (A) the accountant has compared the data which the letter from the chief financial officer specifies as having been derived from the independently audited, year-end financial statements for the latest fiscal year with the amounts in such financial statements; and (B) in connection with that procedure: (i) such amounts were found to be in agreement; or (ii) no matters came to the attention of the accountant which caused them to believe that the specified data should be adjusted; and

(4) a written verification of the current bond rating from the applicable bond rating agency, if the owner or operator is using Alternative II of the letter signed

by the owner's or operator's chief financial officer specified in §37.351 of this title; and

(5) a schedule identifying intangible assets used to calculate tangible net worth.

(d) After the initial submission of items specified in subsection (c) of this section, the owner or operator must send updated information to the executive director within 90 days after the close of each succeeding fiscal year. This information shall consist of all items specified in subsection (c) of this section.

(e) If the owner or operator no longer meets the requirements of subsection (b) of this section, a notice shall be sent to the executive director of intent to establish alternate financial assurance as specified in this subchapter. The notice must be sent by certified mail within 90 days after the end of the fiscal year for which the year-end financial data show that the owner or operator no longer meets the requirements. The owner or operator shall provide the alternate financial assurance within 120 days after the end of such fiscal year.

(f) The executive director may, based on a reasonable belief that the owner or operator may no longer meet the requirements of subsection (b) of this section, require reports of financial condition at any time from the owner or operator in addition to those specified in subsection (c) of this section. If the executive director finds, on the basis of such reports or other information, that the owner or operator no longer meets the requirements of subsection (b) of this section, the owner or operator must provide alternate financial assurance as specified in this subchapter within 30 days after notification of such a finding.

(g) The executive director may disallow use of this test on the basis of qualifications in the opinion expressed in the independent certified public accountant's report on examination of the owner's or operator's financial statements. An adverse opinion or disclaimer of opinion shall be cause for disallowance. The executive director shall evaluate other qualifications on an individual basis. The owner or operator shall provide alternate financial assurance as specified in this subchapter within 30 days after notification of the disallowance.

§37.261. Corporate Guarantee. (172)

(a) An owner or operator may satisfy the requirements of financial assurance for closure, post closure, or corrective action by obtaining a written guarantee, hereafter referred to as "corporate guarantee," which conforms to the requirements of this section, in addition to the requirements as specified in Subchapters A and B of this chapter (relating to General Financial Assurance Requirements and Financial Assurance Requirements for Closure, Post Closure, and Corrective Action).

(b) The guarantor shall be the direct or higher-tier parent corporation of the owner or operator or a corporation with a substantial business relationship with the owner or operator. The guarantor must meet the requirements for owners or operators as specified in §37.251 of this title (relating to Financial Test). The guarantor must comply with the terms of the corporate guarantee.

(c) The wording of the corporate guarantee must be identical to the wording specified in §37.361 of this title (relating to Corporate Guarantee). The corporate guarantee shall accompany the items sent to the executive director as specified in §37.251(c) of this title.

(d) If the guarantor has a substantial business relationship with the owner or operator, in addition to the requirements specified in this chapter for the financial test and corporate guarantee, the guarantor will submit a description of the substantial business relationship and the value received in consideration of the guarantee; an original or certified original copy of the Resolution by the Board of Directors or a certified letter from the chief financial officer, authorizing the corporate guarantee on behalf of the entity; an original or certified original copy of the Resolution by the Board of Directors authorizing the formation or acquisition of the guaranteed entity; an organizational chart which shows the relationship between the two entities; the partnership agreement or other agreements, articles, or bylaws which set out the formation, structure, and operation of the guaranteed entity. After the initial submission of these items to demonstrate a substantial business relationship, if there has been no change in the substantial business relationship, the chief financial officer may submit a letter attesting that there has been no change.

(e) The terms of the corporate guarantee shall provide that:

(1) if the owner or operator fails to perform closure, post closure, or corrective action at the facility(ies) covered by the corporate guarantee in accordance with the permits and other applicable requirements whenever required to do so, the guarantor shall do so or establish a trust fund as specified in §37.201 of this title (relating to Trust Fund) in the name of the owner or operator in the

amount of the current cost estimate;

(2) the corporate guarantee shall remain in force unless the guarantor sends notice of termination by certified mail to the owner or operator and the executive director and the owner or operator has obtained, and the executive director has approved, alternative financial assurance; and

(3) if the owner or operator fails to provide alternate financial assurance as specified in this subchapter and obtain the written approval of such alternate assurance from the executive director Texas Commission on Environmental Quality Page 14 Chapter 37 - Financial Assurance within 90 days after receipt by both the owner or operator and the executive director of a notice of termination of the corporate guarantcc from thc guarantor, the guarantor shall provide such alternate financial assurance in the name of the owner or operator.

§37.271. Local Government Financial Test. (172)

An owner or operator may satisfy the requirements of financial assurance for closure, post closure, or corrective action by establishing a local government financial test or a local government financial test and local government guarantee, which conforms to the requirements of this section, in addition to the requirements specified in Subchapters A and B of this chapter (relating to General Financial Assurance Requirements; and Financial Assurance Requirements for Closure, Post Closure, and Corrective Action). An owner or operator who satisfies the requirements of paragraphs (1) - (3) of this section may demonstrate financial assurance up to the amount specified in paragraph (4) of this section.

(1) In order to satisfy the financial component of the test, the owner or operator must meet the criteria of either subparagraph (A) or (B) of this paragraph and in addition must meet certain general conditions outlined in subparagraph (C) of this paragraph.

(A) The owner or operator must satisfy each of the following financial ratios based on its most recent audited annual financial statement:

(i) a ratio of cash plus marketable securities to total expenditures greater than or equal to 0.05; and

(ii) a ratio of annual debt service to total expenditures less than or equal to 0.20.

(B) If the owner or operator:

(i) of a facility other than a municipal solid waste landfill has outstanding, rated, general obligation bonds that are not secured by insurance, a letter of credit, or other collateral or guarantee, those bonds must have a current rating of Aaa, Aa, A, or Baa, as issued by Moody's, or AAA, AA, A, or BBB, as issued by Standard and Poor's on all such general obligation bonds; or

(ii) of a municipal solid waste landfill subject to Chapter 330 of this title (relating to Municipal Solid Waste) has bonds as defined in Subchapter R of this chapter (relating to Financial Assurance for Municipal Solid Waste Facilities) and those bonds are not secured by insurance, a letter of credit, or other collateral or guarantee, those bonds must have a current rating of Aaa, Aa, A, or Baa, as issued by Moody's, or AAA, AA, A, or BBB, as issued by Standard and Poor's on all such financial obligations.

(C) In addition to meeting the criteria listed under subparagraph (A) or (B) of this paragraph, the following general conditions must be met.

(i) The owner or operator shall prepare its financial statements in conformity with Generally Accepted Accounting Principles for governments and have its financial statements audited by an independent certified public accountant (or appropriate state agency).

(ii) The owner or operator must not have operated at a deficit equal to 5.0% or more of total annual revenue in each of the past two fiscal years.

(iii) The owner or operator must not currently be in default on any outstanding general obligation bonds.

(iv) The owner or operator must not have any outstanding general obligation bonds rated lower than Baa as issued by Moody's or BBB as issued by Standard and Poor's. (v) The owner or operator must not have received an adverse opinion, disclaimer of opinion, or other qualified opinion from the independent certified public accountant (or appropriate state agency) auditing its financial statements as required under clause (i) of this subparagraph. However, the executive director may evaluate qualified opinions on a case-by-case basis and allow use of the financial test in cases where the executive director deems the qualification insufficient to warrant disallowance of use of the test.

(D) The following terms used in this section are defined as follows.

(i) Deficit equals total annual revenues minus total annual expenditures.

(ii) Total revenues is the sum of the following seven items:

(I) "Total Revenues" of the General Fund;

(II) "Total Revenues" of Special Revenue Funds;

(III) "Total Revenues" of the Debt Service Fund;

(IV) "Total Revenues" of Capital Project Funds;

(V) "Total Operating Revenues" of Enterprise Funds;

(VI) if positive, "Total Non-Operating Revenues (Net)" of Enterprise Funds; and

(VII) if positive, "Total Non-Operating Revenues (Net)" of Internal Service Funds.

(iii) Total expenditures is the sum of the following six items:

(I) A Total Expenditures@ of the General Fund;

(II) A Total Expenditures@ of Special Revenue Funds;

(III) A Total Expenditures@ of the Debt Service Fund;

(IV) A Total Operating Expenses Before Depreciation@ of Enterprise Funds;

(V) if negative, A Total Non-Operating Revenues (Net)@ of Enterprise Funds; and

(VI) if negative, A Total Non-Operating Revenues (Net)@ of Internal Service Funds; except if the local government is not using accrual accounting and is not including depreciation in its expenditures, include routine capital outlays and debt repayment as a substitute for depreciation.

(iv) Cash and current investments is the sum of "Cash," "Cash Equivalents" (e.g., bank deposits, very short-term debt securities, money market funds), and "Current Investments" (e.g., interest or dividend bearing securities that are expected to be held for less than one year), in the General Fund, Special Revenue Funds, Debt Service Fund, Enterprise Funds, and Internal Service Funds, as reported on the Comprehensive Annual Financial Report=s (CAFR) Combined Balance Sheet. Note that cash, cash equivalents, and current investments are included in this term even if they are: pooled; with a fiscal agent; or restricted, provided that the assets belong to the General Fund, Special Revenue Funds, Debt Service Fund, Enterprise Funds, and Internal Service Funds. Specifically excluded from this definition are accounts receivable, retirement assets, real property, fixed assets, and other non-current assets, as well as any assets (including cash) in Capital Project Funds.

(v) Debt service is the sum of all amounts in any Debt Service category (including bond principal, other debt principal, interest on bonds, interest on other debt) in the General Fund, Special Revenue Funds, Debt Service Fund, and Capital Projects Funds as reported on the CAFR's Combined Statement of Revenues, Expenditures and Changes in Fund Balances/Equity; plus all principal and interest expense in Enterprise Funds and Internal Service Funds, as reported on the CAFR's Combined Statement of Revenues, Expenses and Changes in Retained Earnings/Fund Balances.

(2) In order to satisfy the public notice component of the test, the local government owner or operator must place a reference to the closure, post closure, or corrective action costs assured through the financial test into its next CAFR after the effective date of this section or prior to the initial receipt of waste at the facility, whichever is later. Disclosure must include the nature and source of closure, post closure, or corrective action requirements; the reported liability at the balance sheet date;

Financial Assurance the estimated total closure or post closure cost remaining to be recognized; the percentage of landfill capacity used to date; and the estimated landfill life in years. A reference to corrective action costs must be placed in the CAFR not later than 120 days after the corrective action remedy has been selected in accordance with the requirements of §330.415 of this title (relating to Implementation of the Corrective Action Program). For the first year the financial test is used to assure costs at a particular facility, the reference may instead be placed in the operating record until issuance of the next available CAFR if timing does not permit the reference to be incorporated into the most recently issued CAFR or budget. For closure and post closure costs, conformance with Government Accounting Standards Board Statement 18 assures compliance with the public notice component.

(3) In order to satisfy the recordkeeping and reporting component of the test, the local government owner or operator must submit the following four items to the executive director:

(A) a letter signed by the local government's chief financial officer worded as specified in §37.371 of this title (relating to Local Government Financial Test) that:

(i) lists all the current cost estimates covered by a financial test as described in paragraph (4) of this section;

(ii) provides evidence and certifies that the local government meets the conditions of either paragraph (1)(A) or (B), and (1)(C) of this section; and

(iii) certifies that the local government meets the conditions of paragraphs (2) and (4) of this section;

(B) the local government's independently audited year-end financial statements for the latest fiscal year, including the unqualified opinion of the auditor. The auditor must be an independent certified public accountant (CPA) or an appropriate state agency that conducts equivalent comprehensive audits;

(C) a report to the local government from the local government=s independent CPA or the appropriate state agency which:

(i) is based on performing an agreed upon procedures engagement relative to the financial ratios required by paragraph (1)(A) of this section, if applicable, and the requirements of paragraph (1)(C)(i), (ii), and (v) of this section; and

(ii) the CPA or state agency=s report states the procedures performed and the CPA or state agency=s findings; and (D) a copy of the CAFR used to comply with paragraph (2) of this section and certification that the requirements of General Accounting Standards Board Statement 18 have been met.

(4) Financial Assurance The portion of the closure, post closure, or corrective action costs for which an owner or operator can assure under this paragraph is determined as follows.

(A) If the local government owner or operator does not assure other environmental obligations through a financial test, it may assure closure, post closure, or corrective action costs that equal up to 43% of the local government=s total annual revenue.

(B) If the local government owner or operator assures other environmental obligations through a financial test, including, but not limited to, those associated with hazardous waste treatment, storage, and disposal facilities under Chapter 335 of this title (relating to Industrial Solid Waste and Municipal Hazardous Waste) and 40 Code of Federal Regulations (CFR) Parts 264 and 265, petroleum underground storage tank facilities under Chapter 334 of this title (relating to Underground and Aboveground Storage Tanks) and 40 CFR Part 280, underground injection control facilities under Chapter 331 of this title (relating to Underground Injection Control) and 40 CFR §144.62, polychlorinated biphenyl storage facilities under 40 CFR Part 761, it must add those costs to the closure, post closure, or corrective action costs it seeks to assure under this paragraph. The total that may be assured must not exceed 43% of the local government=s total annual revenue.

(5) Annual updates of the financial test documentation must be submitted to the executive director within 180 days after the close of each succeeding fiscal year. This information must consist of all the items required under paragraph (3) of this section.

(6) A local government must satisfy the requirements of the financial test at the close of each fiscal year. If the local government owner or operator no longer meets the requirements of paragraphs (1) - (4) of this section, the local government must send notice to the executive director of intent to establish alternate financial assurance. This notice must be sent within 90 days after the end of the fiscal year for which the year-end financial data shows that the local government no longer meets the requirements. The local government must provide alternate financial assurance

within 120 days after the end of such fiscal year.

(7) The local government is no longer required to comply with the requirements of this section when the conditions as specified in §37.61 of this title (relating to Termination of Mechanisms) are met.

(8) The executive director, based on a reasonable belief that the local government owner or operator may no longer meet the requirements of the local government financial test, may require additional reports of financial condition from the local government at any time. If the executive director finds on the basis of such reports or other information, that the local government owner or operator no longer meets the requirements of the financial test, the local government must provide alternate financial assurance as specified in this subchapter within 30 days after notification of such a finding. Adopted August 23, 2006 Effective September 14, 2006 Texas Commission on Environmental Quality Page 19 Chapter 37 –

§37.281. Local Government Guarantee. (172)

An owner or operator may satisfy the requirements of financial assurance for closure, post closure, or corrective action by obtaining a local government guarantee provided by a local government. The local government guarantee must meet the requirements of this section, in addition to the requirements in Subchapters A and B of this chapter (relating to General Financial Assurance Requirements and Financial Assurance Requirements for Closure, Post Closure, and Corrective Action). The local government guarantor must meet the requirements of the local government financial test in §37.271 of this title (relating to Local Government Financial Test) and must comply with the following terms of the local government guarantee.

(1) If the owner or operator fails to perform closure, post closure, or corrective action of a facility covered by the guarantee, the guarantor will: (A) perform, or pay a third party to perform, closure, post closure, or corrective action as required; or (B) establish a fully funded trust fund as specified in §37.201 of this title (relating to Trust Fund) in the name of the owner or operator.

(2) The guarantee will remain in force unless the guarantor sends notice of cancellation by certified mail to the owner or operator and to the executive director. Cancellation may not occur, however, during the 120 days beginning on the date of the receipt of the notice of cancellation by both the owner or operator and the execu-

tive director, as evidenced by the return receipts.

(3) If a guarantee is canceled, the owner or operator must, within 90 days following receipt of the cancellation notice by the owner or operator and the executive director, obtain alternate financial assurance and submit evidence of that alternate financial assurance to the executive director. If the owner or operator fails to provide alternate financial assurance within the 90-day period, the guarantor must provide that alternate assurance within 120 days following the guarantor=s notice of cancellation.

(4) The owner or operator must submit to the executive director an originally signed local government guarantee worded as specified in §37.381 of this title (relating to Local Government Guarantee) along with the items required in §37.271(3) of this title. The items must be updated annually in accordance with the requirements of the local government financial test.

(5) The owner or operator is no longer required to comply with the requirements of this section when the conditions as specified in §37.61 of this title (relating to Termination of Mechanisms) are met.

(6) If a local government guarantor no longer meets the requirements of §37.271 of this title, the owner or operator must, within 90 days, obtain alternate financial assurance, and submit such evidence of the alternate financial assurance to the executive director. If the owner or operator Texas Commission on Environmental Quality Page 20 Chapter 37 - Financial Assurance fails to obtain alternate financial assurance within that 90-day period, the guarantor must provide that alternate financial assurance within the next 30 days.

2. **§37.2015.** Submission of Documents. An owner or operator required to provide financial assurance must submit an originally signed financial assurance mechanism prior to issuance of registration. The signed financial assurance mechanism must be in effect at the time it is submitted.
3. **§37.2021.** Financial Assurance Requirements for Transporters of Used Oil. A used oil transporter must show proof of insurance to the commission in the forms and levels as prescribed by the Texas Department of Transportation or the U.S. Department of Transportation (49 USC §11506). The document issued by the Texas Department of Transportation or the U.S. Department of Transportation

which shows the used oil transporter is currently satisfying department requirements for transporting used oil will be an acceptable form of demonstrating proof of insurance and should be submitted to the commission. If a used oil transporter is not required to be registered as a motor carrier with either of these agencies, then proof of insurance in the form of an original signed certificate of insurance and in levels sufficient to pay for bodily injury and property damage liability caused by the used oil must be submitted to the commission directly by an insurance agent. In all cases, the name of the used oil transporter must be identical to the party named on the applicable insurance form.

4. **Exemption to Financial Responsibility Requirement:**
   Owners or operators which are state or federal government entities whose debts and liabilities are the debts and liabilities of a state or the United States are exempt. Local political subdivisions are not exempt.
5. **Document Submission:**
   An owner or operator who is required to provide financial assurance must submit an originally signed financial assurance mechanism prior to the issuance of a registration by our Registration and Reporting (R&R) Section. The signed financial assurance mechanism must be in effect at the time it is submitted.
   If you become registered and plan to expand your facility or reduce capacity, this will more than likely affect your cost estimates and in turn, your financial assurance obligations.

   Commentary: *The statement above is very important and could result in your submission being rejected so you must be sure any changes, no matter how slight are accounted for, documented and recalculate your financial assurance requirements then contact the TCEQ to verify if a resubmittal is required.*

6. **Financial Assurance Increase and Inflation Accountability**

§37.141 Increase in Current Cost Estimate: addresses any increase to your financial assurance:

Whenever the current cost estimate increases to an amount greater than the amount being provided in the financial assurance mechanism(s), the owner or operator must either cause the amount of the financial assurance to be increased and submit evidence of such increase to the executive director, or obtain additional financial assurance in accordance with this chapter to cover the increase. This adjustment must be made within 60 days after the owner or operator becomes aware, or is notified by the executive director, of the increase.

§37.141 Annual Inflation Adjustments to Closure Cost Estimates: addresses the annual inflation management requirement and determination.

During the active life of the facility, the owner or operator must adjust the current cost estimate for inflation within 60 days prior to the anniversary date of the first establishment of the financial assurance mechanism. For owners or operators using the financial test or corporate guarantee, the current cost estimate must be adjusted for inflation within 30 days after the close of the firm's fiscal year and before submission of updated information to the executive director as specified in this chapter. The adjustment must be made using an inflation factor derived from the most recent annual Implicit Price Deflator for Gross National Product published by the United States Department of Commerce in its Survey of Current Business, as specified in paragraphs (1) and (2) of this section. The inflation factor is the result of dividing the latest published annual Deflator by the Deflator for the previous year.

(1) The first adjustment is made by multiplying the current cost estimate by the inflation factor. The result is the adjusted current cost estimate.

(2) Subsequent adjustments are made by multiplying the latest adjusted cost estimate by the latest inflation factor.

(3) If there is a revised current cost estimate in current dollars approved by the executive director prior to the anniversary date described in this section, then the owner or operator is not required to make an inflation adjustment. However, a revised current cost estimate in current dollars shall be adjusted for inflation for subsequent years as required by this section.

Note: *You can obtain the recent inflation factor by going to the TCEQ Website available from* https://www.tceq.texas.gov/adminservices/financial-administration/revenue/annual_inflation_factors.html

7. Soil Term Utilization:

Our Rules allow the term "soil remediation" to mean the same as "closure" in order to promote consistency among the various financial assurance mechanisms that are utilized among different programs.

8. Mailing and Contact Information:

Mailing Address:

Texas Commission on Environmental Quality
Attn: Financial Assurance Section, MC-184
P.O. Box 13087
Austin, TX 78711-3087

Overnight Mail:
Texas Commission on Environmental Quality
Attn: Financial Assurance Section, MC-184
12100 Park 35 Circle
Austin, TX 78753
Phone: 512/239-6239
FAX: 512/239-6242

PART XV: MATERIAL ANALYSIS PLAN (MAP)

Applicability to Material Analysis Plan

Federal and state regulations require processors of used oil to develop and maintain on site a Material Analysis Plan (MAP). As indicated on the TCEQ website and the TCEQ Used Oil Handlers Handbook **(109),** you are subject to the requirement as a used oil processor in Texas as part of the Used Oil Handler group this guide addresses. Please note that from a reading of 40 CFR §279.50 **(67)** Applicability you may not be a processor but remember you must also comply with state requirements that may be more restrictive, as in this application, if you store used oil more than 35 days you should register as a processor / re-refiner.

"If you are involved in chemical or physical operations designed to (1) produce fuel oils, lubricants, or other products derived from used oil; or (2) make used oil more suitable for production of these products, you should register as a **used oil processor/re-refiner**. Also, if you store used oil more than 35 days, you should register as a used oil processor/re-refiner."

1. **40 CFR §279.50 (67) Applicability. (redacted regulatory cite)**

   (a) The requirements of this subpart apply to owners and operators of facilities that process used oil. Processing means chemical or physical operations designed to produce from used oil, or to make used oil more amenable for production of, fuel oils, lubricants, or other used oil-derived products. Processing includes, but is not limited to: blending used oil with virgin petroleum products, blending used oils to meet the fuel specification, filtration, simple distillation, chemical or physical separation and re-refining. The requirements of this subpart do not apply to:

(1) Transporters that conduct incidental processing operations that occur during the normal course of transportation as provided in Sec. 279.41; or

(2) Burners that conduct incidental processing operations that occur during the normal course of used oil management prior to burning as provided in Sec. 279.61(b).

2. **30 TAC §324.12. Processors and Re-refiners** (applicability analysis plan requirement redacted regulatory cite) **(173)**

The commission incorporates by reference rules for owners and operators of used oil processing and re-refining facilities in 40 Code of Federal Regulations (CFR) Part 279, particularly Subpart F. In addition, the commission requires the following:

(3) Analysis plan. Each facility must prepare an analysis plan. The facility will follow the plan when sampling and analyzing, keeping records, and complying with analytical requirements for documenting that used oil is not listed hazardous and/or the used oil fuel specification has been met. This plan must specify the frequency of sampling and analysis. It must also specify procedures and analysis to assure listed hazardous wastes are not mixed with the used oil received. It must also contain procedures for handling a shipment of contaminated used oil. A facility need not prepare an analysis plan if it:

(A) only processes its own used oil; and

(B) uses adequate process knowledge instead of analysis to prove that the used oil meets rule requirements.

3. **§ 279.55 Analysis plan. (Requirements) (174)**

Owners or operators of used oil processing and re-refining facilities must develop and follow a written analysis plan describing the procedures that will be used to comply with the analysis requirements of §279.53 and, if applicable, §279.72. The owner or operator must keep the plan at the facility.

(a) ***Rebuttable presumption for used oil in § 279.53.*** At a minimum, the plan must specify the following:

(1) Whether sample analyses or knowledge of the halogen content of the used oil will be used to make this determination.

(2) If sample analyses are used to make this determination:

(i) The sampling method used to obtain representative samples to be analyzed. A representative sample may be obtained using either:

(A) One of the sampling methods in appendix I of part 261 of this chapter; or

(B) A method shown to be equivalent under §§ 260.20 and 260.21 of this chapter;

(ii) The frequency of sampling to be performed, and whether the analysis will be performed on-site or off-site; and

(iii) The methods used to analyze used oil for the parameters specified in § 279.53; and

(3) The type of information that will be used to determine the halogen content of the used oil.

(b) ***On-specification used oil fuel in § 279.72.*** At a minimum, the plan must specify the following if § 279.72 is applicable:

(1) Whether sample analyses or other information will be used to make this determination;

(2) If sample analyses are used to make this determination:

(i) The sampling method used to obtain representative samples to be analyzed. A representative sample may be obtained using either:

(A) One of the sampling methods in appendix I of part 261 of this chapter; or

(B) A method shown to be equivalent under §§ 260.20 and 260.21 of this chapter;

(ii) Whether used oil will be sampled and analyzed prior to or after any processing/re-refining;

(iii) The frequency of sampling to be performed, and whether the analysis will be performed on-site or off-site; and

(iv) The methods used to analyze used oil for the parameters specified in §279.72; and

(3) The type of information that will be used to make the on-specification used oil fuel determination.

MAP Example Table Of Contents

An example MAP Table of Contents is provided below, but it is critical that the MAP is site-specific. It must also cover all oil related materials, antifreeze and any other regulated material managed at your site.

Table Of Contents

3.19 Material Acceptance Criteria
3.20 Used Oil/Oily Water-Prior to Unloading
3.21 Used Antifreeze-Prior to Unloading
3.22 Drum Loads-Prior to Unloading

PART XVI: USED OIL HANDLER CLOSURE PLAN REQUIREMENTS

Used Oil Processors/Re-Refiners

Used Oil Handler processors/re-refiners are subject to Closure Plan requirements in Texas; they are also required to establish Financial Assurance. Processors/re-refiners start the establishments of Financial Assurance as identified above; this is used to cover any "soil remediation" required by the facility during operations or after operations stop when closure is required prior to establishing that the site is closed and capable of being used for other purposes. The TCEQ website notes the following regarding used oil handler requirements and financial assurance: "Our Rules allow the term "soil remediation" to mean the same as "closure" in order to promote consistency among the various financial assurance mechanisms that are utilized among different programs." Specific requirements to do closure and associated activities are identified primarily in 40 CFR §279.54 **(174)** Used oil management, and 30 TAC §324.22. **(115)** Soil Remediation Requirements for Used Oil Handlers.

If you do not meet the initial clean-up standards identified and as the State and Federal requirements you must then comply with 40 CFR §265.310 **(176)** Closure and post-closure care, §265.117 **(175)** Post-closure care and use of property, §265.118 **(175)** Post-closure plan; amendment of plan, §265.119 **(175)** Post-closure notices, §265.120 **(175)** Certification of completion of post-closure care, and §265.145 **(175)** Financial assurance for post-closure care. Complying with the Federal and State regulations it is more than a little significant problem and will involve significant TCEQ involvement as well as being a costly process. The closure requirements point out how significant your spill clean-up procedures and quick response to **ANY** release on-site is important to avoid future costs associated with closure.

Redacted sections of the above regulations are provided below with italicized comments. In all situations it is imperative that you document all closure activities that include any analysis, disposal characterization, profiles submitted to any landfill or other disposition facility, Bills of Lading, Uniform Hazardous Waste Manifests, or any other documentation such as a closure report from a contractor / consultant. To close out any Used Oil Handler permit associated with the facility the documentation will be required for review by the TCEQ.

1. **§ 279.54 Used Oil Management. (174)**

   Used oil processor/re-refiners are subject to all applicable Spill Prevention, Control and Countermeasures *(SPCC)* (40 CFR part 112) in addition to the requirements of this subpart. Used oil processors/re-refiners are also subject to the Underground Storage Tank (UST) (40 CFR part 280) standards for used oil stored in underground tanks whether or not the used oil exhibits any characteristics of hazardous waste, in addition to the requirements of this subpart.

   Note: *The UST requirements are not addressed in this Guide, however SPCC requirements are addressed in later sections.*

(h) ***Closure -***

(1) Aboveground tanks. Owners and operators who store or process used oil in aboveground tanks must comply with the following requirements:

(i) At closure of a tank system, the owner or operator must remove or decontaminate used oil residues in tanks, contaminated containment system components, contaminated soils, and structures and equipment contaminated with used oil, and manage them as hazardous waste, unless the materials are not hazardous waste under this chapter.

(ii) If the owner or operator demonstrates that not all contaminated soils can be practicably removed or decontaminated as required in paragraph (h)(1)(i) of this section, then the owner or operator must close the tank system and perform post-closure care in accordance with the closure and post-closure care requirements that apply to hazardous waste landfills (§ 265.310 of this chapter).

***(2) Containers.** Owners and operators who store used oil in containers must comply with the following requirements:*

(i) At closure, containers holding used oils or residues of used oil must be removed from the site;

(ii) The owner or operator must remove or decontaminate used oil residues, contaminated containment system components, contaminated soils, and structures and equipment contaminated with used oil, and manage them as hazardous waste, unless the materials are not hazardous waste under part 261 of this chapter.

2. §324.22. Soil Remediation Requirements for Used Oil Handlers. (115)

***Commentary:** Note: That the Texas regulations identify "It also applies to owners and operators of used oil transfer, processing, re-refining, and off-specification used oil burning facilities referred to as "used oil handlers."" As applicable to Financial Assurance BUT only processors and re-refiners are subject to closure requirements as identified in §279.54 Used oil management identified above as stated in §324.22 (e)*

(a) This section applies to transporters of used oil who are seeking registration under this chapter. It also applies to owners and operators of used oil transfer, processing, re-refining, and off-specification used oil burning facilities referred to as "used oil handlers." It does not apply to a used oil handler which is owned or otherwise effectively controlled by the owners or operators where the used oil is generated.

(e) **As required, processors or re-refiners who store or process used oil in above-ground tanks must, at closure of a tank system, demonstrate financial assurance in the amount of the cost to comply with the closure requirements of 40 CFR §279.54(h). If the used oil handler cannot demonstrate that all contaminated soils are removed or decontaminated as required in 40 CFR §279.54(h), the used oil handler must further demonstrate financial assurance in the amount required to cover the soil and perform post closure in accordance with the closure and post closure care requirements that apply to hazardous waste landfills under 40 CFR §265.310, 265.117-265.120, and 265.145.**

PART XVII: OIL SPILL PREVENTION

As a Texas Used Oil Hander you are subject to the requirements of Title 40, code of Federal Regulations, Part 112, identified as the Spill Prevention, Control, and Countermeasure (SPCC) and Facility Response Plan (FRP) rule.

The EPA Guidance Document

"A Facility Owner/Operator's Guide to Oil Pollution Prevention" **(177)** identifies that "The purpose of the Spill Prevention, Control, and Countermeasure (SPCC) (and FRP) rule is to help facilities prevent a discharge of oil into navigable waters or adjoining shorelines. This rule is part of the U.S. Environmental Protection Agency's oil spill prevention program and was published under the authority of Section 311(j)(1)(C) of the Federal Water Pollution Control Act (Clean Water Act) in 1974. "

Subject to the requirements does not mean you have to comply with all the requirements but you must consider your specific applicability based on your facility activities. This section of the guide is intended to take you through the primary elements of the SPCC / FRP plan for determination of your applicability and what you need to do to be in compliance.

SPCC / FRP Overview

In 2002, the U.S. Environmental Protection Agency (EPA) amended the Oil Pollution Prevention regulation (40 CFR part 112), which includes requirements for specific facilities to prepare, amend, and implement Spill Prevention, Control, and Countermeasure (SPCC) Plans and Facility Response Plans (FRP). The regulation is largely performance-based which allows flexibility in meeting the rule requirements to prevent discharges of oil to navigable waters and adjoining shorelines. It is important to note that if your business handles other materials than used oil, oily water, and antifreeze you will need to refer to the applicable regulations regarding your specific material.

A primary reference for this guide section is the EPA document SPCC Guidance for Regional Inspectors (Dec 2013) **(51)** and copies of this document may be obtained online at www.epa.gov/oilspill. In addition, updates to the document will be available online. This is a significant resource and provides significant more detail on SPCC issues than provided here related to used oil management as the focus of this guide. Also, as always don't hesitate to call the TCEQ regarding SPCC guidance or the US Coast Guard for FRP guidance.

Regulatory Requirement

Identified below in abbreviated form are the applicable TAC regulations and associated 40 CFR Part 279 and the 40 CFR Part 112 application that has the regulatory requirements for the SPCC and FRP plans regarding the various Used Oil Handler types and the regulatory definition of "Oil" that because you handle used oil puts you in the potential regulated group based on your facility type and activities.

A. **40 CFR §112.2 defines oil: (43)** Oil means oil of any kind or in any form, including, but not limited to: fats, oils, or greases of animal, fish, or marine mammal origin; vegetable oils, including oils from seeds, nuts, fruits, or kernels; and, other oils and greases, including petroleum, fuel oil, sludge, synthetic oils, mineral oils, oil refuse, or oil mixed with wastes other than dredged spoil.

B.TAC §324.1 Federal Rule Adoption: (115) Person(s) managing used oil must comply with the requirements in this chapter and the requirements in 40 Code of Federal Regulations (CFR) Part 279, Standards for the Management of Used Oil, as amended through July 14, 2006, at 71 FedReg 40280, which are adopted by reference.

Applicable Facility: (43)

1. **40 CFR §112.2:** Facility means any mobile or fixed, onshore or offshore building, structure, installation, equipment, pipe, or pipeline (other than a vessel or a public vessel) used in oil well drilling operations, oil production, oil refining, oil storage, oil gathering, oil processing, oil transfer, oil distribution, and waste treatment, or in which oil is used, depend on several site-specific factors, including, but not limited to, the ownership or operation of buildings, structures, and equipment on the same site and the types.

It is important to note that if your facility does not fit into one of the determinations below easily or you have special considerations you may want to review the entire 40 CFR §112.2 for more references to possible considerations for your determination as to your applicable facility.

2. **Facility boundaries determination factors for consideration:**
   a. Ownership, management, and operation of the buildings, structures, equipment, installations, pipes, or pipelines on the site;
   b. Similarity in functions, operational characteristics, and types of activities occurring at the site;
   c. Adjacency; or Shared drainage pathways (e.g., same receiving waterbodies).
   d. The facility owner or operator, or a Professional Engineer (PE) on behalf of the facility owner/operator, determines what constitutes the "facility."

   Commentary: *The facility determination for purposes of the SPCC rule should be the same as that used to determine FRP applicability.*

   e. While the facility owner/operator has some discretion in defining the parameters of the facility, the boundaries of a facility should not be drawn to purposely avoid regulation under 40 CFR part 112.
   f. For example, two contiguous operational areas, each with 700 gallons in aboveground storage capacity, that have the same owner, perform similar functions, are attended by the same personnel, and are in other ways indistinguishable from each other, would reasonably be expected to represent a single facility under the SPCC rule, and would therefore be required to have an SPCC Plan, since the capacity of this facility is above the 1,320 gallon aboveground threshold. These two operational areas would not be defined as two separate facilities under the definition of "facility" in §112.2.
3. **TAC §324.12 Processors and Re-refiners: (115)** The commission incorporates by reference rules for owners and operators of used oil processing and re-refining facilities in 40 Code of Federal Regulations (CFR) Part 279, particularly Subpart F.
4. **40 CFR §279.54 Used oil management: (174)** Used oil processor/re-refiners are subject to all applicable Spill Prevention, Control and Countermeasures (40 CFR part 112) in addition to the requirements of this subpart.

5. **TAC §324.11 Transporter and Transfer Facilities: (115)** The commission incorporates by reference rules for used oil transporters and transfer facilities in 40 Code of Federal Regulations (CFR) Part 279, particularly Subpart E.
6. **40 CFR §279.45 Used oil storage at transfer facilities: (69)** Used oil transporters are subject to all applicable Spill Prevention, Control and Countermeasures (40 CFR part 112) in addition to the requirements of this subpart.

This regulatory reference above applies to used oil transfer facilities. Used oil transfer facilities are transportation related facilities including loading docks, parking areas, storage areas, and other areas where shipments of used oil are held for more than 24 hours during the normal course of transportation and not longer than 35 days. Transfer facilities that store used oil for more than 35 days are subject to regulation under subpart F of this part.

7. **TAC §324.7 Collection Centers: (115)** The commission incorporates by reference rules for owners or operators of all "do-it-yourselfer used oil collection centers" and "used oil collection centers" (as defined in 40 Code of Federal Regulations (CFR) §279.1) in 40 CFR Part 279, particularly Subpart D.
8. **40 CFR §279.31 Used oil collection centers: (69) (a) Applicability.** This section applies to owners or operators of used oil collection centers. A used oil collection center is any site or facility that accepts/aggregates and stores used oil collected from used oil generators regulated under subpart C of this part who bring used oil to the collection center in shipments of no more than 55 gallons under the provisions of §279.24(a). Owners or operators of all used oil collection centers must: (1) Comply with the generator standards in subpart C of this part;
9. **40 CFR §279.30 (69) Do-it-yourselfer used oil collection centers**: (a) *Applicability.* This section applies to owners or operators of all do-it-yourselfer (DIY) used oil collection centers. A DIY used oil collection center is any site or facility that accepts/aggregates and stores used oil collected only from household do-it-yourselfers. (b) *DIY used oil collection center requirements.* Owners or operators of all DIY used oil collection centers must comply with the generator standards in subpart C of this part.
10. **40 CFR §279.32 (69)** Used Oil aggregation points owned by the generator: (a) Applicability. This section applies to owners or operators of all used oil aggregation points. A used oil aggregation point is any site or facility that accepts, aggregates,

and/or stores used oil collected only from other used oil generation sites owned or operated by the owner or operator of the aggregation point, from which used oil is transported to the aggregation point in shipments of no more than 55 gallons under the provisions of §279.24(b). (b) Used oil aggregation point requirements. Owners or operators of all used oil aggregation points must comply with the generator standards in subpart C of this part.

11. **TAC §324.6 Generators: (115)** The commission incorporates by reference rules for used oil generators in 40 Code of Federal Regulations Part 279, particularly Subpart C. A person or entity that services equipment involving removal of used oil or changes used oil at a customer's home or business and transports the used oil from the site in quantities less than or equal to 55 gallons may choose to be the generator. If the service company removing the used oil from equipment does not assume generator responsibility, the site owner or operator will remain the generator.
12. **40 CFR §279.22 Used oil storage: (61)** Used oil generators are subject to all applicable Spill Prevention, Control and Countermeasures (40 CFR part 112) in addition to the requirements of this Subpart. Used oil generators are also subject to the Underground Storage Tank (40 CFR part 280) standards for used oil stored in underground tanks whether or not the used oil exhibits any characteristics of hazardous waste, in addition to the requirements of this subpart.
13. **TAC §324.14 Marketers of Used Oil Fuel: (115)** The commission incorporates by reference rules for marketers of used oil which will be burned for energy recovery. These rules are found in 40 Code of Federal Regulations Part 279, Subpart H.

Potential to Discharge Into "Navigable Waters"

With an assumption that you are a designated facility above and meet the oil storage requirements of at least 1,300 gallons in aboveground storage (specific guidance on storage to follow) or have more than 42,000 gallons of oil storage underground making you subject to SPCC and possibly FRP requirements. You next determination is your potential to discharge into "Navigable Waters" that further identifies your compliance requirements.

Threshold Determination as to Applicability Associated With Facility Potential to Discharge Into "Navigable Waters"

Commentary: *This is an often-debated issue related to Used Oil Facilities as it appears by many to be a more subjective than objective decision by regulators. A discussion of the definition follows but from a practical matter if you store used oil in the quantities identified above in a facility identified within the regulations you are going to be applicable from the TCEQ perspective. I have included the full text of the EPA SPCC Regional Inspectors Guidance (2013)* ***(51)*** *for reference and as you review it you will notice how it states that even though it is your responsibility as an owner or operator of a "facility" it is your determination but also states that "The rule does not define the term "reasonably be expected." This puts you as a Used Oil Handler in a very difficult position because you cannot be wrong but seems a contradiction by the regulators. To their defense however, many fact patterns related to operations and facility proximity to navigable waters would be difficult to define by regulation. It is easy for anyone with a facility sitting next to or close to a waterway, lake or shoreline but that is not the case for most facilities.*

a. Section 112.2 provides the SPCC rule's definition of "navigable waters." This definition has been revised on several occasions, most recently in 2008. The current definition of navigable waters for the SPCC rule is the definition promulgated byEPA in 1973 (73 FR 71941, November 26 2008).

   EPA and the U.S. Army Corps of Engineers have issued guidance on implementing Supreme Court decisions that affect CWA jurisdiction over navigable waters. A copy of the "EPA and Army Corps rule to define "Waters of the United States" by Claudia Copeland specialist in Resources and Environmental Policy (2017) can be found at https://fas.org/sgp/crs/misc/R43455.pdf .
b. With the above consideration the regulatory citation below and applicability to a facility is summarized for SPCC and FRP application is identified.
c. **40 CFR §112.2** "Navigable waters" of the United States means "navigable waters" as defined in section 502(7) of the FWPCA, and includes:

(1) All navigable waters of the United States, as defined in judicial decisions prior

to the passage of the 1972 Amendments to the FWPCA (Pub. L. 92-500), and tributaries of such waters;

(2) Interstate waters;

(3) Intrastate lakes, rivers, and streams which are utilized by interstate travelers for recreational or other purposes; and

(4) Intrastate lakes, rivers, and streams from which fish or shellfish are taken and sold in interstate commerce.

d. SPCC Applicability: A facility is covered by the SPCC rule if it has an aggregate aboveground oil storage capacity greater than 1,320 U.S. gallons or a completely buried storage capacity greater than 42,000 U.S. gallons and there is a reasonable expectation of an oil discharge into or upon navigable waters of the U.S. or adjoining shorelines.

e. FRP Applicability: The FRP requirement for onshore facilities applies to any facility that, "because of its location, could reasonably be expected to cause substantial harm to the environment by discharging into or on the navigable waters, adjoining shorelines, or the exclusive economic zone.

f. It could happen, even though unlikely, after consideration of all the issues of reasonable expectation of an oil discharge to navigable waters based on If an owner or operator determination that, due to the location, the facility cannot reasonably be expected to discharge oil as described in §112.1(b), he should be prepared to provide the rationale and any supporting documentation to an EPA (TCEQ) inspector that explains why the facility does not have an SPCC Plan.

Reasonable Expectation of Discharge

An SPCC or FRP requirement are dependent on your applicability to one of the two definitions associated with potential discharge. First, for SPCC applicability your potential must represent a "Reasonable Expectation of Discharge," which applies only to facilities that, due to their location, can *reasonably be expected to discharge oil* as described in §112.1(b). The rule does not define the term "reasonably be expected." The owner or operator of each facility must determine the potential for a discharge from his/her facility. According to §112.1(d)(1)(i), this determination must be based solely upon consideration of the geographical and locational aspects of the facility. An owner or operator should consider the location of the facility in relation to a stream, ditch, gully, or storm sewer;

the volume of material likely to be spilled; drainage patterns; and soil conditions. An owner or operator may not consider constructed features, such as dikes, equipment, or other manmade structures that prevent, contain, hinder, or restrain a discharge as described in §112.1(b), when making this determination.

The SPCC rule applies only to facilities that, due to their location, can reasonably be expected to discharge oil as described in §112.1(b). The rule does not define the term "reasonably be expected." The owner or operator of each facility must determine the potential for a discharge from his/her facility. According to §112.1(d)(1)(i), this determination must be based solely upon consideration of the geographical and locational aspects of the facility. An owner or operator should consider the location of the facility in relation to a stream, ditch, gully, or storm sewer; the volume of material likely to be spilled; drainage patterns; and soil conditions. An owner or operator may not consider constructed features, such as dikes, equipment, or other manmade structures that prevent, contain, hinder, or restrain a discharge as described in §112.1(b), when making this determination.

A facility owner or operator, however, should consider the presence of manmade structures that may serve to transport discharged oil to navigable waters, such as sanitary or storm water drainage systems, even if they lead to a publicly owned treatment work (POTW) prior to ultimate discharge into navigable waters. The presence of a treatment system such as a POTW cannot be used to determine that the facility is not reasonably expected to discharge to navigable waters or adjoining shorelines. POTWs can fail to contain oil. They are not designed to handle oil discharges and are on occasion forced to bypass to receiving waterbodies during extreme weather events or when upsets occur in the treatment system.

The following factors may be useful to consider in determining whether there is a reasonable expectation of a discharge:

1. ***Past discharges*** of oil from the facility or a neighboring facility that reached a navigable water or adjoining shoreline may indicate that another could be reasonably expected;
2. ***Facility location*** relative to navigable waters, a watercourse and/or intervening natural drainage could cause a discharge to the navigable waters to be reasonably expected;
3. ***On-site conduits and certain underground features***, such as sewer lines, storm

sewers, power or cable lines, or groundwater could facilitate the transport of discharged oil off-site to navigable waters;

4. ***Unique geological or geographic features*** could facilitate the transport of discharged oil off-site to navigable waters;
5. ***Precipitation runoff*** could transport oil into navigable waters; and
6. ***Quantity and nature of oil stored.***

SPCC Applicability Determination

With an assumption that you are a designated applicable facility and you have a potential for discharge into navigable waters next step would be to determine volumes of oil applicable and exemptions that may apply.

1. Exemptions and Exclusions
    a. 40 CFR **§112.1** establishes the general applicability of the SPCC rule by describing both the facilities, activities, and equipment that are subject to the rule and those that are excluded.
        1. In general, SPCC-regulated facilities are non-transportation related, have aboveground oil storage capacity of more than 1,320 gallons on site, and could reasonably be expected to discharge oil to navigable waters or adjoining shorelines in quantities that may be harmful.
        2. Facilities owned and operated by federal government agencies are subject to the regulation to the same extent as any other facility (although the federal government is not subject to civil penalties).
        3. Likewise, facilities owned and operated by state and local governments are subject to the regulation.

    b. Section 112.1(d) describes the facilities, activities, and equipment excluded from the rule based on jurisdiction or through exemptions or exclusions from storage capacity calculations.

        1. Exemptions pertain to whether a facility or part thereof is included in the SPCC-regulated universe, and exclusions from storage capacity determine which containers count when determining a facility's total oil storage capacity. In addition to facilities that are excluded from the SPCC rule because they are not subject to EPA's jurisdiction, §112.1(d) ex-

empts:

a. Minerals Management Service

b. Any offshore oil drilling, production, or workover facility that is subject to the federal notices.

c. Completely buried storage tanks.

d. Connected underground piping

e. Underground ancillary equipment, and containment systems, when such tanks are subject to all of the technical requirements of 40 CFR part 280 or a state program approved under 40 CFR part 281 (also known as the Underground Storage Tank regulations).

1. Although these tanks are exempt from the SPCC requirements, they must still be marked on the facility diagram if the facility is otherwise subject to the SPCC rule (§112.7(a)(3)).
2. Any tanks that are not completely buried are considered aboveground storage tanks and subject to the SPCC rule.

f. The following are either excluded from the definition of UST or are exempt from the UST regulations at 40 CFR part 280 (and therefore may be subject to the SPCC rule, if the completely buried tanks contain oil):

1. Tanks with a capacity of 110 gallons or less;
2. Farm or residential tanks with a capacity of 1,100 gallons or less used for storing motor fuel for non-commercial purposes;
3. Tanks used for storing heating oil for consumptive use on the premises where stored; Tanks storing non-petroleum oils, such as animal fat or vegetable oil;
4. Tanks on or above the floor of underground areas (e.g., basements or tunnels);
5. Septic tanks and systems for collecting storm water and wastewater;
6. Flow-through process tanks;
7. Emergency spill and overfill tanks that are expeditiously emptied after use;
8. Surface impoundments, pits, ponds, or lagoons; Any UST system holding RCRA hazardous waste;
9. Any equipment or machinery that contains regulated substances for operational purposes;

10. Liquid trap or associated gathering lines directly related to oil or gas production or gathering operations;
11. Pipeline facilities regulated under the Natural Gas Pipeline Safety Act of 1968, the Hazardous Liquid Pipeline Safety Act of 1979, or intrastate pipelines regulated under state laws comparable to the provisions of above laws;
12. Any UST system that contains de minimis concentration of regulated substances. The following are examples of deferrals from the UST regulations (and therefore may be subject to the SPCC rule):
13. Wastewater treatment tank systems;
14. Any UST systems containing radioactive materials that are regulated under the Atomic Energy Act of 1954;
15. UST systems that are part of emergency generator systems at nuclear power generation facilities;

Airport hydrant fuel distribution systems;

18. UST systems with field-constructed tanks.

c. **Applicability overview is provided from the EPA SPCC Guidance (50) (Table 1) document below:**

Commentary: *You can see by the referenced table that Used Oil Management activities as they relate to storage, transfer facilities, processing, re-refining would be possibly applicable to the requirements of an SPCC plan and based on location also to an FRP as well.*

Table 6 – EPA, Examples of Oil Related Activities That May Be Regulated Under 40 CFR Part 112. Source EPA SPCC Guidance (2005)

Activity	Examples of Oil-related Regulated Activities
Drilling	Drilling a well to extract crude oil or natural gas and associated products (such as wet natural gas) from a subsurface field.
Producing	Extracting product from a well and separating the crude oil and/or gas from other associated products (e.g., water, sediment).
Gathering	Collecting oil from numerous wells, tank batteries, or platforms and transporting it to a main storage facility, processing plant, or shipping point.
Storing	Storing oil in containers prior to use, while being used, or prior to further distribution in commerce.
Processing	Treating oil using a series of processes to prepare the oil for commercial use, consumption, further refining, manufacturing, or distribution.
Refining	Separating crude oil into different types of hydrocarbons through distillation, cracking, reforming, and other processes; separating animal fats and vegetable oils from free fatty acids and other impurities.
Transferring	Transferring oil between containers, such as between a railcar or tank truck and a bulk storage container, or between stock tanks and manufacturing equipment.
Distributing	Selling or marketing oil for further commerce or moving oil using equipment such as highway vehicles, railroad cars, or pipeline systems. Note that businesses commonly referred to as oil distributors and retailers commonly are also "storing" oil, as described above.
Using	Using oil for mechanical or operational purposes in a manner that does not significantly reduce the quantity of oil, such as using oil to lubricate moving parts, provide insulation, or for other purposes in electrical equipment, electrical transformers, and hydraulic equipment.
Consuming	Consuming oil in a manner that reduces the amount of oil, such as burning as fuel in a generator.

1. **Storage Determinations – with an assumption that you have met one or more**

of the applicable regulated activities the next step would be to identify the volume of oil that 40 CFR §112.1 (43) identifies that the SPCC rule applies to a facility that has more than 42,000 gallons of completely buried oil storage capacity or more than 1,320 gallons of aggregate aboveground oil storage capacity, provided it meets the other applicable criteria set forth in §112.1. **(43)**

a. **Don't Count:**

1. Any facility where the storage capacity of completely buried storage tanks and associated piping and equipment does not exceed 42,000 gallons.
2. The aggregate aboveground storage capacity does not exceed 1,320 gallons;
3. Any container with a storage capacity less than 55 gallons at a facility, whether or not subject to the requirements of the SPCC rule;
4. Any facility or part thereof used exclusively for wastewater treatment.
5. Storage containers used exclusively in wastewater treatment.
6. Completely buried tanks and associated piping and equipment that are subject to all of the technical requirements under 40 CFR part 280 or 281;
7. The capacity of any "permanently closed" aboveground storage container.

Note: *This does not mean out of service for maintenance or not being used due to inventory issues but must meet the 40 CFR §112.2 (43) Permanently* ***closed*** *requirements identified below.*

a. **§112.2 Permanently** *closed* means any container or facility for which (1) All liquid and sludge has been removed from each container and connecting line; and (2) All connecting lines and piping have been disconnected from the container and blanked off, all valves (except ventilation valves) have been closed and locked, and conspicuous signs have been posted on each container stating that it is a permanently closed container and noting the date of closure.

2. Oil Water Separators are often a debated even though often not counted and so included here, however it may still be difficult to determine their applicability, I have included a table (Table 6) that is from the EPA SPCC Guidance document identified above for your reference. Separators are not counted in many facilities after

identifying applicability however you must be sure. Follow the reference table I have included a reference flowchart (Flowchart 5) from the same guidance document for your reference. SPCC.

Table 7 – SPCC Rule Applicability for Various Uses of Oil/Water Separators, Source, EPA SPCC Guidance (2005)

Wastewater Treatment	Secondary Containment	Oil Production
Separators are exempt from all SPCC requirements in accordance with §112.1(d)(6) and do not count toward facility storage capacity.	Separators that are used as part of a secondary containment system and are not intended for oil storage or use do not themselves require secondary containment, and do not count toward facility storage capacity. However, they are subject to the design specifications (e.g., capacity) for the secondary containment requirements with which they are designed to comply.	Separators that are bulk storage containers, subject to the provisions of §§112.9(c) or 112.11(b) and (d), are not exempt and count toward the facility storage capacity.

The following statement from the EPA Guidance **(51)** identifies additional issues for your consideration to help insure your separator is categorized correctly. Section 112.1(d)(6) addresses oil/water separators used for wastewater treatment. Facilities or equipment used exclusively for wastewater treatment, and which do not satisfy any requirements of the SPCC rule, are exempt from the SPCC rule requirements. These oil/water separators do not count toward facility storage capacity. Whether a wastewater treatment facility or part thereof is used exclusively for wastewater treatment or used to satisfy an SPCC requirement will often be a facility-specific determination based upon the activities carried out at the facility and upon its configuration. Drainage systems that satisfy the secondary containment requirements of the SPCC rule may use oil/water separators to recover oil and return it to the facility. Examples of oil/water separators that are used to meet SPCC requirements include oil/water separators used to satisfy the secondary containment requirements of §§112.7(c), 112.7(h)(1), 112.8(c)(2), 112.8(c)(11), 112.12(c)(2), and/or 112.12(c)(11). Additionally, the drainage provisions in §§112.8(b) and 112.9(b) set forth design specifications for secondary containment at a facility. Oil/water separators may be used as part of a facility drainage system to meet the secondary containment requirements of the rule. Oil/water separators used to satisfy these rule requirements are subject to applicable secondary containment requirements, but they do not count toward storage capacity. From the same guidance flowchart 4 below provides additional reference for separator applicability.

Flowchart 4 – TCEQ Texas Hazardous and Nonhazardous Wastes

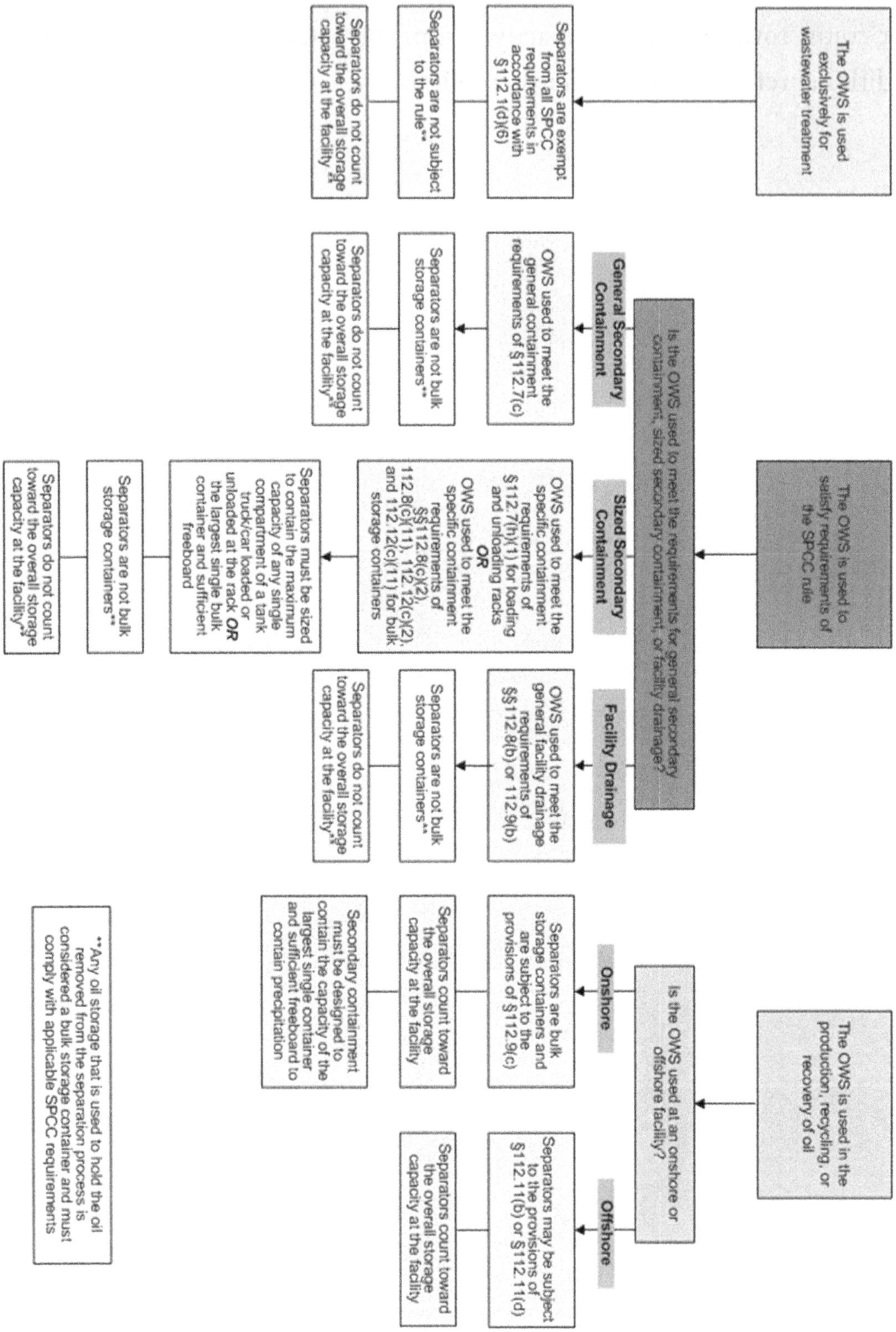

3. What Is Counted:
   a. Capacity of containers (e.g., bulk storage containers, oil-filled equipment, mobile/portable containers) with a capacity of 55 gallons or greater. Note that any container brought on to a facility that has never stored oil is not subject to the SPCC rule, nor is it counted toward the facility capacity until it stores oil. Any other container that at one time stored oil but no longer contains oil or sludge, which is brought on to a facility and meets the definition of permanently closed, is not subject to the SPCC rule nor is it counted toward the facility capacity until it stores oil. Practically, unless you have cleaned and labeled as empty with a date cleaned a 55 gallon container or larger that once contained oil it must be counted.
   b. ***Remember***: Under the SPCC rule, if a container has the requisite capacity, it does not matter whether the container is actually filled to that capacity. The storage capacity of a container is defined as the shell capacity of the container. If a certain portion of a container is incapable of storing oil because of its integral design (e.g., mechanical equipment or other interior components take up space), then the shell capacity of the container is reduced to the volume the container might hold (67 FR 47081). Generally, the shell capacity is the rated design capacity rather than the working/operational capacity. This means that used oil drums of 55 gallons or greater on your property, currently filled or not, are to be counted for SPCC applicability unless they meet the 67 FR 47081 noted above.
4. **Oil and water mixtures no matter how small the quantity of Oil** with water the water mixture containers are subject to the SPCC rule. A mixture of wastewater and oil is "oil" under the statutory and regulatory definition of the term (33 U.S.C. 1321(a)(1) and 40 CFR 110.2 and 112.2). A discharge of wastewater containing oil to navigable waters or adjoining shorelines in a "harmful quantity" (40 CFR part 110) is prohibited (see July 17, 2002, 67 FR 47069). One example of an oil and water mixture is produced water.
5. An SPCC applicability flowchart from the EPA Guidance Document "SPCC Guidance For Regional Inspectors (2013) (51) is provided below for your reference:

Flowchart 5 – EPA SPCC Applicability - Reference EPA Guidance Document

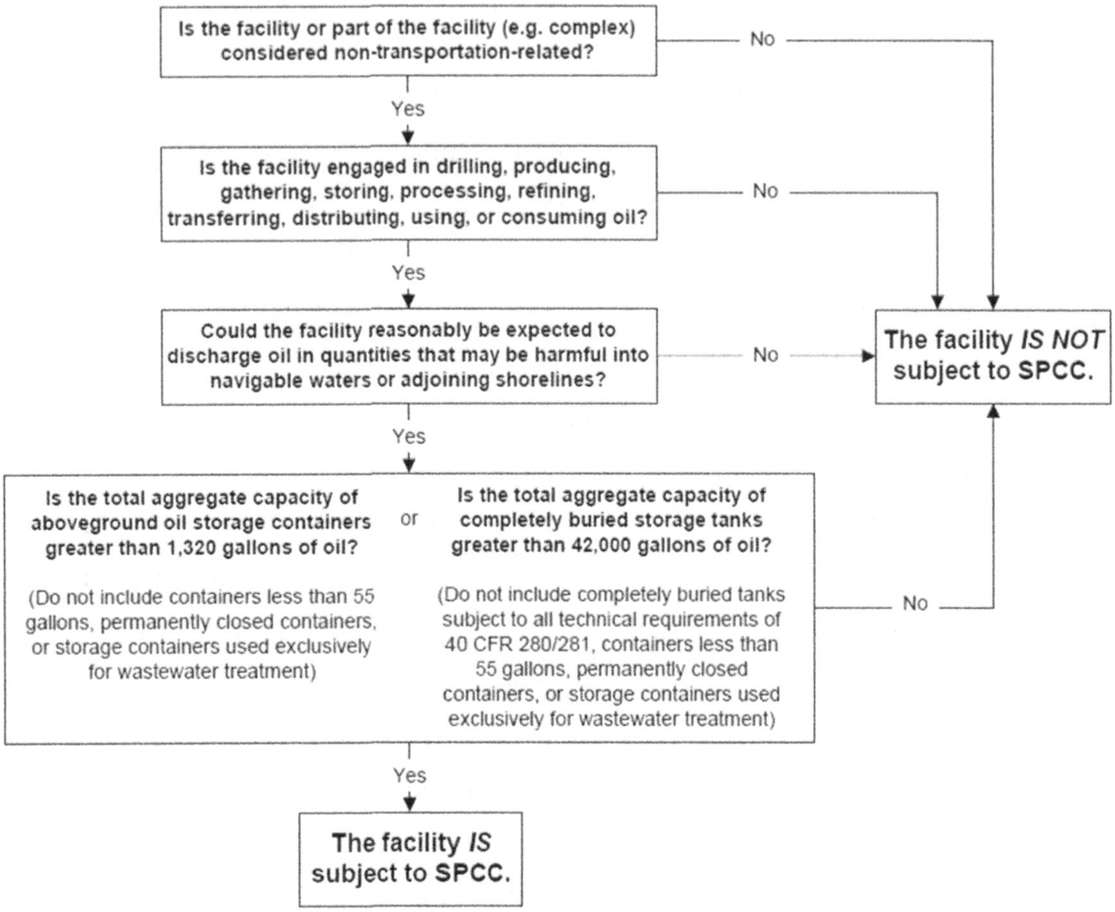

6. Second, for the FRP requirements to apply in addition to the SPCC requirement your non-transportation-related facility is SPCC-regulated, and an oil discharge from your facility could reasonably be expected to cause "Substantial Harm" to the environment from a discharge to navigable waters of the U.S. or the adjoining shoreline, you must prepare an FRP and submit it to the EPA region. The Regional Administrator will then determine whether your facility is a "significant and substantial harm" facility, which would require the RA to review and approve your FRP. If your facility does not meet the substantial harm criteria discussed in the sections below, you must complete a certification form and maintain it at the facility for review by EPA during facility inspections along with your SPCC plan. I have provided a flowchart from the 40 CFR §112 Appendix C below to assist you in your determination of substantial harm.

Flowchart 6 – EPA, FRP Applicability, Reference Title 40 Chapter I Subchapter D Part 112 Appendix C

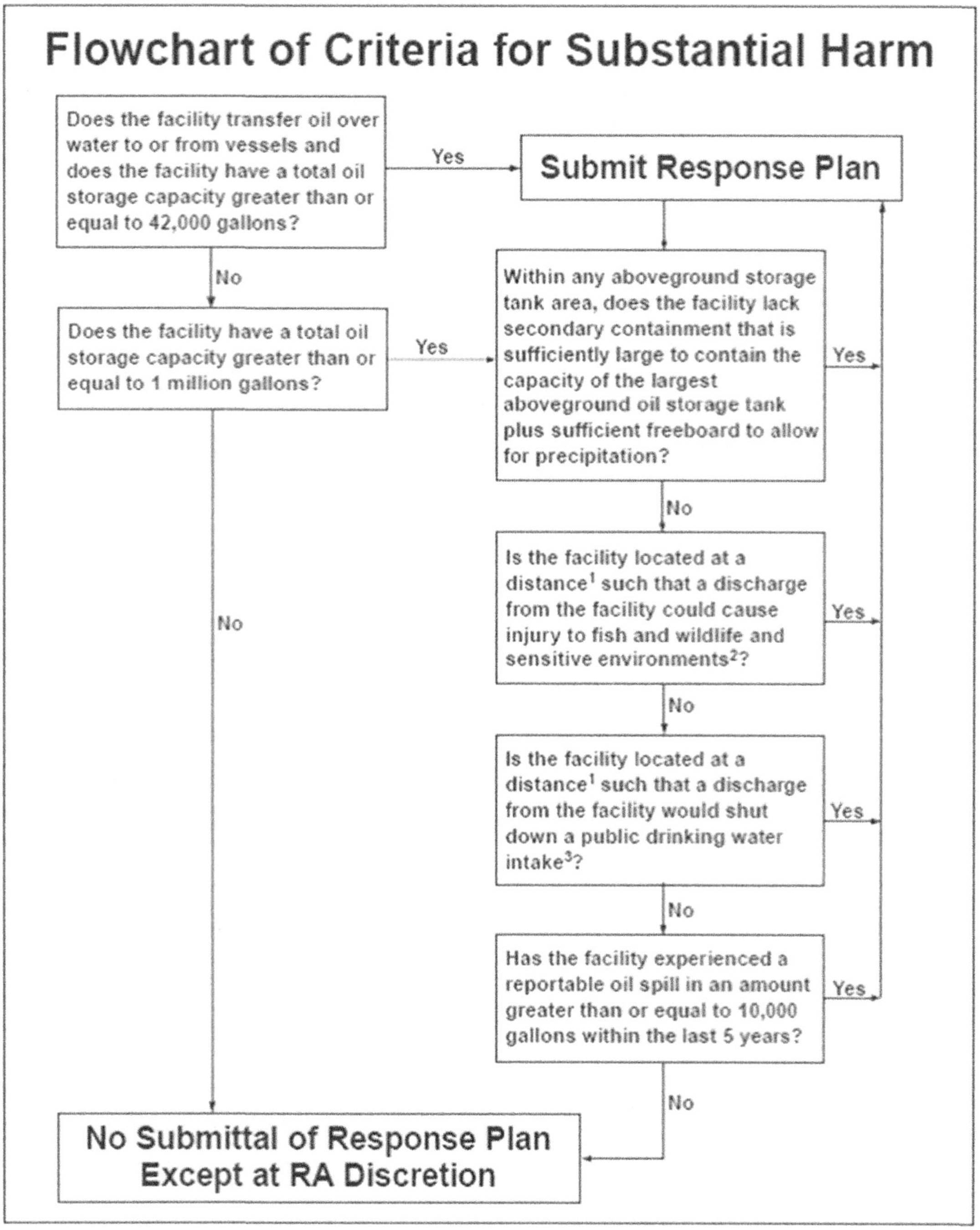

Further specifics as to FRP specific requirements are provided below following the SPCC requirements.

C. Overall Purpose and Scope: The following content represent a more detailed overview of the SPCC and FRP references, rule intent and delegation, Statutory Framework, Regulatory History and Implementation, Regional Admin-

istrator (RA) issues, Clean Water Act (CWA) references, United States Coast Guard and Texas Government Land Office involvement in the FRP

1. SPCC and FRP Intent and Delegated Authority
   a. Subparts A through C of part 112 are often referred to as the "SPCC rule." Focusing on oil spill prevention, preparedness, and response.
   b. The SPCC / FRP rule is designed to protect public health, public welfare, and the environment from potential harmful effects of oil discharges to navigable waters and adjoining shorelines.
   c. The rule requires facilities that could reasonably be expected to discharge oil in quantities that may be harmful into navigable waters of the United States and adjoining shorelines to develop and implement SPCC / FRP Plans.
   d. The SPCC / FRP Plans ensure that these facilities put in place containment and countermeasures that will prevent oil discharges. The requirement to develop, implement, Revise the SPCC / FRP plan, Train employees to carry it out, Part 112.
   e. Facility Response Plans (FRPs). EPA has established requirements that define who must prepare and submit an FRP and what must be included in the Plan. These requirements are found in Subpart D of 40 CFR part 112, which is often referred to as the "FRP rule." Although the SPCC and FRP rules are related, and certain SPCC-regulated facilities must also comply with FRP requirements The SPCC / FRP rule carries out EPA's authority under CWA §311.
   f. EPA was delegated the authority to regulate non-transportation-related onshore and offshore facilities that could reasonably be expected to discharge oil into navigable waters of the United States or adjoining shorelines (35 FR 11677, July 22, 1970). Executive Order 11548 was superseded by Executive Orders 11735 and 12777, respectively (38 FR 21243, August 7, 1973; 56 FR 54757, October 22, 1991).
   g. The U.S. Department of Transportation (DOT) was delegated authority over transportation-related onshore facilities, deepwater ports, and vessels. A Memorandum of Understanding (MOU) between the Secretary of Transportation and the EPA Administrator, dated November 24, 1971 (36 FR 24080, December 18, 1971), defines non transportation related facilities (34) and

transportation-related facilities. In addition, the U.S. Department of the Interior (DOI) regulates specific offshore facilities, including associated pipelines. The jurisdictional responsibilities of EPA, DOT, and DOI in relation to offshore facilities are further discussed in another Memorandum of Understanding, dated November 8, 1993. (This MOU is included as Appendix B to 40 CFR part 112.)

2. Statutory Framework:
   a. The Federal Water Pollution Control Act of 1972, (178) as amended, or Clean Water Act, is the principal federal statute for protecting navigable waters, adjoining shorelines, and the waters of the contiguous zone from pollution. Section 311 of the CWA addresses the control of oil and hazardous substance discharges, and provides the authority for a program to prevent, prepare for, and respond to such discharges.
   b. Specifically, §311(j)(1)(C) mandates regulations establishing procedures, methods, equipment, and other requirements to prevent discharges of oil from vessels and facilities and to contain such discharges.
   c. Under CWA §311(a)(1), "oil" is defined to mean "oil of any kind or in any form but not limited to, petroleum, fuel oil, sludge, oil refuse, and oil mixed with wastes other than dredged spoil." Petroleum oils include crude and refined petroleum products, asphalt, gasoline, fuel oils, mineral oils, naphtha, sludge, oil refuse, and oil mixed with wastes other than dredged spoil (67 FR 47075). The "Oil" definition also includes animal fats and vegetable oils and synthetic Oils.
   d. The U.S. Coast Guard (USCG) compiled a list of substances it considers oil, based on the CWA definition. The list is available on the USCG Website. Note, however, that the USCG list is not comprehensive and does not define "oil" for purposes of 40 CFR part 112. EPA may determine that a substance, chemical, material, or mixture is an oil even if it is not on the USCG list.
   e. In 1975, EPA published a notice on the applicability of the SPCC rule to non-petroleum oils. The notice confirmed that all facilities processing and storing non-petroleum oils in the quantities and under circumstances set out in 40 CFR part 112 are required to prepare and implement an SPCC Plan in accordance with that part (40 FR 28849, July 9, 1975).EPA stated that the broad

and comprehensive definition of "oil" in the CWA is consistent with the expressed congressional intent to strengthen federal law for the prevention, control, and cleanup of oil spilled in the aquatic environment.

f. Both EPA and the U.S. Coast Guard consistently interpreted and administered §311 as applicable to spills of non-petroleum-based oils, particularly because of the common physical and chemical properties of animal and vegetable oils and petroleum oils, and their common potential for adverse environmental impact when discharged into water.
g. The Oil Pollution Act of 1990 (OPA) streamlined and strengthened EPA's ability to prevent, prepare for, and respond to catastrophic oil discharges. Specifically, OPA expands prevention and preparedness activities, improves response capabilities, ensures that shippers and owners or operators of facilities that handle oil pay the costs of discharges that do occur, expands research and development programs.
h. Establishes an Oil Spill Liability Trust Fund.
i. OPA §4202(a)(6) amended CWA §311(j) to require regulations to be promulgated that require owners or operators of certain vessels and facilities to prepare and submit Facility Response Plans (FRPs) for responding to a worst case discharge of oil and to a substantial threat of such a discharge (CWA §311(j)(5)). EPA published the FRP rule on July 1, 1994, as an amendment to 40 CFR part 112.

3. **Regulatory History and Implementation: (150)** The SPCC rule was initially promulgated in 1973, with modifications to the SPCC requirements proposed for public comment on several occasions in the 1990s. Incorporating many aspects of the earlier proposals, final revisions to the rule were published in the *Federal Register* (FR) in July 2002 and became effective on August 16, 2002. Even though the original final version gave some extensions for facility compliance at the time that are now expired so all of the final rule requirements now in the regulations are to be complied with by applicable facilities.
4. **SPCC Regional Administrator (RA) is TCEQ:**
   a. Regional Administrator (RA) – TCEQ overriding authority:
      1. First a practical compliance issue about applicability to SPCC requirements is that the Regional Administrator (RA), in Texas it is the TCEQ can de-

termine on its own that you are applicable despite any specific criteria you may or may not be applicable too. 40 CFR Section 112.1(f) allows the Regional Administrator (RA) to require preparation of an SPCC Plan or applicable part by the owner or operator of an otherwise exempted facility that is subject to EPA jurisdiction, this means basically every used oil handler, under CWA §311(j) of the CWA. This provision is designed to address gaps in other regulatory regimes that might best be remedied by requiring a facility to have an SPCC Plan. For example, a facility may be exempted from the SPCC rule because its storage capacity is below the regulatory threshold, but the facility may have been the cause of repeated discharges as described in §112.1(b).

a. Factors the RA may consider in making a determination to require that a facility prepare an SPCC Plan include, but are not limited to:
    1. Physical characteristics of the facility;
    2. The presence of secondary containment;
    3. The discharge history of the facility;
    4. The proximity of the facility to sensitive environmental areas such as wetlands, parks, or wildlife refuges.
b. The RA might require an entire Plan, or might require only a partial Plan addressing secondary containment, for example, to prevent future discharges. Sections 112.1(f)(1) through (5) describe the process for an RA to determine applicability. The process includes specific time deadlines for both the RA and the facility owner or operator, as well as requirements for the type of information and delivery method. The requirements lists the deadlines and responsibilities of the RA and the facility owner or operator to appeal the RA determination that he/she must prepare an SPCC Plan.

SPCC Plan Contents Example Plan

"Unified Oil Company, Ltd. SAMPLE Spill Prevention, Control, and Countermeasure (SPCC) Plan" from "Appendix D A Sample Bulk Storage Facility" within the document associated with a fictitious Oil storage Facility and is taken directly from the EPA document SPCC Guidance for Regional Inspectors (Dec 2013) **(51)** and copies of this document may be obtained online at www.epa.gov/oilspill. In addition, updates to the document will be available online. As identified above this is a significant resource and provides significant more detail on SPCC issues than provided here related to used oil management as the focus of this guide. Also, as always don't hesitate to call the TCEQ regarding SPCC guidance or the US Coast Guard for FRP guidance. To be able to generate your SPCC plan it is critical that you refer to this appendix or decide to have a third party develop a plan on your behalf then this content listing can be used as a reference to insure your third party is addressing each applicable section.

General content descriptions are identified below, again refer to the guidance document referenced above for significantly more detail.

Overview of the Requirements of a Written SPCC Plan

The SPCC plan must describe all measures taken at your facility to prevent and control a release of oil or petroleum products, including but not limited to:

1. Facility operations, staffing, site security, spill history and documentation of annual training for oil-handling employees;
2. Oil release scenarios that include possible volume of the spill and the direction of flow;
3. Notification procedures, which must include an emergency call list and should include notification of nearby facilities that may be in the direction of flow and could be impacted by the release;
5. A facility site plan showing areas of oil storage and transfer;
6. A description of containment structures or equipment used to prevent releases;
7. A description of the procedures to stop, contain and clean up any released materials, including procedures for managing collected rain water;
8. Oil storage inspection procedures;
9. A written commitment of manpower, equipment and materials to expeditiously control and remove any amount of oil that may be spilled; and
10. Written procedures for integrity and leak testing of tanks, containers, valves and piping
11. See Section L (4) below for required Professional Engineer (PE) Certification requirement and possible qualification as a "Qualified Facility" under the Applicability criteria related to the exemption to be able to do a self certification. Initial

criteria is that you can only have a maximum total of 10,000 gallon total used oil storage with certain other criteria.

TABLE OF CONTENTS

Part 5: Discharge Response

List of Tables

Appendices

-iii- Version 1.0, 11/28/2005

K. With an SPCC plan requirement comes a mandatory training requirement as addressed by 40 CFR 112.7(f). (43) Note: Hazwoper Training IS NOT SPCC TRAINING. Also with training as an additional element you must also make a determination as to facility substantial harm determination.

40 CFR 112.7(f)

(f) Personnel, training, and discharge prevention procedures.

(1) At a minimum, train your oil-handling personnel in the operation and maintenance of equipment to prevent discharges; discharge procedure protocols; applicable pollution control laws, rules, and regulations; general facility operations; and, the contents of the facility SPCC Plan.

(2) Designate a person at each applicable facility who is accountable for discharge prevention and who reports to facility management.

(3) Schedule and conduct discharge prevention briefings for your oil-handling personnel at least once a year to assure adequate understanding of the SPCC Plan for that facility. Such briefings must highlight and describe known discharges as described in § 112.1(b) or failures, malfunctioning components, and any recently developed precautionary measures.

"Discharge" includes, but is not limited to, any spilling, leaking, pumping, pouring, emitting, emptying, or dumping of oil, but excludes discharges in compliance with a permit under section 402 of the CWA; discharges resulting from circumstances identified, reviewed, and made a part of the public record with respect to a permit issued or modified under section 402 of the CWA, and subject to a condition in such permit; or continuous or anticipated intermittent discharges from a point source, identified in a permit or permit application under section 402 of the CWA, that are caused by events occurring within the scope of relevant operating or treatment systems. For purposes of this part, the term discharge shall not include any discharge of oil that is authorized by a permit issued under section 13 of the River and Harbor Act of 1899 (33 U.S.C. 407).

112.1(b)

(b) Except as provided in paragraph (d) of this section, this part applies to any owner or operator of a non-transportation-related onshore or offshore facility engaged in drilling, producing, gathering, storing, processing, refining, transferring, distributing, using, or consuming oil and oil products, which due to its location, could reasonably be expected to discharge oil in quantities that may be harmful, as described in part 110 of this chapter, into or upon the navigable waters of the United States or adjoining shorelines, or into or upon the waters of the contiguous zone, or in connection with activities under the Outer Continental Shelf Lands Act or the Deepwater Port Act of 1974, or that may affect natural resources belonging to, appertaining to, or under the exclusive management authority of the United States (including resources under the Magnuson Fishery Conservation and Management Act) that has oil in:

(1) Any aboveground container;

(2) Any completely buried tank as defined in § 112.2;

Completely buried tank means any container completely below grade and covered with earth, sand, gravel, asphalt, or other material. Containers in vaults, bunkered tanks, or partially buried tanks are considered aboveground storage containers for purposes of this part.

(3) Any container that is used for standby storage, for seasonal storage, or for temporary storage, or not otherwise "permanently closed" as defined in § 112.2;

"Permanently closed" means any container or facility for which:

(1) All liquid and sludge has been removed from each container and connecting line; and

(2) All connecting lines and piping have been disconnected from the container and blanked off, all valves (except for ventilation valves) have been closed and locked, and conspicuous signs have been posted on each container stating that it is a permanently closed container and noting the date of closure.

(4) Any "bunkered tank" or "partially buried tank" as defined in § 112.2, or any container in a vault, each of which is considered an aboveground storage container for purposes of this part.

"Bunkered tank" means a container constructed or placed in the ground by cutting the earth and re-covering the container in a manner that breaks the surrounding natural grade, or that lies above grade, and is covered with earth, sand, gravel, asphalt, or other material. A bunkered tank is considered an aboveground storage container for purposes of this part.

"Partially buried tank" means a storage container that is partially inserted or constructed in the ground, but not entirely below grade, and not completely covered with earth, sand, gravel, asphalt, or other material. A par-

tially buried tank is considered an aboveground storage container for purposes of this part.

1. Training records must document who received SPCC training, the dates of the training, and the topics covered. These documents must be maintained for a period of three (3) years. The model plan suggests you maintain your training records in section 3.8.
2. From the SPCC Guidance (51) identified above related to Unified Oil example plan section 3.8 provides an overview of Personnel, Training, and Discharge Prevention Procedures associated with 40 CFR 112.7(f) compliance and provides typical activities it performs in compliance.
   a. The Facility Manager is the facility designee and is responsible for oil discharge prevention, control, and response preparedness activities at this facility.
   b. Unified Oil management has instructed oil-handling facility personnel in the operation and maintenance of oil pollution prevention equipment, discharge procedure protocols, applicable pollution control laws, rules and regulations, general facility operations, and the content of this SPCC Plan.
   c. Any new facility personnel with oil-handling responsibilities are provided with this same training prior to being involved in any oil operation.
   d. Annual discharge prevention briefings are held by the Facility Manager for all facility personnel involved in oil operations. The briefings are aimed at ensuring continued understanding and adherence to the discharge prevention procedures presented in the SPCC Plan. The briefings also highlight and describe known discharge events or failures, malfunctioning components, and recently implemented precautionary measures and best practices.
   e. Facility operators and other personnel will have the opportunity during the briefings to share recommendations concerning health, safety, and environmental issues encountered during facility operations.
   f. A simulation of an on-site vehicular discharge has been conducted,

and future training exercises will be periodically held to prepare for possible discharge responses.

g. Records of the briefings and discharge prevention training are kept on the form shown in Appendix E (as provided in the example SPCC Plan) and maintained with this SPCC Plan for a period of three years.

3. Substantial Harm Criteria

With your plan you must include the form or equivalent Applicability To Substantial Harm Criteria assessment that identifies your need to also to an Facility Response Plan (FRP) and include it in your SPCC plan – Typically as Appendix B as identified in the model plan identified above. Below is an example completed Applicability To Substantial Harm Criteria.

Within your completed plan you must provide a completed Applicability of Substantial Harm Criteria form, example provided below. The following provides the criterial used to determine Substantial Harm from EPA guidance. Please note when reviewing the guidance that even if you determine you do not meet the criteria the RA can still require an FRP based on its review. In your regulatory and guidance review you will also see the substantial harm criteria also identified as "Significant and Substantial Harm" as an attempt to indicate the significance by regulators to qualify the term but no guidance could be found to qualify any difference in the statements as to compliance requirements.

a. **A facility may pose "substantial harm" according to the FRP rule if it: (179)**

1. Has a total oil storage capacity greater than or equal to 42,000 gallons and it transfers oil over water to/from vessels; or
2. Has a total oil storage capacity greater than or equal to 1 million gallons and meets one of the following conditions:
   a. Does not have sufficient secondary containment for each aboveground storage area.
   b. Is located at a distance such that a discharge from the facility

could cause "injury" to fish, wildlife, and sensitive environments.

   c. Is located at a distance such that a discharge from the facility would shut down a public drinking water intake.
   d. Has had, within the past five years, a reportable discharge greater than or equal to 10,000 gallons.
3. If the facility does not meet the criteria (either under 1 or 2), then the facility is not subject to the FRP rule via self-identification.
4. The criteria presented here represents the self-identification criteria to be used by the facility owner/operator to determine if the facility is subject to the FRP rule. If the facility meets the criteria, an FRP is required to be prepared and submitted to the appropriate EPA Regional Office. For more information on self-selection, please see: Section 112.20 of the Oil Pollution Prevention regulation and Attachment C-II Certification of the Applicability of the Substantial Harm Criteria.

a. Once the certification form and FRP are submitted to the region, the RA will review and determine if the facility should be classified as a significant and substantial harm facility. If the RA determines that the facility could cause significant and substantial harm, the FRP requires approval by the RA. An RA determines if a facility could, because of its location cause significant and substantial harm to the environment by discharging oil into or on the navigable waters and adjoining shorelines. This is determined by factors similar to the substantial harm criteria, as well as:
   - the age of tanks;
   - type of transfer operations;
   - oil storage capacity;
   - lack of secondary containment;
   - proximity to fish, wildlife, and sensitive environments or drinking-water intakes;
   - spill history and frequency of past discharges; or
   - other information, including local impacts on public health.

Facilities that pose significant and substantial harm must have their plans reviewed and approved by EPA.

5. With completing your SPCC you identify as the form indicates below the requirement for your facility to also have a Facility Response Plan (FRP). The next sections cover the requirement overview associated with the FRP in Texas.

APPLICABILITY OF SUBSTANTIAL HARM CRITERIA

Does the facility transfer oil over-water to or from vessels and does the facility have a total oil storage capacity greater than or equal to 42,000 gallons?

Yes No X

Does the facility have a total oil storage capacity greater than or equal to 1 million gallons and does the facility lack secondary containment that is sufficiently large to contain the capacity of the largest aboveground oil storage tank plus sufficient freeboard to allow for precipitation within any aboveground oil storage tank area?

Yes No X

Does the facility have a total oil storage capacity greater than or equal to 1 million gallons and is the facility located at a distance (as calculated using the appropriate formula in Appendix C or a comparable formula) such that a discharge from the facility could cause injury to fish and wildlife and sensitive environments?

Yes X No

Does the facility have a total oil storage capacity greater than or equal to 1 million gallons and is the facility located at a distance (as calculated using the appropriate formula in Appendix C or a comparable formula) such that a discharge from the facility would shut down a public drinking water intake?

Yes No X

Does the facility have a total oil storage capacity greater than or equal to 1 million gallons and has the facility experienced a reportable oil spill in an amount greater than or equal to 10,000 gallons within the last 5 years?

Yes No X

CERTIFICATION

I certify under penalty of law that I have personally examined and am familiar with the information submitted in this document, and that based on my inquiry of those individuals responsible for obtaining this information, I believe that the submitted information is true, accurate, and complete.

Dave Foulke
TT&M HES&S Manager

6/9/2008
Date

L. Facility Response Plans:

With completing your SPCC you identify as the form indicates above the requirement for your facility to also have a Facility Response Plan (FRP). The next sections cover the requirements overview associated with the FRP in Texas.

1. Overview of the requirements begins with a short history of the regulatory development and reasons for its requirement based on an overview provided by the EPA guidance documents and the RA role in Texas that includes the Government Land Office (GLO)

FRP History: (22) The Oil Pollution Act (OPA) of 1990 streamlined and strengthened EPA's ability to prevent and respond to catastrophic oil spills. A trust fund financed by a tax on oil is available to clean up spills when the responsible party is incapable or unwilling to do so. The OPA requires oil storage facilities and vessels to submit to the Federal government plans detailing how they will respond to large discharges. EPA has published regulations for aboveground storage facilities; the Coast Guard has done so for oil tankers. The OPA also requires the development of Area Contingency Plans to prepare and plan for oil spill response on a regional scale. In 1990, the Oil Pollution Act amended the Clean Water Act to require some oil storage facilities to prepare Facility Response Plans. On July 1, 1994, EPA finalized the revisions that direct facility owners or operators to prepare and submit plans for responding to a worst-case discharge of oil (Subpart D).

Facility Response Plan (FRP) Applicability: Facilities that could reasonably be expected to cause "substantial harm" to the environment by discharging oil into or on navigable waters are required to prepare and submit Facility Response Plans (FRPs). Facilities that could cause "significant and substantial harm" are required to have their plans approved by an EPA Regional Administrator (RA). As identified above the requirement is triggered when a facility is identified as posing substantial harm either through a self-identification process, or by the determination of an RA.

In Texas the RA role is more complicated than many other states as the role is shared by the TCEQ and the Texas General Land Office (GLO) : Starting with TCEQ statement from their Website : "The TCEQ and the Texas General Land Office (GLO) jointly serve as the state's primary members of the federal Regional Response Team for the U.S. Environmental Protection Agency's Region VI, and the TCEQ represents the state in matters related to inland spills or releases of oil, and inland or coastal releases of hazardous substances or other pollutants." However due to the issue of a release into any navigable waterway that can cause substantial harm the US Coast Guard also has jurisdiction over your facility and must approve your FRP. Your Regional Administrator (RA) for FRP is shared by the EPA Region IV and the US Coast Guard that based on federal authorities has the final say regarding your FRP. However as the GLO has authority as the State representative it also has RA authority though the EPA based on the TCEQ authorities but cannot overrule the US Coast Guard. I indicated it is complicated!

In Texas you would submit your FRP to the US Coast Guard, EPA Region IV and the Texas GLO to receive certification (discussed below) and approval by Region IV who typically delegates approval to the US Coast Guard in your area.

All the authorities above have been established through various Memorandum of Understanding (MOU) that establishes the jurisdictional responsibilities as well as Presidential Executive Orders that further identify delegations. For detail information on these MOUs and Executive Orders please see Appendix B to Part 112—Memorandum of Understanding Among the Secretary of the Interior, Secretary of Transportation, and Administrator of the Environmental Protection Agency.

2. FRP overview, contents and GLO requirements.

The following overview identifies content requirements for Facility Response Plans (FRPs) from the EPA website (180) and following the content summary is the GLO certification requirement then an example outline for an FRP for a Texas facility. The example FRP outline is not to be relied on and you should refer to the Model Plan identified in Appendix F of the Oil Pollution Prevention regulation (40 CFR 112) indicated above that discusses content for each element.

Below are general overview issues that must be addressed within your FRP.

a. Be consistent with the National Contingency Plan and applicable Area Contingency Plans;

b. Identify a qualified individual having full authority to implement removal actions, and require immediate communication between that person and the appropriate federal authorities and responders;

c. Identify and ensure availability of resources to remove, to the maximum extent practicable, a worst-case discharge (see: Appendix E of the Oil Pollution Prevention regulation (40 CFR 112));

d. Describe training, testing, unannounced drills, and response actions of persons on the vessel or at the facility;

e. Be updated periodically;

f. Be resubmitted to an EPA Regional Office for approval of each significant change.

3. **Key elements include:**

a. Emergency Response Action Plan, which serves as both a planning and action document, should be maintained as an easily accessible, stand-alone section of the overall plan

b. Facility information, including its name, type, location, owner, operator information

c. Emergency notification, equipment, personnel, and evacuation information

d. Identification and analysis of potential spill hazards and previous spills

e. Discussion of small, medium, and worst-case discharge scenarios and response actions

f. Description of discharge detection procedures and equipment

g. Detailed implementation plan for response, containment, and disposal

h. Description and records of self-inspections, drills and exercises, and response training

i. Diagrams of facility site plan, drainage, and evacuation plan

j. Security (e.g., fences, lighting, alarms, guards, emergency cut-off valves and locks, etc.)

k. Response plan coversheet

4. **Facility Response Plan Format:**

From the EPA Website the following is a listing of content elements required. A model plan is available from Appendix F of the Oil Pollution Prevention regulation (40 CFR 112) located at

https://www.ecfr.gov/cgi-bin/textidx?SID=934dd060745fb05152cd5fea2224d24b&mc=true&node=ap40.22.112_121.f&rgn=div9 (181)

The model plan provides significant information as to specific content requirements and needs to be referenced if you have any intent in doing your own plan. From a practical standpoint it is rare for any small facility to do their own plan but even if you don't do your own plan to better understand the requirements spending time reviewing the requirements beyond the regulations is strongly recommended as you will be responsible for plan implementation at your site. I have included two reference statements 1.0 and 1.1 from the model plan for your understanding of the requirements and the type of guidance provided within the Appendix F.

Appendix F to Part 112—Facility-Specific Response Plan

Table of Contents

1.0 Model Facility-Specific Response Plan

(A) Owners or operators of facilities regulated under this part which pose a threat of substantial harm to the environment by discharging oil into or on navigable waters or adjoining shorelines are required to prepare and submit facility-specific response plans to EPA in accordance with the provisions in this appendix. This appendix further describes the required elements in §112.20(h).

(B) Response plans must be sent to the appropriate EPA Regional office. (*Table 7A provided below lists each EPA Regional office and the address where owners or operators must submit their response plans as of 2018. It is recommended you call the contact listed for each region to see if the region FRP coordinator would prefer an electronic version for review rather than sending a hardcopy.*)Those facilities deemed by the Regional Administrator (RA) to pose a threat of significant and substantial harm to the environment will have their plans reviewed and approved by EPA. In certain cases, information required in the model response plan is similar to information currently maintained in the facility's Spill Prevention, Control, and Countermeasures (SPCC) Plan as required by 40 CFR 112.3. In these cases, owners or operators may reproduce the information and include a photocopy in the response plan.

(C) A complex may develop a single response plan with a set of core elements for all regulating agencies and separate sections for the non-transportation-related and transportation-related components, as described in §112.20(h). Owners or operators of large facilities that handle, store, or transport oil at more than one geographically distinct location (e.g., oil storage areas at opposite ends of a single, continuous parcel of property) shall, as appropriate, develop separate sections of the response plan for each storage area.

Table 8 – FRP Contacts And Mailing Addresses

USA EPA REGION / AREA OF OVERSIGHT	CONTACT NAME	CONTACT PHONE	CONTACT EMAIL	FRP MAILING ADDRESS
US EPA Region 1 (CT, ME, MA, NH, RI, and VT)	Cosmo Caterino	(617) 918-1264	Caterino.Cosmo@epa.gov	US EPA Region 1 Attn: Cosmo Caterino FRP Coordinator Mail Code OSRR02-2 5 Post Office Square Boston, MA 02109
US EPA Region 2 (NJ, NY, Puerto Rico, and the U.S. Virgin Islands)	Francesco Maimone	(732) 321-4483	maimone.francesco@epa.gov	US EPA Region 2 Attn: FRP Coordinator 2890 Woodbridge Ave MS-211 Edison, NJ 08837
US EPA Region 3 (DE, DC, MD, PA, VA, and WV)	Arlin Galarza-Hernandez	(215) 814-3223	galarza-hernandez.arlin@epa.gov	US EPA Region 3 Attn: Arlin Galarza-Hernandez SPCC/FRP Coordinator Office of Enforcement, Oil and Prevention Branch (3HS61) 1650 Arch Street Philadelphia, PA 19103-2029
US EPA Region 4 (AL, FL, GA, KY, MS, NC, SC, and TN)	Ted Walden	(404) 562-8752	walden.ted@epa.gov	US EPA Region 4 Attn: Ted Walden SPCC/FRP Coordinator 61 Forsyth Street SW Atlanta, GA 30303
US EPA Region 5 (IL, IN, MI, MN, OH, and WI)	Alex Tzallas	(312) 886-0622	tzallas.alexander@epa.gov	US EPA Region 5 Attn: FRP Coordinator 77 W. Jackson Blvd. SE-5J Chicago, IL 60604
US EPA Region 6 (AR, LA, NM,	Bryant	(214) 665-	smalley.bryant@epa.gov	US EPA Region 6

USA EPA REGION / AREA OF OVERSIGHT	CONTACT NAME	CONTACT PHONE	CONTACT EMAIL	FRP MAILING ADDRESS
OK, and TX)	Smalley	7368		Attn: FRP Coordinator 1445 Ross Avenue, Suite 1200 Mail Code: 6SF-EO Dallas, TX 75202-2733
	Chris Perry	(214) 665-6702	perry.chris@epa.gov	
US EPA Region 7 (IA, KS, MO, and NE)	Susan Fisher	(913) 551-7919	Fisher.Susan@epa.gov	US EPA Region 7 Attn: Susan Fisher 11201 Renner Blvd Lenexa, KS 66219
US EPA Region 8 (CO, MT, ND, SD, UT, and WY)	Bre Bockstahler	(303) 312-6034	bockstahler.breann@epa.gov	US EPA Region 8 Attn: Bre Bockstahler FRP Coordinator Emergency Response Program (8EPR-ER) 1595 Wynkoop Street Denver, Colorado 80202
US EPA Region 9 (AZ, CA, HI, NV, American Samoa, Commonwealth of the Northern Mariana Islands, Federated States of Micronesia, Guam, Marshall Islands, and Republic of Palau)	Donn Zuroski	(415) 972-3035	zuroski.donn@epa.gov	US EPA Region 9 Attn: Donn Zuroski Federal On-Scene Coordinator Emergency Response Program (8EPR-ER) Mail Code (SFD-9-4) 75 Hawthorne Street, San Francisco CA 94105
US EPA Region 10 (AK, ID, OR, WA and 271 native tribes)	**Alaska,** Bob Whittier	(907) 271-3247	whittier.robert@epa.gov	**ALASKA ONLY** US EPA Region 10 On-Scene Coordinator Mail Code: AOO/A 222 West 7th Avenue #19 Anchorage, AK 99513-7588

USA EPA REGION / AREA OF OVERSIGHT	CONTACT NAME	CONTACT PHONE	CONTACT EMAIL	FRP MAILING ADDRESS
	Washington and Oregon, Richard Franklin	(503) 326-2917	franklin.richard@epa.gov	**WASHINGTON AND OREGON** US EPA Region 10 On-Scene Coordinator Mail Code: OOO 805 SW Broadway, Suite 500 Portland, OR 97205
	Idaho, Stephen Ball	(208) 378-6510	ball.stephen@epa.gov	**IDAHO** US EPA Region 10 On-Scene Coordinator 950 W. Bannock, Suite 900 Boise, ID 83702
				NON-SPECIFIC U.S. EPA, Region 10 Oil Program 1200 6th Avenue, Suite 900 ECL-116 Seattle, WA 98101-3140

1.1 Emergency Response Action Plan

Several sections of the response plan shall be co-located for easy access by response personnel during an actual emergency or oil discharge. This collection of sections shall be called the Emergency Response Action Plan. The Agency intends that the Action Plan contain only as much information as is necessary to combat the discharge and be arranged so response actions are not delayed. The Action Plan may be arranged in a number of ways. For example, the sections of the Emergency Response Action Plan may be photocopies or condensed versions of the forms included in the associated sections of the response plan. Each Emergency Response Action Plan section may be tabbed for quick reference. The Action Plan shall be maintained in the front of the same binder that contains the complete response plan or it shall be contained in a separate binder. In the latter case, both binders shall be kept together so that the entire plan can be accessed by the qualified individual and appropriate spill response personnel. The Emergency Response Action

Plan shall be made up of the following sections:

1. Qualified Individual Information (Section 1.2) partial
2. Emergency Notification Phone List (Section 1.3.1) partial
3. Spill Response Notification Form (Section 1.3.1) partial
4. Response Equipment List and Location (Section 1.3.2) complete
5. Response Equipment Testing and Deployment (Section 1.3.3) complete
6. Facility Response Team (Section 1.3.4) partial
7. Evacuation Plan (Section 1.3.5) condensed
8. Immediate Actions (Section 1.7.1) complete
9. Facility Diagram (Section 1.9) complete

5. §112.20 Facility response plans. Below is a redacted excerpt from 40 CFR §112.20 addressing FRP management that discusses RA interactions and responsibilities as well as your plan submittal requirements.

(a) The owner or operator of any non-transportation-related onshore facility that, because of its location, could reasonably be expected to cause substantial harm to the environment by discharging oil into or on the navigable waters or adjoining shorelines shall prepare and submit a facility response plan to the Regional Administrator, according to the following provisions:

(1) For the owner or operator of a facility in operation on or before February 18, 1993 who is required to prepare and submit a response plan under 33 U.S.C. 1321(j)(5), the Oil Pollution Act of 1990 (Pub. L. 101-380, 33 U.S.C. 2701 *et seq.*) requires the submission of a response plan that satisfies the requirements of 33 U.S.C. 1321(j)(5) no later than February 18, 1993.

(i) The owner or operator of an existing facility that was in operation on or before February 18, 1993 who submitted a response plan by February 18, 1993 shall revise the response plan to satisfy the requirements of this section and resubmit the response plan or updated portions of the response plan to the Regional Administrator by February 18, 1995.

(ii) The owner or operator of an existing facility in operation on or before February 18, 1993 who failed to submit a response plan by February 18, 1993 shall prepare and submit a response plan that satisfies the requirements of this section to the Regional Administrator before August 30, 1994.

(2) The owner or operator of a facility in operation on or after August 30, 1994 that satisfies the criteria in paragraph (f)(1) of this section or that is notified by the Regional Administrator pursuant to paragraph (b) of this section shall prepare and submit a facility response plan that satisfies the requirements of this section to the Regional Administrator.

(i) For a facility that commenced operations after February 18, 1993 but prior to August 30, 1994, and is required to prepare and submit a response plan based on the criteria in paragraph (f)(1) of this section, the owner or operator shall submit the response plan or updated portions of the response plan, along with a completed version of the response plan cover sheet contained in appendix F to this part, to the Regional Administrator prior to August 30, 1994.

(ii) For a newly constructed facility that commences operation after August 30, 1994, and is required to prepare and submit a response plan based on the criteria in paragraph (f)(1) of this section, the owner or operator shall submit the response plan, along with a completed version of the response plan cover sheet contained in appendix F to this part, to the Regional Administrator prior to the start of operations (adjustments to the response plan to reflect changes that occur at the facility during the start-up phase of operations must be submitted to the Regional Administrator after an operational trial period of 60 days).

(iii) For a facility required to prepare and submit a response plan after August 30, 1994, as a result of a planned change in design, construction, operation, or maintenance that renders the facility subject to the criteria in paragraph (f)(1) of this section, the owner or operator shall submit the response plan, along with a completed version of the response plan cover sheet contained in appendix F to this part, to the Regional Administrator before the portion of the facility undergoing change commences operations (adjustments to the response plan to reflect changes that occur at the facility during the start-up phase of operations must be submitted to the Regional Administrator after an operational trial period of 60 days).

(iv) For a facility required to prepare and submit a response plan after August 30, 1994, as a result of an unplanned event or change in facility characteristics that renders the facility subject to the criteria in paragraph (f)(1) of this section, the owner or operator shall submit the response plan, along with a completed version of the response plan cover sheet contained in appendix F to this part, to the Regional Administrator within six months of the unplanned event or change.

(3) In the event the owner or operator of a facility that is required to prepare and submit a response plan uses an alternative formula that is comparable to one contained in appendix C to this part to evaluate the criterion in paragraph (f)(1)(ii)(B) or (f)(1)(ii)(C) of this section, the owner or operator shall attach documentation to the response plan cover sheet contained in appendix F to this part that demonstrates the reliability and analytical soundness of the alternative formula.

(4) *Preparation and submission of response plans—Animal fat and vegetable oil facilities.* The owner or operator of any non-transportation-related facility that handles, stores, or transports animal fats and vegetable oils must prepare and submit a facility response plan as follows:

(i) *Facilities with approved plans.* The owner or operator of a facility with a facility response plan that has been approved under paragraph (c) of this section by July 31, 2000 need not prepare or submit a revised plan except as otherwise required by paragraphs (b), (c), or (d) of this section.

(ii) Facilities with plans that have been submitted to the Regional Administrator. Except for facilities with approved plans as provided in paragraph (a)(4)(i) of this section, the owner or operator of a facility that has submitted a response plan to the Regional Administrator prior to July 31, 2000 must review the plan to deter-

mine if it meets or exceeds the applicable provisions of this part. An owner or operator need not prepare or submit a new plan if the existing plan meets or exceeds the applicable provisions of this part. If the plan does not meet or exceed the applicable provisions of this part, the owner or operator must prepare and submit a new plan by September 28, 2000.

(iii) Newly regulated facilities. The owner or operator of a newly constructed facility that commences operation after July 31, 2000 must prepare and submit a plan to the Regional Administrator in accordance with paragraph (a)(2)(ii) of this section. The plan must meet or exceed the applicable provisions of this part. The owner or operator of an existing facility that must prepare and submit a plan after July 31, 2000 as a result of a planned or unplanned change in facility characteristics that causes the facility to become regulated under paragraph (f)(1) of this section, must prepare and submit a plan to the Regional Administrator in accordance with paragraph (a)(2)(iii) or (iv) of this section, as appropriate. The plan must meet or exceed the applicable provisions of this part.

(iv) Facilities amending existing plans. The owner or operator of a facility submitting an amended plan in accordance with paragraph (d) of this section after July 31, 2000, including plans that had been previously approved, must also review the plan to determine if it meets or exceeds the applicable provisions of this part. If the plan does not meet or exceed the applicable provisions of this part, the owner or operator must revise and resubmit revised portions of an amended plan to the Regional Administrator in accordance with paragraph (d) of this section, as appropriate. The plan must meet or exceed the applicable provisions of this part.

(b)(1) The Regional Administrator may at any time require the owner or operator of any non-transportation-related onshore facility to prepare and submit a facility response plan under this section after considering the factors in paragraph (f)(2) of this section. If such a determination is made, the Regional Administrator shall notify the facility owner or operator in writing and shall provide a basis for the determination. If the Regional Administrator notifies the owner or operator in writing of the requirement to prepare and submit a response plan under this section, the owner or operator of the facility shall submit the response plan to the Regional Administrator within six months of receipt of such written notification.

(2) The Regional Administrator shall review plans submitted by such facilities to determine whether the facility could, because of its location, reasonably be expected to cause significant and substantial harm to the environment by discharging oil into or on the navigable waters or adjoining shorelines.

(c) The Regional Administrator shall determine whether a facility could, because of its location, reasonably be expected to cause significant and substantial harm to the environment by discharging oil into or on the navigable waters or adjoining shorelines, based on the factors in paragraph (f)(3) of this section. If such a determination is made, the Regional Administrator shall notify the owner or operator of the facility in writing and:

(1) Promptly review the facility response plan;

(2) Require amendments to any response plan that does not meet the requirements of this section;

(3) Approve any response plan that meets the requirements of this section; and

(4) Review each response plan periodically thereafter on a schedule established by the Regional Administrator provided that the period between plan reviews does not exceed five years.

(d)(1) The owner or operator of a facility for which a response plan is required under this part shall revise and resubmit revised portions of the response plan within 60 days of each facility change that materially may affect the response to a worst case discharge, including:

(i) A change in the facility's configuration that materially alters the information included in the response plan;

(ii) A change in the type of oil handled, stored, or transferred that materially alters the required response resources;

(iii) A material change in capabilities of the oil spill removal organization(s) that provide equipment and personnel to respond to discharges of oil described in paragraph (h)(5) of this section;

(iv) A material change in the facility's spill prevention and response equipment or emergency response procedures; and

(v) Any other changes that materially affect the implementation of the response plan.

(2) Except as provided in paragraph (d)(1) of this section, amendments to personnel and telephone number lists included in the response plan and a change in the oil spill removal organization(s) that does not result in a material change in support capabilities do not require approval by the Regional Administrator. Facility owners or operators shall provide a copy of such changes to the Regional Administrator as the revisions occur.

(3) The owner or operator of a facility that submits changes to a response plan as provided in paragraph (d)(1) or (d)(2) of this section shall provide the EPA-issued facility identification number (where one has been assigned) with the changes.

(4) The Regional Administrator shall review for approval changes to a response plan submitted pursuant to paragraph (d)(1) of this section for a facility determined pursuant to paragraph (f)(3) of this section to have the potential to cause significant and substantial harm to the environment.

(e) If the owner or operator of a facility determines pursuant to paragraph (a)(2) of this section that the facility could not, because of its location, reasonably be expected to cause substantial harm to the environment by discharging oil into or on the navigable waters or adjoining shorelines, the owner or operator shall complete and maintain at the facility the certification form contained in appendix C to this part and, in the event an alternative formula that is comparable to one contained in appendix C to this part is used to evaluate the criterion in paragraph (f)(1)(ii)(B) or (f)(1)(ii)(C) of this section, the owner or operator shall attach documentation to the certification form that demonstrates the reliability and analytical soundness of the comparable formula and shall notify the Regional Administrator in writing that an alternative formula was used.

(f)(1) A facility could, because of its location, reasonably be expected to cause substantial harm to the environment by discharging oil into or on the navigable waters or adjoining shorelines pursuant to paragraph (a)(2) of this section, if it meets any of the following criteria applied in accordance with the flowchart contained in attachment C-I to appendix C to this part:

(i) The facility transfers oil over water to or from vessels and has a total oil storage capacity greater than or equal to 42,000 gallons; or

(ii) The facility's total oil storage capacity is greater than or equal to 1 million gallons, and one of the following is true:

(A) The facility does not have secondary containment for each aboveground storage area sufficiently large to contain the capacity of the largest aboveground oil storage tank within each storage area plus sufficient freeboard to allow for precipitation;

(B) The facility is located at a distance (as calculated using the appropriate formula in appendix C to this part or a comparable formula) such that a discharge from the facility could cause injury to fish and wildlife and sensitive environments. For further description of fish and wildlife and sensitive environments, see Appendices I, II, and III of the "Guidance for Facility and Vessel Response Plans: Fish and Wildlife and Sensitive Environments" (see appendix E to this part, section 13, for availability) and the applicable Area Contingency Plan prepared pursuant to section 311(j)(4) of the Clean Water Act;

(C) The facility is located at a distance (as calculated using the appropriate formula in appendix C to this part or a comparable formula) such that a discharge from the facility would shut down a public drinking water intake; or

(D) The facility has had a reportable oil discharge in an amount greater than or equal to 10,000 gallons within the last 5 years.

(2)(i) To determine whether a facility could, because of its location, reasonably be expected to cause substantial harm to the environment by discharging oil into or on the navigable waters or adjoining shorelines pursuant to paragraph (b) of this section, the Regional Administrator shall consider the following:

(A) Type of transfer operation;

(B) Oil storage capacity;

(C) Lack of secondary containment;

(D) Proximity to fish and wildlife and sensitive environments and other areas determined by the Regional Administrator to possess ecological value;

(E) Proximity to drinking water intakes;

(F) Spill history; and

(G) Other site-specific characteristics and environmental factors that the Regional Administrator determines to be relevant to protecting the environment from harm by discharges of oil into or on navigable waters or adjoining shorelines.

(ii) Any person, including a member of the public or any representative from a Federal, State, or local agency who believes that a facility subject to this section could, because of its location, reasonably be expected to cause substantial harm to the environment by discharging oil into or on the navigable waters or adjoining shorelines may petition the Regional Administrator to determine whether the facility meets the criteria in paragraph (f)(2)(i) of this section. Such petition shall include a discussion of how the factors in paragraph (f)(2)(i) of this section apply to the facility in question. The RA shall consider such petitions and respond in an appropriate amount of time.

(3) To determine whether a facility could, because of its location, reasonably be expected to cause significant and substantial harm to the environment by discharging oil into or on the navigable waters or adjoining shorelines, the Regional Administrator may consider the factors in paragraph (f)(2) of this section as well as the following:

(i) Frequency of past discharges;

(ii) Proximity to navigable waters;

(iii) Age of oil storage tanks; and

(iv) Other facility-specific and Region-specific information, including local impacts on public health.

(g)(1) All facility response plans shall be consistent with the requirements of the National Oil and Hazardous Substance Pollution Contingency Plan (40 CFR part 300) and applicable Area Contingency Plans prepared pursuant to section 311(j)(4) of the Clean Water Act. The facility response plan should be coordinated with the local emergency response plan developed by the local emergency planning committee under section 303 of Title III of the Superfund Amendments and Reauthorization Act of 1986 (42 U.S.C. 11001 et seq.). Upon request, the owner or operator should provide a copy of the facility response plan to the local emergency planning committee or State emergency response commission.

(2) The owner or operator shall review relevant portions of the National Oil and Hazardous Substances Pollution Contingency Plan and applicable Area Contingency Plan annually and, if necessary, revise the facility response plan to ensure consistency with these plans.

(3) The owner or operator shall review and update the facility response plan periodically to reflect changes at the facility.

(h) A response plan shall follow the format of the model facility-specific response plan included in appendix F to this part, unless you have prepared an equivalent response plan acceptable to the Regional Administrator to meet State or other Federal requirements. A response plan that does not follow the specified format in appendix F to this part shall have an emergency response action plan as specified in paragraphs (h)(1) of this section and be supplemented with a cross-reference section to identify the location of the elements listed in paragraphs (h)(2) through (h)(10) of this section. To meet the requirements of this part, a response plan shall address the following elements, as further described in appendix F to this part:

6. GLO Requirements in Texas:

Texas General Land Office (GLO) Certification: Your FRP can be used to meet the GLO requirements and receive the certification required but you must insure all GLO requirements are contained within your FRP. It is theoretically possible but highly unlikely that your facility on the water front may not require an FRP under the federal rules but may still be subject to the GLO requirements that would require your development of a plan meeting GLO requirements and requiring you to obtain a certification absent a waiver as identified in §19.4 or exemption identified in §19.12 **(80)**.

1. TAC Title 31 - NATURAL RESOURCES AND CONSERVATION, Part 1 GENERAL LAND OFFICE, and Chapter 19 - OIL SPILL PREVENTION AND RESPONSE address GLO requirements and applicability. Title 31 §19.11 Classification of Waterfront and Offshore "Facilities Waterfront and offshore facilities are classified based on their capacity to transfer or store oil. Oil that is integral to equipment, such as oil in transformers or lubricating oil in machinery, is not included in determining storage or transfer capacity." **(80)**

(1) Small–A facility that transfers oil through pipelines, flow lines, gathering lines, or trunk lines with a line diameter of four inches or less or that has the capacity to store 1,320 gallons or less of oil.

(2) Intermediate–A facility that transfers oil through pipelines, flow lines, gathering lines, or trunk lines with a line diameter of greater than four inches up to and including twelve inches or that has the capacity to store more than 1,320 gallons up to and including 250,000 gallons of oil.

(3) Large–A facility that transfers oil through pipelines, flow lines, gathering lines, or trunk lines with a line diameter greater than 12 inches or that has the capacity to store more than 250,000 gallons of oil."

"Title 31 §19.12 Facility Certification Requirements **(80)**

(b) Applicability. This section applies to any person who operates a waterfront or offshore facility. If an operator controls part of a facility which is waterfront or offshore, the entire facility in which oil is handled under the control of that operator must be covered by the discharge prevention and response certificate. Pipelines, flowlines, gathering lines, or transmission lines that transfer oil across an area of coastal waters are considered facilities. A combination of interrelated or adjacent tanks, impoundments, pipelines, gathering lines, flow lines, separator or treatment facilities, and other structures, equipment, or devices under common ownership or operation will be considered a single facility under OSPRA. Interrelated means the devices are all an integral part of one commercial or industrial operation or are managed and controlled by a single entity. The term includes facilities owned by units of federal, state, or local government, as well as privately owned facilities.

(c) Current certificate required to operate. No entity may operate a waterfront or offshore facility without a current discharge prevention and response certificate issued by the GLO. This requirement does not apply, however, to an entity that operates a facility and has obtained a waiver from the facility certification requirement pursuant to §19.4 of this title (relating to Waiver) or if an exemption applies to the facility.

(d) Certificate void when operator changes or facility classification level increases. A discharge prevention and response certificate is issued to a specific operator and for a particular facility classification level. When the operator of a facility changes, the discharge prevention and response certificate is void. The new operator of the facility will need to submit an application for a certificate to the GLO before beginning to operate the facility. A certificate is also void when the facility changes its operations in a manner that increases its facility classification level. If an operator increases storage capacity or installs new oil transfer lines at a facility, causing the facility classification to change from small to intermediate or large or from intermediate to large, the operator will need to apply for a new certificate.

(e) Obtaining certificate application forms. The operator of a facility must apply for a discharge prevention and response certificate by submitting a completed application form to the GLO. Application forms are available from the General Land Office, Oil Spill Prevention and Response Program, 1700 North Congress Avenue, Austin, Texas 78701-1495 or from any regional office of the GLO. The application form can also be downloaded from the GLO's Oil Spill Prevention and Response Program website, www.glo.state.tx.us/oilspill.

(f) Signature requirements. The certificate application must be signed by a representative of the facility operator who has approved the facility's discharge prevention and response plan and has the authority to commit the necessary resources to implement the plan.

(g) Facility inspections. After the GLO determines the application is administratively complete, the GLO will contact the facility operator to schedule an on-site inspection and review of the facility's discharge prevention and response plan. The inspection and plan review will cover the following elements:

(1) the facility's compliance with applicable regulations;

(2) whether the discharge prevention and response plan adequately addresses all the elements required by §19.13;

(3) whether the discharge prevention and response plan specifically addresses the worst case unauthorized discharge and demonstrates the facility can adequately respond to the worst case unauthorized discharge from the facility; and whether the discharge prevention and response plan has been implemented.

(h) Additional information. After the on-site inspection, the GLO may require an applicant to submit additional information to resolve any issues related to the applicant's discharge prevention and response preparedness. The GLO may also require an applicant to develop and implement additional measures to prevent and respond to unauthorized discharges of oil.

(i) Notification that certification requirements have been met. When the GLO determines the facility has adequately addressed its discharge prevention and response requirements and has submitted sufficient information in its application, the GLO will notify the facility operator that the certification requirements have been met. The operator will then be informed of the facility classification level (small, intermediate, or large).

(j) Certification fee. A fee of $25 will be assessed for every facility to be certified, but the fee should not be submitted with the completed application form. The facility operator will be instructed to submit the fee to the GLO after the GLO determines a certificate will be issued to the operator.

(k) Term for certificates. The GLO will issue certificates with a term of five years from the date of issuance. Each certificate will be assigned an identification number, which will allow the facility operator to review and amend the facility information on the GLO's Oil Spill Prevention and Response Program interactive website. The identification number will be sent to the person who signed the application form, along with instructions on how to update and renew the certificate.

(l) Discretionary submittal of discharge prevention and response plan. After a certificate is issued to a facility, the GLO may require the facility operator to submit to the GLO a complete copy of its discharge prevention and response plan for review. Submittal of the plan for review may be required if the GLO determines the facility may not be adequately implementing its plan to prevent and respond to unauthorized discharges of oil.

(m) Exemptions. The following facilities that handle oil do not need to apply to the GLO for a discharge prevention and response certificate:

(1) Mobile or portable oil-handling equipment, such as a mobile offshore drilling unit, when it is fixed in place for less than 90 days.

(2) A farm, ranch, or residential property that stores up to and including 1,320 gallons of oil for farming, ranching, or residential purposes.

- A facility that stores oil exclusively in underground tanks and does not transfer oil to vessels in the water.
- A facility that stores or transfers oil only in containers with a volume of 55 gallons or less.

(n) Effect of certificate on other violations. Issuance of a certificate does not estop the state in an action brought under OSPRA, or any other law, from alleging a violation of any such law, other than failure to have a certificate.

2. Part of the Texas requirements under Title 31 §19.14 **(80)** require an Annual Updating of Application Information; Renewal and Suspension of Certificates. ***Note: You are requested to use the Internet link to update and check status of your application as identified in (a)(1) below.***

(a) Annual review of application information. Facility operators are required to report annually any changes in the information submitted to the GLO in their applications for certificates. Changes must be reported by the anniversary of the date the certificate was issued, but operators are encouraged to update the information more frequently. Facility operators can update information on file with the GLO in the following ways:

(1) Internet. The GLO has established a link on its website (www.glo.texas.gov)

to allow facility operators to review and amend application information on file with the GLO. Facility operators can use the identification number, which is issued with the certificate, to access this interactive link. To minimize the GLO's administrative expense of updating information, the GLO encourages certificate holders to use the Internet to revise facility information on file with the GLO.

(2) Mail or facsimile. If a facility operator cannot update application information over the Internet, updated information can be sent by mail or facsimile to the appropriate GLO regional office. Addresses and facsimile numbers for the regional office covering a particular facility can be obtained by calling the main oil spill program office in Austin at (512) 475-1575.

(b) Renewing certificates. Operators are responsible for ensuring that certificates are renewed by their expiration dates. The GLO will not send expiration notices to operators.

(1) All certificates, which will be issued for a period of five years, will specify the date of expiration. To renew a certificate, certificate holders must complete and submit to the GLO a new application form. To give the GLO sufficient time to review the application, it must be submitted to the GLO at least 15 days before the expiration date.

(2) In reviewing the application to renew a certificate, the GLO may conduct an on-site review of the facility's discharge prevention and response plan. The GLO may require the applicant to amend its plan if the GLO determines the plan does not adequately address the elements required by §19.13.

(3) A fee of $25 will be assessed for renewal of a certificate. The GLO will inform the certificate holder that the fee is being assessed after the application is reviewed and a determination has been made that the certificate will be renewed.

(c) Notification to GLO when facility closes or is shut-in. A facility operator is required to notify the GLO when the facility closes or when the facility is shut-in and no longer handling oil.

(d) Certificate suspension. Suspension of a certificate requires the facility owner or operator to apply for a new certificate. The GLO may suspend a certificate if the facility operator violates a provision of OSPRA or rules or orders adopted under authority of OSPRA. A certificate may also be suspended if the GLO determines the facility has not adequately implemented its discharge prevention and response plan or the facility's response to an unauthorized discharge of oil was inadequate. Before suspending a certificate, the GLO will inform the certificate holder in writing that suspension is being considered. The reasons for the proposed suspension will be specified, and the certificate holder will be afforded an opportunity to address the problems. If the GLO ultimately determines the certificate holder has not adequately addressed the facility's problems and suspension of the certificate is appropriate, the facility operator may request and is entitled to a hearing on the suspension in the same manner provided under Chapter 2 of this Title, relating to Rules of Practice and Procedure for contested case hearings before the GLO.

3. Below is the Texas Title 31 GLO requirements which are substantially the same as the Federal requirements and you may submit your FRP to meet the requirements but you must insure you have addressed all the GLO requirements as identified.

Title 31 – Natural Resources And Conservation, Part 1 General Land Office, Chapter 19 – Oil Spill Prevention And Response, Subpart B, Spill Prevention and Preparedness §19.13

(a) Applicability. This section applies to any person who operates a waterfront or offshore facility and must obtain a discharge prevention and response certificate.

(b) Implementation of plans. An operator of any facility that requires certification must develop and implement a written discharge prevention and response plan. Before issuing a certificate, the GLO will conduct an on-site review of the plan. The GLO will determine whether the facility's plan contains all the information required by this section and has been fully implemented.

(c) Required elements of discharge prevention and response plans for all facility classifications. Operators of all facilities that require certification must prepare discharge prevention and response plans which include the following information:

(1) the owner and operator of the facility;

(2) the person or persons in charge of the facility, as required by §19.16 of this title (relating to Person in Charge);

(3) the name and address (both physical and mailing) of the facility;

(4) a description of the facility, including:

(A) the location of the facility by latitude and longitude;
(B) the facility's primary activity;
(C) the types of oil handled, whether material safety data sheets (MSDS) have been prepared for them, and the location where the MSDS are maintained;
(D) the storage capacity of each tank used for storing oil;
(E) the diameter of all lines through which oil is transferred;
(F) the average daily throughput of oil at the facility; and
(G) the dimensions and capacity in barrels of the largest oil-handling vessel which docks at the facility.

(5) for a facility which normally does not have personnel on-site, a commitment to maintain in a prominent location a sign or placard which states that the GLO and National Response Center are to be notified of an oil spill and gives the 24-hour phone numbers for notifying the GLO and National Response Center;
(6) a general description of measures taken by the facility to prevent unauthorized discharges of oil;
(7) a plan to conduct an annual oil spill drill that entails notifying the GLO and National Response Center and keeping a log at the facility which documents when the notification drill was conducted and facility personnel who participated in it;
(8) if oil is transferred at the facility, emergency transfer procedures to be implemented if an actual or threatened unauthorized discharge of oil occurs at the facility;
(9) strategic plans to contain and clean up unauthorized discharges of oil from the facility;
(10) a statement that all facility personnel who might be involved in an oil spill response have been informed that detergents or other surfactants are prohibited from being used on an oil spill in the water, and that dispersants can only be used with the approval of the Regional Response Team, the interagency group composed of federal and state agency representatives that coordinates oil spill responses; and
(11) a description of any secondary containment or diversionary structures or

equipment at the facility to prevent discharged oil from reaching coastal waters, including the methodology for determining that the structures or equipment are adequate to prevent oil from reaching coastal waters.

(d) Additional requirements for facilities classified as intermediate. In addition to the requirements in §19.13(c), operators of intermediate facilities must prepare written discharge prevention and response plans which include the following information:

(1) a description of the worst case unauthorized discharge of oil reasonably likely to occur at the facility and the rationale used to determine the worst case unauthorized discharge;

(2) a description and map of environmentally sensitive areas that would be impacted by the worst case unauthorized discharge and plans for protecting these areas if an oil spill occurs at the facility;

(3) a description of the facility's response strategies to contain and clean up the worst case unauthorized discharge;

(4) a description of discharge prevention procedures implemented at the facility, including procedures to prevent discharges from transfers of oil;

(5) a plan to conduct an annual oil spill drill that includes the following elements:

(A) notifying the GLO and National Response Center;

(B) notifying any third parties, such as discharge cleanup organizations, which have agreed to respond to an oil spill and confirming they would be able to respond to an oil spill at the facility on the day of the drill;

(C) if the facility has spill response equipment stored on-site, deployment of a representative portion of the equipment which would be used to respond to the type of discharge most likely to occur at the facility; and

(D) a log documenting when the annual drill was conducted and the facility personnel who participated in it; and

(6) if the operator has entered into any oil spill response or cleanup contracts or basic ordering agreements with a discharge cleanup organization, copies of the contracts or agreements or a narrative description of their terms.

(e) Additional requirements for facilities classified as large. In addition to the requirements in §19.13(c), operators of large facilities must prepare written discharge prevention and response plans which include the following information:

(1) maps showing vehicular access to the facility, pipelines to and from the facility, and nearby residential or other populous areas;

(2) a site plan of the facility showing:

(A) the location of all structures in which oil is stored;

(B) the location of all areas where oil is transferred at the facility; and

(C) drainage and diversion systems at the facility, such as sewers, outfalls, catchment or containment systems or basins, sumps, and all watercourses into which surface runoff from the facility drains (all of which may be shown on the site plan or maps);

(3) a plan to conduct an annual oil spill drill that includes the following elements:

(A) notifying the GLO and National Response Center;

(B) notifying any third parties, such as discharge cleanup organizations, which have agreed to respond to an oil spill and confirming they would be able to respond to an oil spill at the facility on the day of the drill;

(C) if the facility has spill response equipment stored on-site, deployment of a representative portion of the equipment which would be used to respond to the type of discharge most likely to occur at the facility; and

(D) a log documenting when the annual drill was conducted and the facility personnel who participated in it;

(4) a detailed description of the facility's discharge prevention and response capability, including:

(A) leak detection and safety systems to prevent accidental discharges of oil, including a description of equipment and procedures;

(B) schedules, methods, and procedures for testing, maintaining, and inspecting storage tanks, pipelines, and other equipment used for handling oil;

(C) schedules, methods, and procedures for conducting accidental discharge response drills;

(D) whether the facility's oil spill response capability will primarily be based on contracts or agreements with third parties or on the facility's own personnel and equipment;

(E) planned response actions, the chain of command, lines of communication, and procedures for notifying the GLO, emergency response and public safety entities, other agencies, and neighboring facilities in the event of an unauthorized discharge of oil;

(F) oil spill response equipment and supplies located at the facility, their ownership and location, and the time required to deploy them;

(G) if the facility owns and maintains oil spill response equipment, the schedules, methods, and procedures for maintaining the equipment in a state of constant readiness for deployment;

(H) if the operator has entered into any oil spill response or cleanup contracts or basic ordering agreements with a discharge cleanup organization, copies of the contracts or agreements or a narrative description of their terms;

(I) the worst case unauthorized discharge of oil reasonably likely to occur at the facility and the rationale used to determine the worst case unauthorized discharge;

(J) a description and map of environmentally sensitive areas that would be impacted by the worst case unauthorized discharge and plans for protecting these areas if an oil spill occurs at the facility;

(K) a description of response strategies that would be implemented to contain and clean up the worst case unauthorized discharge;

(L) information on the facility's program for training facility personnel on accidental discharge prevention and response;

(M) information on facility personnel who have been specifically designated to respond to an oil spill, including any training they have received and where the training records are maintained;

(N) plans for transferring oil during an emergency; plans for recovering, storing, separating, transporting, and disposing of oily waste materials generated during an oil spill response; and

(O) plans for providing emergency medical treatment, site safety, and security during an oil spill.

4. With your FRP requirement as with your SPCC you have a very specific training requirement that must be accomplished along with your SPCC training which is a separate requirement even though technically it is possible to combine them both it is rarely ever done. Note that the training like the SPCC training must be site specific. Note

the reference in §112.21 (c) for where to get guidance as to the training specifics. WEB reference for the Appendix E section 13 (36) is retrieved from https://www.ecfr.gov/cgi-bin/text-idx?SID=1c51d40600f1ee4665809e1befd53271&mc=true&node=ap40.24.112_121.e&rgn=div9

Also included below for your reference from the Model Plan are 40 CFR Appendix F to Part 112—Facility-Specific Response Plan sections 1.8.2 Facility Drills/Exercises, 1.8.2.1 Qualified Individual Notification Drill Logs, 1.8.2.2 Spill Management Team Tabletop Exercise Logs, 1.8.3 Response Training, 1.8.3.1 Personnel Response Training Logs, and 1.8.3.2 Discharge Prevention Meetings Logs.

§112.21 Facility response training and drills/exercises. (37)

(a) The owner or operator of any facility required to prepare a facility response plan under §112.20 shall develop and implement a facility response training program and a drill/exercise program that satisfy the requirements of this section. The owner or operator shall describe the programs in the response plan as provided in §112.20(h)(8).

(b) The facility owner or operator shall develop a facility response training program to train those personnel involved in oil spill response activities. It is recommended that the training program be based on the USCG's Training Elements for Oil Spill Response, as applicable to facility operations. An alternative program can also be acceptable subject to approval by the Regional Administrator.

   (1) The owner or operator shall be responsible for the proper instruction of facility personnel in the procedures to respond to discharges of oil and in applicable oil spill response laws, rules, and regulations.

   (2) Training shall be functional in nature according to job tasks for both supervisory and non-supervisory operational personnel.

   (3) Trainers shall develop specific lesson plans on subject areas relevant to facility personnel involved in oil spill response and cleanup.

(c) The facility owner or operator shall develop a program of facility response drills/exercises, including evaluation procedures. A program that follows the National Preparedness for Response Exercise Program (PREP) (see appendix E to this part, section 13, for availability) will be deemed satisfactory for purposes of this section. An alternative program can also be acceptable subject to approval by the Regional Administrator.

[59 FR 34101, July 1, 1994, as amended at 65 FR 40798, June 30, 2000]

1.8.2 Facility Drills/Exercises

(A) CWA section 311(j)(5), as amended by OPA, requires the response plan to contain a description of facility drills/exercises. According to 40 CFR 112.21(c), the facility owner or operator shall develop a program of facility response drills/exercises, including evaluation procedures. Following the PREP guidelines (see appendix E to this part, section 13, for availability) would satisfy a facility's requirements for drills/exercises under this part. Alternately, under §112.21(c), a facility owner or operator may develop a program that is not based on the PREP guidelines. Such a program is subject to approval by the Regional Administrator based on the description of the program provided in the response plan.

(B) The PREP Guidelines specify that the facility conduct internal and external drills/exercises. The internal exercises include: qualified individual notification drills, spill management team tabletop exercises, equipment deployment exercises, and unannounced exercises. External exercises include Area Exercises. Credit for an Area or Facility-specific Exercise will be given to the facility for an actual response to a discharge in the area if the plan was utilized for response to the discharge and the objectives of the Exercise were met and were properly evaluated, documented, and self-certified.

(C) Section 112.20(h)(8)(ii) requires the facility owner or operator to provide a description of the drill/exercise program to be carried out under the response plan. Qualified Individual Notification Drill and Spill Management Team Tabletop Drill logs shall be provided in sections 1.8.2.1 and 1.8.2.2, respectively. These logs may be included in the facility response plan or kept as an annex to the facility response plan. See section 1.3.3 of this appendix for Equipment Deployment Drill Logs.

1.8.2.1 Qualified Individual Notification Drill Logs

Qualified Individual Notification Drill Log
Date: ___ / ___ / _____
Company: ____
Qualified Individual(s):
____

Emergency Scenario: ____
____
____

Evaluation:
____
____
____

Changes to be Implemented:
____
____

Time Table for Implementation:
____

1.8.2.2 Spill Management Team Tabletop Exercise Logs

Spill Management Team Tabletop Exercise Log

Date: ___ / ___ / ______

Company:

____

Qualified Individual(s):

____

Emergency Scenario:

____

____

____

Evaluation:

____

____

____

Changes to be Implemented:

____

____

Time Table for Implementation:

____

1.8.3 Response Training

Section 112.21(a) requires facility owners or operators to develop programs for facility response training. Facility owners or operators are required by §112.20(h)(8)(iii) to provide a description of the response training program to be carried out under the response plan. A facility's training program can be based on the USCG's Training Elements for Oil Spill Response, to the extent applicable to facility operations, or another response training program acceptable to the RA. The training elements are available from the USCG Office of Response (G-MOR) at 302) 267-0518 or fax (202) 267-4085. Personnel response training logs and discharge prevention meeting logs shall be included in sections 1.8.3.1 and 1.8.3.2 of the response plan respectively. These logs may be included in the facility response plan or kept as an annex to the facility response plan.

1.8.3.1 Personnel Response Training Logs

Personnel Response Training Log

Name	Response training/date and number of hours	Prevention training/date and number of hours

1.8.3.2 Discharge Prevention Meetings Logs

Discharge Prevention Meeting Log

Date: ___ / ___ / ______

Attendees:

____

____

____

____

Subject/issue identified	Required action	Implementation date

M. Additional consideration factors associated with your SPCC and FRP development. Below are topics associated with your plan that are offered to clarify or provide additional reference to issues covered within the guide.

1. CONTAINMENT ISSUE:

Loading/Unloading Activities: (50) DOT regulates equipment used for the fueling of locomotive units, as well as the rights-of-way on which they operate. EPA regulates the activity of loading or unloading oil in bulk into storage containers (such as those on tank trucks or railroad cars), as well as all equipment involved in this activity (e.g., a hose or loading arm attached to a storage tank system). Different requirements apply to oil transfer areas and to loading/unloading racks at a regulated facility. A transfer area is any area of a facility where oil is transferred between bulk storage containers and tank trucks or railroad cars. These areas are subject to the general secondary containment requirements in §112.7(c). If a "loading/unloading rack" (as defined in §112.2) is present, the requirements of §112.7(h) apply to the loading/unloading rack area. For more information, refer to *Chapter 4: Secondary Containment and Impracticability* which discusses secondary containment requirements for loading/unloading areas and racks.

2. Marine Terminals (50)

A marine terminal is an example of a "complex" subject to both U.S. Coast Guard (USCG) and EPA jurisdiction. The jurisdictional boundary of a complex facility for both USCG and EPA is defined in 33 CFR part 154, *Facilities Transferring Oil or Hazardous Material in Bulk* under the definition of a marine transportation-related facility (MTR facility) in §154.1020. The USCG regulates the pier structures, transfer hoses, hose-piping connection, containment, controls, and transfer piping associated with the transfer of oil between a vessel and an onshore facility. EPA regulates the tanks, internal piping, loading racks, and vehicle/rail operations that are completely within the non-transportation portion of the facility. EPA jurisdiction begins at the first valve inside secondary containment. If there is no secondary containment, EPA jurisdiction begins at the valve or manifold adjacent to the storage tank. *Appendix H* includes drawings that show EPA's regulatory jurisdiction at complexes, including an example of a marine terminal.

3. Vessels (Ships/Barges)(50)

The U.S. Coast Guard regulates the loading or unloading of oil in bulk from a vessel to an onshore facility, as well as the oil-carrying vessel and the connecting piping (33 CFR part 155, Oil or Hazardous Material Pollution Prevention Regulations for Vessels). In this scenario, a vessel is a ship or a barge. The oil passes from the USCG's jurisdiction to that of the EPA when it passes the first valve inside the secondary containment for the storage container at an otherwise regulated facility. If there is no secondary containment, EPA's jurisdiction begins at the first valve or manifold closest to the storage container. Storage tanks and appurtenances for the reception of oily ballast water or tank washings from vessels are under USCG jurisdiction.

Vessels themselves are specifically exempt from 40 CFR part 112 under §112.1(d)(1)(iii). EPA also clarified that barges or other watercraft that store oil, and have been determined by the Coast Guard to be permanently moored, are no longer vessels, but storage containers that are part of an offshore facility (67 FR 47075, July 17, 2002).

4. SPCC Plan Professional Engineer (PE) Certification requirement and "Qualified Facility Status" application. The EPA and Texas require an SPCC applicable facility to have your plan certified by a PE as indicated in the SPCC outline information above but also provides for Self Certification under certain criteria as identified below. The reference for this section is EPA Guidance "Spill Prevention Control and Countermeasure (SPCC) Plan Qualified Facilities Applicability" and it is suggested you review if you meet the criteria identified below as your plan format must comply with the 40 CFR §112 Subparts as identified that is not addressed in the guide sections above and is available from https://www.epa.gov/sites/production/files/2013-08/documents/qf_app_guidance_0.pdf

The exception rule provides simplified spill prevention requirements for qualified facilities. The owner or operator of a qualified facility can self-certify the facility's SPCC Plan. There are two types of qualified facilities. To determine if your facility is a qualified facility and what type it is, you'll need the following information:

- The total capacity of aboveground oil storage containers at the facility and
- Information on oil spills from the facility for the past three years.

As noted above you are applicable to the SPCC requirements if your used oil facility has a the total aggregate aboveground oil storage capacity greater than 1,320 gallons of oil; or is the total aggregate capacity of completely buried storage tanks greater than 42,000 gallons of oil?

Professional Engineer (PE) Certified SPCC Plans

If your facility does not qualify as either a Tier I or Tier II qualified facility, that is addressed below, then you must follow the requirements of 40 CFR 112.7 and subparts B or C of the rule, and have the SPCC Plan certified by a PE.

A PE must also be involved when you deviate from any of the SPCC requirements (i.e., either by providing environmentally equivalent alternatives or a contingency plan instead of secondary containment). You can still self-certify the SPCC Plan, if your facility is a Tier I or II qualified facility but the deviation must be certified by a PE.

The facility owner or operator is responsible for preparation of the SPCC Plan. This does not mean that you cannot have a third party (consultant) prepare or assist you in the plan preparation but you are responsible for the content from a compliance enforcement position, but it must be certified by a PE who will confirm that:

- The PE is familiar with the requirements of the rule;
- The PE or his agent has visited and examined the facility;
- The SPCC Plan has been prepared in accordance with good engineering practices, including consideration of applicable industry standards, and with the requirements of the rule;
- Procedures for required inspections and testing have been established; and
- The SPCC Plan is adequate for the facility.

6. Qualified Facilities

If your facility meets the criteria described in the table below, primary criteria is that **the facility total aboveground oil storage capacity is 10,000 gallons or less.** If you meet the primary criteria then you may fall into either a Tier I or Tier II category. As identified in the guidance you then determine if each of your aboveground oil storage containers has a capacity of 5,000 gallons or less, then your facility is a Tier I Qualified Facility and you can choose to complete and certify the SPCC Plan template found in Appendix G of the SPCC rule. A copy of this template can be found on the SPCC page at https://www.epa.gov/oil-spills-prevention-and-preparedness-regulations/tier-i-qualified-facility-spcc-plan-template#Tier I SPCC Template

If your facility is not eligible to use the template, you may still qualify to self-certify your SPCC Plan as a Tier II Qualified Facility. The requirements for a Tier II Qualified Facility Plan are similar to a PE-certified Plan (i.e., the Plan must follow the requirements of 40 CFR 112.7 and subparts B or C of the rule; however, the owner or operator certifies the Plan).

An owner/operator that certifies a facility's SPCC Plan attests that he/she is familiar with the SPCC requirements and has visited and examined the facility. The owner/operator also certifies that:

- The Plan has been prepared in accordance with accepted and sound industry practices and standards and with the rule requirements;
- Procedures for required inspections and testing have been established;
- The Plan is being fully implemented;
- The facility meets the qualifying criteria;
- The Plan does not deviate from rule requirements except as allowed and as certified by a PE; and
- Management approves the Plan and has committed resources to implement

Follow the table below to determine if you can self-certify your SPCC Plan:

Table 9 – Qualified Facility Applicability, EPA SPPC Guidance (2005)

<table>
<tr><th colspan="3">If the facility total aboveground oil storage capacity is 10,000 gallons or less ...</th></tr>
<tr><th>And...</th><th>And the facility has...</th><th>Then the facility is a:</th></tr>
<tr><td rowspan="2">Within three years prior to the Plan certification date, or since becoming subject to the SPCC rule if in operation for less than three years, the facility has not had:
A single discharge of oil to navigable waters or adjoining shorelines exceeding 1,000 gallons, or
Two discharges of oil to navigable waters or adjoining shorelines each exceeding 42 gallons within any 12- month period.</td><td>No individual aboveground oil containers greater than 5,000 gallons;</td><td>Tier I Qualified Facility: Complete and self-certify Plan template (Appendix G to 40 CFR part 112) in lieu of a full PE-certified Plan or other self-certified SPCC Plan.</td></tr>
<tr><td>Any individual aboveground oil container greater than 5,000 gallons;</td><td>Tier II Qualified Facility: Prepare a self-certified Plan in accordance with all applicable requirements of §112.7 and subparts B or C of the rule, in lieu of a PE- certified Plan.</td></tr>
</table>

PART XVIII: AST TANK MANAGEMENT

TANK MANAGEMENT IN TEXAS:

Texas tank management involves several compliance issues related to Aboveground Storage Tanks (AST) that are associated with your storage of Used Oil in Texas. This Guide will only address AST requirements and not Underground Storage Tanks (AST). Tank Management issues for Used Oil Handlers are associated with your SPCC and FRP requirements addressed above that requires you to have a management process that includes inspection, testing and record keeping that must be addressed in your SPCC and or FRP.

The SPCC requirements are also applicable to your FRP but you are allowed to use the same information in both plans. The initial general requirements are identified in Appendix F to Part 112—Facility-Specific Response Plan – Model FRP requirements Plan Section 1.8.1.1. Tank Inspection "The tank inspection checklist presented below has been included as guidance during inspections and monitoring. Similar requirements exist in 40 CFR part 112, subparts A through C. Duplicate information from the SPCC Plan may be photocopied and inserted in this section. "

40 CFR Part 112 identifies requirements associated with Tank Management cited below and the Texas AST management requirements are identified in TAC Title 30 Part 1 Chapter 334 referenced below. Additional reference for the information below are based on the EPA "Spill Prevention, Control and Countermeasure Plan (SPCC) Program - Bulk Storage Container Inspection Fact Sheet" (55). *Additional comments are provide italicized. It should be noted that not all specific requirements are identified and you may need to reference applicable regulations as identified in the summary provided for a more detailed understanding of requirements associated with your specific facility.*

SPCC REQUIREMENTS

§112.7 General Requirements for Spill Prevention, Control, and Countermeasure

Plans (SPCC)

(e) Inspections, tests, and records. Conduct inspections and tests required by this part in accordance with written procedures that you or the certifying engineer develop for the facility. You must keep these written procedures and a record of the inspections and tests, signed by the appropriate supervisor or inspector, with the SPCC Plan for a period of three years. Records of inspections and tests kept under usual and customary business practices will suffice for purposes of this paragraph.

Fact Sheet Information (55)

§112.8(c)(6), §112.12(c)(6)(i) Test or inspect each aboveground container for integrity on a regular schedule and whenever you make material repairs. You must determine, in accordance with industry standards, the appropriate qualifications for personnel performing tests and inspections, the frequency and type of testing and inspections, which take into account container size, configuration, and design (such as containers that are: shop-built, field-erected, skid-mounted, elevated, equipped with a liner, double-walled, or partially buried). Examples of these integrity tests include, but are not limited to: visual inspection, hydrostatic testing, radiographic testing, ultrasonic testing, acoustic emissions testing, or other systems of non-destructive testing. You must keep comparison records and you must also inspect the container's supports and foundations. In addition, you must frequently inspect the outside of the container for signs of deterioration, discharges, or accumulation of oil inside diked areas. Records of inspections and tests kept under usual and customary business practices satisfy the recordkeeping requirements of this paragraph. Note: The above text is an excerpt of the SPCC rule. Refer to the full text of 40 CFR part 112.

> ***Commentary:*** *The detail offered as to type of testing requirement examples and the requirement for comparison records and all other associated records. With tank issues as well as other equipment you need to have an understanding of all historical conditions and be able to show that documentation when requested.*

> ***Commentary:*** *As identified above you must apply "industry standards" that you document as to which standard you applied and often presents a chal-*

lenge to most operators but can result in a significant non-compliance if not complied with and often requires third party support to accomplish.

"What are industry standards? Industry standards are technical guidelines created by experts in a particular industry for use throughout that industry. Standards-developing organizations use a consensus process to establish the minimum accepted industry practice. The SPCC rule requires that the Plan be prepared in accordance with good engineering practice. Standards play a role in determining good engineering practice when developing spill prevention procedures and an inspection program for an SPCC regulated facility.

Implementing an inspection program based on a particular industry standard is ultimately up to the owner/operator. When an owner/operator indicates in the SPCC Plan that he intends to use a standard to comply with a particular rule requirement (e.g. integrity testing), then it is mandatory to implement the relevant portions of the standard (i.e. those that address integrity testing of the container). The American Petroleum Institute (API) Standard 653, "Tank Inspection, Repair, Alteration, and Reconstruction" and the Steel Tank Institute (STI) "SP001 Standard for the Inspection of Aboveground Storage Tanks" (STI SP001) are two commonly used inspection standards for aboveground bulk storage.

More information on industry standards: API Standard 653 (API-653)–Tank Inspection, Repair, Alteration, and Reconstruction API-653 covers steel storage tanks built to design specifications in the API 650 standard and its predecessor API 12. It provides minimum requirements for maintaining the integrity of tanks after they have been placed in service and addresses inspection, repair, alteration, relocation, and reconstruction. This standard is typically used to establish an integrity testing program for field-erected tanks. Go to the API website for more information on their standards: http://www.api.org/ STI Standard SP001 (STI SP001)–Standard for the Inspection of Aboveground Storage Tanks This standard focuses primarily on inspection of welded, metal, shop-fabricated and small field-erected tanks. Also included is the inspection of smaller, portable containers such as 55-gallon drums, intermediate bulk containers (IBCs) and other such containers that may be of metal or plastic construction. Go to the STI website for more information on the SP001 standard: http://www.steeltank.com/"

Tank Inspection and Facility Inspection statements from the 40 CFR Model Plan Guidance (181)

1.8.1 Facility Self-Inspection: Under 40 CFR 112.7(e), you must include the written procedures and records of inspections for each facility in the SPCC Plan. You must include the inspection records for each container, secondary containment, and item of response equipment at the facility. You must cross-reference the records of inspections of each container and secondary containment required by 40 CFR 112.7(e) in the facility response plan. The inspection record of response equipment is a new requirement in this plan. Facility self-inspection requires two-steps: (1) a checklist of things to inspect; and (2) a method of recording the actual inspection and its findings. You must note the date of each inspection. You must keep facility response plan records for five years. You must keep SPCC records for three years.

1.8.1.1. Tank Inspection: The tank inspection checklist presented below has been included as guidance during inspections and monitoring. Similar requirements exist in 40 CFR part 112, subparts A through C. Duplicate information from the SPCC Plan may be photocopied and inserted in this section. The inspection checklist consists of the following items:

TAC Title 30 Part 1 Chapter 334 (112) Redacted References for Requirements.

a. §334.1 Purpose and Applicability

(a) Purpose. The purposes of this chapter are to:

(1) provide a comprehensive regulatory program for hazardous substance and petroleum substance underground storage tank (UST) systems, and a limited regulatory program for petroleum product aboveground storage tanks (ASTs), as prescribed by the Texas Water Code, Chapter 26, Subchapter I and Subchapter K;

(2) establish minimum standards and procedures to reasonably protect and maintain the quality of the state's groundwater and surface water resources from environmental contamination that could result from any releases of harmful substances stored in such tanks;

(3) provide for the use of risk-based corrective action; and

(4) generally provide for the protection of human health and safety, as well as the

protection of the overall environment of the state.

(c) Applicability to ASTs.

(1) An AST is subject to the applicable regulations in this chapter only when such tank;

(A) meets the definition of "aboveground storage tank" in §334.2 of this title;

(B) contains, has contained, or will contain a "petroleum product" as defined in §334.2 of this title;

(C) is not exempted from regulation under §334.123 of this title (relating to Exemptions for Aboveground Storage Tanks (ASTs)); and

(D) is not excluded from regulation under §334.124 of this title (relating to Exclusions for Aboveground Storage Tanks (ASTs)).

(2) The requirements and provisions in this chapter apply to regulated ASTs, and to the registration, installation notification, reporting, recordkeeping, release reporting and corrective action (including risk-based corrective action), fee assessment, and other applicable requirements associated with such tanks, as more fully described in this chapter.

(3) The applicable requirements and provisions in this chapter shall apply equally to all owners and operators of regulated ASTs, including individuals, trusts, firms, joint-stock companies, corporations, governmental corporations, partnerships, associations (including nonprofit and charity organizations), states, municipalities, commissions, political subdivisions of a state, interstate bodies, consortiums, joint ventures, commercial and noncommercial entities, and the United States government (including all of its departments), except as otherwise provided in this chapter.

(4) The following types of ASTs are subject to the applicable regulations in this chapter if they meet the general qualifications for an AST in paragraph (1) of this subsection:

(A) compartmental tanks, when at least one of the compartments is used to store petroleum products; and

(B) dual-use or multiple-use tanks which alternately store two or more substances when at least one of the stored substances is a petroleum product;

(5) If a storage tank containing a petroleum product technically meets the defini-

tions of both an AST and an UST under this chapter, then the tank will be considered an UST, and must conform with all applicable requirements for USTs in this chapter.

(6) Consistent with the exemption for heating oil tanks in §334.123(a)(2) of this title, an AST storing a petroleum product (such as kerosene or diesel) which is primarily used as a heating oil substitute for heating purposes on the premises where stored, and which is secondarily used as a motor fuel for the operation of internal combustion engines, is exempt from the regulations of this chapter.

(d) Applicability of specific sections to USTs and ASTs.

(1) USTs are subject to all the applicable provisions of this chapter, except Subchapter F of this chapter (relating to Aboveground Storage Tanks). Underground petroleum storage tanks are also subject to all applicable provisions of Chapter 37, Subchapter I of this title.

(2) ASTs are subject to all the applicable provisions of this chapter, except:

(A) §334.3 of this title (relating to Exemptions for Underground Storage Tanks (USTs) and UST Systems), §334.4 of this title (relating to Exclusions for Underground Storage Tanks (USTs) and UST Systems), §334.5 of this title (relating to General Prohibitions for Underground Storage Tanks (USTs) and UST Systems), §334.6 of this title (relating to Construction Notification for Underground Storage Tanks (USTs) and UST Systems), §334.7 of this title (relating to Registration for Underground Storage Tanks (USTs) and UST Systems), §334.8 of this title (relating to Certification for Underground Storage Tanks (USTs) and UST Systems), and §334.9 of this title (relating to Seller's Disclosure), and §334.10 of this title (relating to Reporting and Recordkeeping);

(B) Subchapter B of this chapter (relating to Underground Storage Tank Fees);

(C) Subchapter C of this chapter (relating to UST Technical Standards); and

(D) Subchapter I of this chapter (relating to Underground Storage Tank Contractor Registration and Installer Licensing).

§334.121 Purpose and Applicability for Aboveground Storage Tanks (ASTs) (89)

(a) Purpose. The purpose of this subchapter is to provide a regulatory program for aboveground storage tanks (ASTs) storing petroleum products, as prescribed by the Texas Water Code, Chapter 26, Subchapter I, to maintain and protect the quality of groundwater and surface water resources in the state from certain substances in ASTs that may pollute such groundwater and surface water resources, and to provide for the protection of human health and safety, as well as the protection of the overall environment of the state.

(b) Applicability.

(1) An AST shall be subject to the regulations in this subchapter only when such tank:

(A) meets the definition of "aboveground storage tank" in §334.122 of this title (relating to Definitions for Aboveground Storage Tanks (ASTs));

(B) contains, has contained, or will contain a "petroleum product" as defined in §334.2 of this title (relating to Definitions);

(C) is not exempted from regulation in §334.123 of this title (relating to Exemptions for Aboveground Storage Tanks (ASTs)); and

(D) is not excluded from regulation in §334.124 of this title (relating to Exclusions for Aboveground Storage Tanks (ASTs)).

(2) The requirements and provisions in this subchapter are applicable to regulated ASTs, and to the registration, installation notification, reporting, recordkeeping, release reporting and corrective action, fee assessment, and other requirements associated with such tanks, as more fully described in this subchapter.

(3) The requirements and provisions in this subchapter apply equally to all owners and operators of regulated ASTs, including individuals, trusts, firms, joint-stock companies, corporations, governmental corporations, partnerships, associations (including nonprofit and charity organizations), states, municipalities, commissions, political subdivisions of a state, interstate bodies, consortiums, joint ventures, commercial and noncommercial entities, and the United States government (including all of its departments), except as otherwise provided in this subchapter.

(4) The following types of aboveground tanks are subject to the regulations in this

subchapter if they meet the general qualifications for an AST in paragraph (1) of this subsection:

(A) compartmental tanks, when at least one of the compartments is used to store petroleum products; and

(B) dual-use or multiple-use tanks which alternately store two or more substances, when at least one of the stored substances is a petroleum product.

(5) If a storage tank containing a petroleum product technically meets the definitions of both an AST and an underground storage tank (UST) under this chapter, then the tank will be considered an UST, and must conform with all applicable requirements for UST in this chapter.

(6) Consistent with the exemption for heating oil tanks in §334.123(a)(2) of this title, an AST storing a petroleum product (such as kerosene or diesel) which is primarily used as a heating oil substitute for heating purposes on the premises where stored, and which is secondarily used as a motor fuel for the operation of internal combustion engines, is exempt from the regulations of this subchapter.

6. §334.122. Definitions for Aboveground Storage Tanks (ASTs) (89)

Except as provided in subsection (b) of this section, the words and terms used in this subchapter have the meanings prescribed in §334.2 of this title (relating to Definitions), unless the context clearly indicates otherwise.

The following words and terms, as used in this subchapter only, shall have the following meanings, unless the context clearly indicates otherwise.

Aboveground storage tank--A non-vehicular device (including any associated piping) that is made of nonearthen materials; located on or above the surface of the ground, or on or above the surface of the floor of a structure below ground, such as a mine working, basement, or vault; and designed to contain an accumulation of petroleum products.

Associated piping--All underground pipes or aboveground pipes (including related valves, elbows, joints, flanges, connectors, and other fittings) which are directly associated with an AST storing petroleum products, which are located at the AST facility, and through which petroleum products flow or in which petroleum products are contained or stored.

Bulk facility--A facility, including pipeline terminals, refinery terminals, rail and barge terminals, and associated underground and ASTs, connected or separate, from which petroleum products are withdrawn from bulk and delivered into a cargo tank or barge used to transport those products.

Electric generating facility--A plant or facility where on-site equipment is used to generate electrical power on a continuous or regular basis. This term does not include emergency generators, as defined in §334.2 of this title.

Nonearthen materials--Man-made materials designed to provide structural support and containment, such as concrete, steel, or plastic.

Non-vehicular device--A fixed, stationary, or moveable storage vessel which is not affixed or mounted to any self-propelled, towable, or pushable vehicle (e.g., wagon, truck, trailer, railcar, aircraft, boat, or barge).

7. §334.123 Exemptions for Aboveground Storage Tanks (ASTs), (89)

Commentary: *Pay close attention to listing of exemptions as many may apply to your facility especially (2)- (6) and (9). Also note the (c) requirement that you may be requested to provide documentation on your determination that a tank is exempt.*

(a) The following aboveground storage tanks (ASTs) are exempt from regulation under this subchapter:

(1) a farm or residential tank with a capacity of 1,100 gallons or less used for storing motor fuel for noncommercial purposes;

(2) a tank used for storing heating oil for consumptive use on the premises where stored;

(3) a septic tank;

(4) a surface impoundment, pit, pond, or lagoon;

(5) a stormwater or wastewater collection system;

(6) a flow-through process tank;

(7) a tank, liquid trap, gathering line, or other facility used in connection with an activity associated with the exploration, development, or production of oil, gas, or geothermal resources, or any other activity regulated by the Railroad Commission of Texas pursuant to the Natural Resources Code, §91.101;

(8) a tank located on or above the surface of the floor of an underground area,

such as a basement, cellar, mine working, drift, shaft, or tunnel, if the sole or principal substance in the tank is a hazardous substance; and

(9) a tank that is located at or is part of a petrochemical plant, a petroleum refinery, an electric generating facility, or a bulk facility.

- Upon request by the agency, the owner and operator of a tank claimed to be exempted under this section must provide appropriate documentation or other information in a timely manner to support that claim.

8. §334.124 Exclusions for Aboveground Storage Tanks (ASTs). (89)

Commentary: *Section (3) can imply that a used oil and water tank of significant dilution may be excluded but arguing the non-applicability of (3)(B) would be a significant challenge when any SPCC or FRP requirement exists at your facility. It would be advisable to contact the TCEQ if you believed that you had a tank that meets the (3)(B) exclusion for verification. Also as in the exemptions you may requested to provide documentation supporting your exclusion.*

(a) Except as provided in subsection (b) of this section, the following aboveground storage tanks (AST's) are excluded from regulation under this subchapter:

(1) any tank with a capacity of 1,100 gallons or less;

(2) any emergency spill protection or emergency overflow containment tank, including any sump or secondary containment system, which is used solely for the temporary storage or containment of petroleum products resulting from a leak, spill, overfill, or other unplanned release of petroleum products from any source, and where the petroleum products are routinely removed within 48 hours of the discovery of the release, provided that this tank must be inspected for a release no less than once every month;

(3) any tank that contains petroleum products at such dilute concentrations that:

(A) the mixture is not capable of being used as a fuel for the propulsion of a motor vehicle or aircraft; and

(B) any release would not pose any significant threat to human health and safety or the environment;

(4) a transformer or other electrical equipment that is used in the transmission of electricity.

(b) Notwithstanding the exemptions in subsection (a) of this section, any AST containing petroleum products located at a retail service station is subject to the construction notification requirements of §334.126 of the this title (relating to Installation Notification for Aboveground Storage Tanks (ASTs)).

(c) Upon request by the agency, the owner and operator of a tank claimed to be excluded under this section must provide appropriate documentation or other information in a timely manner to support that claim.

9. §334.125 General Prohibitions and Requirements for Aboveground Storage (89) Tanks (ASTs).

Commentary: *This can be a challenging section to comply with when you receive product from common carriers as Tank Registration Certificates for tanks they are putting material into are too made available to them on request.*

Delivery prohibition. Except as provided in paragraph (1) of this subsection, on or after the effective date of this subchapter, no common carrier (as defined in §334.2 of this title (relating to Definitions) shall deposit any petroleum products into an aboveground storage tank (AST) unless he observes that the owner or operator has a valid, current registration certificate, issued by the agency in accordance with §334.127 of this title (relating to Registration of Aboveground Storage Tanks (ASTs)).

(1) For new or replacement AST systems, only during the initial period ending 90 days after that petroleum product is first deposited into such system(s), a common carrier may accept, as adequate to meet this requirement, documentation that the owner or operator has a "temporary delivery authorization" (as defined at §334.127(h) of this title) issued by the agency for the facility at which the new or replacement AST system(s) exists.

(2) A common carrier delivering petroleum product into an AST system may observe a valid, current, original registration certificate (or temporary delivery authorization, if applicable), or a legible copy of the same.

(b) Owner/Operator requirements. The owner and operator of ASTs regulated under this section must make available to a common carrier a valid, current Texas Natural Resource Conservation Commission (TNRCC) (now the TCEQ) tank registration certificate (or TNRCC temporary delivery authorization, as applicable) before delivery of a petroleum product(s) into the AST(s) can be accepted. The bill of lading for the first delivery of petroleum product into any new or replacement AST system at the facility must be attached to the temporary delivery authorization for that facility.

10. §334.126 Installation Notification for Aboveground Storage Tanks (ASTs) (89)

(a) Except as provided in subsection (b) of this section, any person who intends to install a new or replacement aboveground storage tank (AST) must comply with the notification requirements of this section prior to initiating such activity.

(1) Installation notifications shall be submitted to the agency at least 30 days prior to initiating the activity.

(A) The notification may be provided either to the agency's central office in Austin or to the agency's appropriate regional office. The official date of notification shall be the date on which the notification is first received in an agency office.

Commentary: *The TCEQ Underground & Aboveground Storage Tank Construction Notification Form is located below following the AST Registration, Registration Form Instructions and AST Guidance.*

(B) Notification may be provided by the owner or operator, or an authorized representative of the owner or operator, (e.g., the contractor or consultant retained for the activity). Notifications filed by unauthorized persons shall be null and void.

(C) Notifications shall be submitted on the agency's authorized form. The form shall be filled out completely. Upon completion, the form shall be dated and signed by the owner or operator (or authorized representative) and shall be filed in accordance with this paragraph.

(D) When appropriate, installation notifications for ASTs (as required under this section) may be filed together with construction notifications for underground storage tank (UST) activities at the same facility (as required by §334.6 of this title (relating to Construction Notification for Underground Storage Tank (USTs) and UST Systems)), provided that complete and accurate explanation of the activities is included.

(E) Between 24 and 72 hours prior to the scheduled time of initiation of the installation, the owner or operator (or authorized representative) shall contact the agency's appropriate regional office to confirm the time of the initiation of the installation activities.

(F) The requirements and procedures for rescheduling, waiver requests, and expiration as related to installation notifications for ASTs shall be in conformance with the procedures for construction notifications applicable to USTs in §334.6(b)(3)-(5) of this title.

(2) When requested by the agency, any person who intends to install a new or replacement AST shall also submit additional supporting information to assure that the activity is in compliance with applicable statutes and regulations.

(3) In addition to the installation notification requirements of this section, the owner or operator of a proposed AST that is to be located in the designated recharge zone or transition zone of the Edwards Aquifer shall also secure the requisite approval from the executive director prior to initiating any installation or replacement activities, as prescribed in Chapter 313 of this title (relating to Edwards Aquifer).

(4) When an existing UST is to be removed from the ground and is to be subsequently converted to an AST, the person must comply with the applicable technical requirements under §334.55 of this title (relating to Permanent Removal from Service) and with the notification requirements of §334.6 and §334.126 of this title (relating to Construction Notification for Aboveground Storage Tanks (ASTs) and Installation Notification for Aboveground Storage Tanks (ASTs)).

(b) The following ASTs shall not be subject to the installation notification requirements of this section:

- ASTs which are exempt from regulation under §334.123 of this title (relating to Exemptions for Aboveground Storage Tanks (ASTs));
- ASTs which are excluded from regulation under §334.124 of this title (relating to

Exclusions for Aboveground Storage Tanks (ASTs));

(3) movable or mobile ASTs (e.g., skid tanks) which are moved from one location to another on a regular basis, which are not permanently part of any particular facility, and which are otherwise in compliance with the provisions of §334.127(f) of this title (relating to Registration for Aboveground Storage Tanks (ASTs)); except that any movable or mobile ASTs proposed for installation or placement at a retail service station shall remain subject to the installation notification requirements of this section.

11. §334.127, Registration for Aboveground Storage Tanks (ASTs). (89)

Commentary: *This section is critical as a Used Oil Handler regarding your compliance activities as failure to register tanks can lead to significant non-compliance. As indicated above it is also connected to the Prohibitions identified above as applied to delivery by common carriers. Also please be sure you understand requirements for portable tanks such as "Frac" tanks that are often utilized as part of used oil collection and storage activities and have typical volumes of approximately 21,000 gallons and would be applicable to tank management regulations.*

(a) General provisions.

(1) All aboveground storage tanks (ASTs) in existence on or after September 1, 1989, must be registered with the agency on authorized agency forms in accordance with subsection (e) of this section, except for those tanks which:

(A) are exempt from regulation under §334.123 of this title (relating to Exemptions for Aboveground Storage Tanks (ASTs)); or

(B) are excluded from regulation under §334.124 of this title (relating to Exclusions for Aboveground Storage Tanks (ASTs)).

(2) The owner and operator of an AST are responsible for compliance with the tank registration requirements of this section. An owner or operator may designate an authorized representative to complete and submit the required registration information; however, the owner and operator remain responsible for compliance with the provisions of this section.

(3) All ASTs subject to the registration requirements of this section are also subject to the fee provisions in §334.128 of this title (relating to Annual Facility Fees for Aboveground Storage Tanks (ASTs)), except where specifically exempted from such fee provisions. The failure by a tank owner or operator to properly or timely register any tanks shall not exempt the owner from such fee assessment and payment provisions.

(4) Proper completion of the specified agency tank registration form will result in the agency's issuance of a registration certificate for the tanks at the facility covered by that registration. This certificate is tied to the delivery prohibitions detailed in §334.125 of this title General Prohibitions and Requirements for Aboveground Storage Tanks (ASTs).

(b) Existing tanks. Any person who owns or operates an AST subject to the provisions of this section that was in existence on September 1, 1989, shall register such tank with the agency not later than March 1, 1990, on an authorized agency form.

(c) New or replacement tanks. Any person who owns or operates a new or replacement AST subject to the provisions of this section that is placed into service on or after September 1, 1989, must register the tank with the agency on an authorized agency form no later than March 1, 1990, or within 30 days from the date any petroleum product is first placed into the tank, whichever is later.

(d) Changes or additional information. An owner or operator of an AST subject to the provisions of this section must provide written notice to the agency of any changes or additional information concerning the status of any regulated tanks, including, but not limited to, information regarding the operational status, condition, substance stored, ownership, location of records, and number of tanks. This notice must be submitted on an authorized agency form which has been completed in accordance with subsection (e) of this section. This form must be properly completed and signed, and shall include the Texas Natural Resource Conservation Commission (TNRCC) facility identification number in the appropriate space on the form. Notice of any change or additional information must be filed with the agency within 30 days of the occurrence of the change or addition, or within 30 days from the date on which the owner or operator first became aware of the change or addition, as applicable.

(e) Required form for providing AST registration information. The TCEQ Registra-

tion Form and Instructions are provided below for your reference. Also the Fillable PDF you can fill out on screen is available at
https://www.tceq.texas.gov/assets/public/permitting/rrr/forms/0659.pdf (182)
Reference the TCEQ Petroleum Storage Tanks (PSTs) site at
https://www.tceq.texas.gov/permitting/pst_cert.html

Also included below the form and instructions is the current TCEQ AST Guide For Owners and Operators of ASTs

(1) Any AST owner or operator required to submit tank registration information under subsections (a)-(d) of this section must provide all the information indicated on the agency's authorized form for each regulated AST owned.

(2) The tank registration form must be filled out completely and accurately. Upon completion, the form must be dated and signed by the owner, operator, or an authorized representative of the owner or operator, and must be filed with the agency within the time frames specified in this section.

(3) All AST owners or operators required to submit AST registration information under subsections (a)-(d) of this section must provide the registration information for all ASTs located at a particular facility on the same registration form.

(4) All AST owners or operators who own or operate ASTs located at more than one facility must complete and file a separate registration form for each facility where regulated ASTs are located, unless otherwise allowed under subsection (f) of this section.

(5) If additional documents are submitted with new or revised registration data, the specific facility identification information (including the facility identification number, if known) must be conspicuously indicated on each document, and all such documents must be securely attached to and filed with the registration form.

(f) Registration requirements for movable ASTs. Movable or mobile ASTs which are regularly used to store petroleum products (e.g., skid tanks) must also be registered by the owner or operator in accordance with the provisions of this section. When such tanks are intended to be moved from one location to another on a regular basis and are not permanently part of any particular facility, then an owner or operator may register the tanks in accordance with the following procedures:

(1) For the purposes of completing the tank registration form, the owner or operator must identify the facility location for such movable tanks as the owner's or operator's principal business address or location;

(2) The owner or operator must continuously maintain complete and accurate records of the specific location, operational status, condition, and type of petroleum products stored at the owner's or operator's principal business address or location. At any given time, the records must include the required tank information for at least the preceding five years. Such records must be readily accessible and available for inspection upon request by agency personnel; and

(3) Any movable or mobile tank which is registered at the owner's or operator's business address or location, rather than at the actual facility location, must be permanently and legibly labeled with the agency's designated identification number for such tank by painting, decals, tags, or other permanent identification method.

(g) Inadequate information. When any of the required AST registration information submitted to the agency is determined to be inaccurate, unclear, illegible, incomplete, or otherwise inadequate, the agency may require the owner and/or operator to submit additional information. An owner and/or operator must submit any such additional information within 30 days of receipt of such request.

(h) Temporary delivery authorization.

(1) Upon receipt of a TNRCC construction notification form indicating pending installation of a new or replacement AST system(s), the agency will issue a temporary delivery authorization for that tank system(s).

(2) The temporary delivery authorization is valid for no more than 90 days after the first delivery of petroleum product into the new or replacement AST system.

(3) The AST owner and operator are responsible for maintaining complete and accurate records of the date of the first deposit of petroleum product into a new or replacement AST, as well as the date that the initial 90-day period expires. The bill of lading for the first delivery of regulated substance into any new or replacement AST at the facility must be attached to the temporary delivery authorization for that facility.

Form 7 – TCEQ Aboveground Storage Tank Registration Form (TCEQ-0659)

PST _	_ RE_	_ AST
	For internal use only	

Customer No.: CN **Regulated Entity No.: RN**

TCEQ – ABOVEGROUND STORAGE TANK REGISTRATION FORM

For Use in TEXAS — Texas Commission on Environmental Quality

• Please mail completed form to:
Petroleum Storage Tank Registration Program (MC-138)
Texas Commission on Environmental Quality
P. O. Box 13087 **Fax (512) 239-3398**
Austin, Texas 78711-3087

TCEQ Facility ID No.:
TCEQ Owner ID No.:
Federal Tax ID No. :

1. TANK OWNER INFORMATION

The Customer Name submitted here may be updated automatically based on what is current and active with the Texas Secretary of State (SOS) or Texas Comptroller of Public Accounts (CPA) If the Owner Name below is a new Owner, enter previous Owner Name:

TANK OWNER BUSINESS OR LAST NAME: / TANK OWNER FIRST NAME	TYPE OF TANK OWNER ☐ Individual ☐ Corporation ☐ Sole Proprietorship ☐ Federal Gov't ☐ State Gov't ☐ County Gov't ☐ City Gov't ☐ Local Gov't ☐ General Partnership ☐ Limited Partnership ☐ Other
OWNER MAILING ADDRESS:	
CITY: STATE: ZIP CODE: ZIP +4:	LOCATION OF RECORDS: ☐ At facility ☐ Offsite at:
COUNTRY (OUTSIDE USA) / E-MAIL ADDRESS	OFFSITE RECORDS LOCATION ADDRESS CITY STATE
OWNER'S AUTHORIZED REPRESENTATIVE TITLE: TELEPHONE NO/Ext.	RECORDS CUSTODIAN/CONTACT PERSON: TELEPHONE NO.
TX State TAX ID (11 Digits) / TX SOS/CPA Filing NO	FAX NO: / INDEPENDENTLY OWNED & OPERATED ☐ YES ☐ NO
FEDERAL TAX ID (9 Digits) / DUNs NO	NUMBER OF EMPLOYEES ☐ 0-20 ☐ 21-100 ☐ 101-250 ☐ 251-500 ☐ 501 & HIGHER

2. FACILITY INFORMATION

The Regulated Entity Name submitted may be updated in order to meet the TCEQ Agency Data Standards (removal of organizational endings such as Inc, LP, or LLC).

FACILITY NAME:	TYPE OF FACILITY: ☐ Emergency Generator ☐ Wholesale ☐ Retail ☐ Farm or Residential ☐ Fleet Refueling ☐ Aircraft Refueling ☐ Indian Land ☐ Watercraft Fueling ☐ Industrial/Manufacturing/Chemical Plant
Facility(RE) Address or PHYSICAL Location Description: (No P.O. Boxes)	Number of regulated *USTs at this facility: ____ *Underground Storage Tanks (USTs)
CITY:(Nearest if Physical Loc) / STATE: TX / ZIP CODE (Nearest if Physical Loc) / COUNTY:	Number of regulated *ASTs at this facility: ____ *Aboveground Storage Tanks (ASTs)
Facility Mailing Address	☐ New Regulated Entity ☐ Update to Regulated Entity Name ☐ Update to Regulated Entity Information
CITY: STATE: ZIP CODE: ZIP +4:	What is the primary Business of this Entity? (Do not repeat SIC or NAICS description)
ON-SITE CONTACT PERSON / TITLE: / TELEPHONE NO/Ext:	PRIMARY SIC CODE / SECONDARY SIC CODE
E MAIL ADDRESS: / FAX NUMBER	PRIMARY NAICS CODE / SECONDARY NAICS CODE
LATITUDE Degrees / Minutes / Seconds	LONGITUDE Degrees / Minutes / Seconds

3. TANK OPERATOR*INFORMATION ☐ (mark here if same as owner)

The Customer Name submitted here may be updated automatically based on what is current and active with the Texas Secretary of State (SOS) or Texas Comptroller of Public Accounts (CPA)

"Operator" means any person in day-to-day control of and having responsibility for, the daily operation of the AST system. If the Operator Name below is a new Operator, enter previous Operator Name:__________ CN__________

TANK OPERATOR NAME: *(DO NOT LIST EMPLOYEES OF OPERATOR)*	TYPE OF TANK OPERATOR: ☐ Individual ☐ Corporation ☐ Sole Proprietorship ☐ Federal Gov't ☐ State Gov't ☐ County Gov't ☐ City Gov't ☐ Local Gov't ☐ General Partnership ☐ Limited Partnership ☐ Other
MAILING ADDRESS:	
CITY: STATE: ZIP CODE: ZIP +4:	Date listed person became operator:
COUNTRY (OUTSIDE USA) / E-MAIL ADDRESS	FAX NO: / INDEPENDENTLY OWNED & OPERATED ☐ YES ☐ NO
OPERATOR'S AUTHORIZED REPRESENTATIVE: TITLE: TELEPHONE NO/Ext:	NUMBER OF EMPLOYEES ☐ 0-20 ☐ 21-100 ☐ 101-250 ☐ 251-500 ☐ 501 & HIGHER
TX State TAX ID / TX SOS/CPA Filing No	FEDERAL TAX ID / DUNs NO

TCEQ Facility ID No

TCEQ - AST REGISTRATION FORM

4. REGISTRATION STATUS

REASON FOR SUBMITTING FORM *(Mark all that apply)*:

1 ☐ Initial Registration 2 ☐ AST Ownership Change *(New Owner indicate effective date:)*... ☐

3 ☐ Amendment of: A ☐ Owner Information Update B ☐ Operator Information Update C ☐ Facility Information Update

D ☐ Tank Information Update

4. ☐ Other *(specify)*:

If this update will affect other program areas affiliated to this customer/RN, indicate all affected programs in Section 6 of this form.

Effective Date for Customer Information Updates (mm/dd/yyyy): _______________

5. OWNER CERTIFICATION

I certify under penalty of law that I have personally examined and am familiar with the information submitted in this and all attached documents, and based on my current knowledge and understanding, the submitted information is true, accurate, and complete and that I have signature authority to submit this form on behalf of the entity in Section 1 and/or as required for the updates to the ID number identified in Section 6.

PRINTED NAME OF OWNER/OPERATOR (OR AUTHORIZED REPRESENTATIVE)	TITLE
SIGNATURE OF OWNER/OPERATOR (OR AUTHORIZED REPRESENTATIVE)	DATE OF SIGNATURE (mm/dd/yyyy)

Individuals are entitled to request and review their personal information that the agency gathers on its forms. They may also have any errors in their information corrected. To review such information, contact us at 512/239-2160.

TCEQ ABOVEGROUND (AST) GENERAL INFORMATION

Who Must Register? Registration is required by Title 30 TAC, Chapter 334, Subchapter F, Section 334.123 for all aboveground storage tanks (ASTs) that have been used to store a regulated petroleum product that are in existence as of September 1, 1989, or that are brought into use after September 1, 1989. The deadline for registering existing tanks was March 1, 1990. Owners who put tanks into use after March 1, 1990 must register their tanks with the Texas Commission on Environmental Quality within 30 days from the date any regulated product is placed into the tank.

Which Tanks Are Regulated? AST is defined as a non-vehicular device that is made of non-earthen materials; located on or above the surface of the ground or on or above the surface of the floor of a structure below ground, such as a mine working, basement, or vault; and designed to contain an accumulation of petroleum. Only AST's with a capacity greater than 1100 gallons are regulated.

What Petroleum Products Are Regulated? Petroleum product means a product that is obtained from distilling and processing crude oil and that is capable of being used as a fuel for the propulsion of a motor vehicle or aircraft, including:

1.motor gasoline; 4. kerosene;
2.gasohol, and other blended fuels; 5.distillate fuel oil; and
3.aviation gasoline; 6.diesel #1 and #2.

The term does not include naphtha-type jet fuel, kerosene-type jet fuel, waste oil, or a petroleum product destined for use in chemical manufacturing or feedstock of that manufacturing.

Which Tanks Are Exempt? The following ASTs are exempt from regulation under this program:

1. a farm or residential AST with a capacity of 1,100 gallons or less used for motor fuel for noncommercial purposes;
2. a tank used for storing heating oil for consumptive use on the premises where stored;
3. a septic tank;
4. a surface impoundment, pit, pond, or lagoon;
5. a stormwater or wastewater collection system;
6. a flow-through process tank;
7. a tank, liquid trap, gathering line, or other facility used in connection with the exploration, development, or production of oil, gas, or geothermal resources, or any other activity regulated by the Railroad Commission of Texas pursuant to the Natural Resources Code, §91.101;
8. a transformer or other electrical equipment that contains a regulated substance and that is used in the transmission of electricity, to the extent that such a transformer or equipment is exempted by the United States Environmental Protection Agency under 40 C.F.R. Part 280;
9. an AST is exempt from regulation under this chapter if the sole or principal substance in the tank is a hazardous substance;
10. an interstate pipeline facility, including gathering lines, or an AST connected to such a facility is exempt from regulation under this chapter if the pipeline facility is regulated under:
    (1) the Natural Gas Pipeline Safety Act of 1968 (49 United States Code, §1671 et seq.)
    (2) the Hazardous Liquid Pipeline Safety Act of 1979 (49 United States Code, §2001 et seq.);
11. an intrastate pipeline facility or aboveground storage tank connected to such a facility is exempt from regulation under this chapter if the pipeline facility is regulated under one of the following state laws:
    (1)Natural Resources Code, Chapters 111 and 117
    (2)Texas Civil Statutes, Articles 6053-1 and 6053-2; and
12. an AST that is located at or is part of a petrochemical plant, a petroleum refinery, an electric generating facility, or a bulk facility as that term is defined by §26.3574(a) of the Water Code is exempt from regulation under this chapter but is not exempt for purposes of the fee for delivery of certain petroleum products authorized under $26.3574 of the Water Code.

Amended Registration: An owner of a regulated AST is required to provide written notice to the TCEQ of any changes or additional information concerning the status of any regulated tank, including, but not limited to, operational status condition, product stored, and ownership. When filing an amended registration form, please mark the appropriate box in Section IV. Notice must be filed with the Commission within 30 days from the date of occurrence or knowledge of the status change.

AST Fees: Annual facility fees shall cease to be assessed effective September 1, 2007, and shall not be assessed until such time as reinstated by the commission at an amount determined appropriate by the commission. (Title 30 TAC, Chapter 334, Subchapter F, Section 334.128)

Penalties: Any owner who knowingly fails to register their ASTs or submits false information may be subject to a civil penalty not to exceed $10,000 per day for each violation.

TCEQ Facility ID No

TCEQ - AST REGISTRATION FORM

6. TCEQ PROGRAMS IN WHICH THIS REGULATED ENTITY PARTICIPATES

Check all Programs and write in the permits/registration numbers that will be affected by the updates submitted on this form or the updates may not be made. If the program is not listed, check other and write it in. This identification will help ensure this form will go to the correct Program Areas and that the appropriate permits / registrations are updated.

☐ Animal Feeding Operation	☐ Dam Safety	☐ Districts	☐ PWS
☐ Industrial & Hazardous Waste	☐ Municipal Solid Waste	☐ New Source Review - Air	☐ Edwards Aquifer
☐ OSSF	☐ Petroleum Storage Tank	☐ Sludge	☐ Emmission Inventory Air
☐ Stormwater	☐ Tires	☐ Title V - Air	☐ Used Oil
☐ Utilities	☐ Voluntary Cleanup Program	☐ Wastewater Agriculture	
☐ Wastewater Permit	☐ Water Districts	☐ Water Rights	
☐ Water Utilities	☐ Other ____	☐ Other ____	
☐ Unknown	☐ Licensing - Type(s)	____ ____ ____	

7. DESCRIPTION OF ABOVEGROUND STORAGE TANKS

Tank ID (e.g. 1,2,3 or A, B, C)				
Tank Installation Date *(Month/day/year)*				
Tank Capacity *(U.S. gallons)(must be >1100 gallons)*				
Tank Status 1-In Use *(includes tanks that are inactive but contain product)* 2-Out of Use *(tanks that are inactive and do not contain product)*. Indicate date taken out of use *(mo/day/yr)*.	1- ☐ 2- ____	1- ☐ 2- ____	1- ☐ 2- ____	1- ☐ 2- ____
Product Stored Mark all that apply 1-Gasoline 2-Diesel 3-Kerosene 4-Alcohol Blended Fuel 5-Aviation Gasoline 6-Distillate Fuel Oil	1-☐ 2-☐ 3-☐ 4-☐ 5-☐ 6-☐	1-☐ 2-☐ 3-☐ 4-☐ 5-☐ 6-☐	1-☐ 2-☐ 3-☐ 4-☐ 5-☐ 6-☐	1☐ 2-☐ 3-☐ 4-☐ 5-☐ 6-☐
Material of Construction Mark all that apply 1-Steel 2-Fiberglass 3-Aluminum 4-Corrugated Metal 5-Concrete	1-☐ 2-☐ 3-☐ 4-☐ 5-☐	1-☐ 2-☐ 3-☐ 4-☐ 5-☐	1-☐ 2-☐ 3-☐ 4-☐ 5-☐	1-☐ 2-☐ 3-☐ 4-☐ 5-☐
Containment Mark all that apply 1-Earthen Dike 2-Containment Liner 3-Concrete 4-None	1-☐ 2-☐ 3-☐ 4-☐	1-☐ 2-☐ 3-☐ 4-☐	1-☐ 2-☐ 3-☐ 4-☐	1-☐ 2-☐ 3-☐ 4-☐
Stage I Vapor Recovery ** See rule & location exemption information.* 1-Stage I (AST to tanker truck): Installation date: • Type: 1a-Stage 1 two-point system 1b-Stage 1 coaxial system • Exempt by: 1c-TCEQ Rule*	1- ____ 1a-☐ 1b-☐ 1c-☐	1- ____ 1a-☐ 1b-☐ 1c-☐	1- ____ 1a-☐ 1b-☐ 1c-☐	1- ____ 1a-☐ 1b-☐ 1c-☐

* STAGE I VAPOR RECOVERY - Please indicate whether your system has Stage I vapor recovery equipment and the installation date of the equipment. Applicable requirements may be found in 30 TAC, §115.221-229 and §115.241-249. If your AST system is not located in a non-attainment county or one of the 95 covered attainment counties, completion of this section is not necessary. For a complete list of covered attainment counties, please refer to 30 TAC, §115.10.

1. Stage I - system used to capture vapors from the AST during deliveries. Stage I is required in non-attainment counties and in the 95 covered attainment counties if throughput is greater than 125,000 gallons.

If you have questions on how to fill out this form or regarding the PST program, please contact us at 512/239-2160.

Individuals are entitled to request and review their personal information that the agency gathers on its forms. They may also have any errors in their information corrected. If you wish to review such information, contact us at 512/239-2160.

For data verification purposes, please check our IWR (Integrated web reporting) web page:
www15.tceq.texas.gov/crpub/index.cfm?fuseaction=regent.RNSearch

TCEQ-0659 (Rev. 01/01/2017) For more information, visit our web site at www.tceq.texas.gov. Page 3 of 3

Texas Commission on Environmental Quality
Petroleum Storage Tank Program

Aboveground Storage Tank Registration Form Instructions

WHAT IS THIS FORM? This form (form 0659) provides two functions for the agency: to report Aboveground Storage Tank (AST) registration information and to provide core data for the TCEQ's Central Registry Program. The Central Registry maintains basic information about the people, businesses, and institutions that the TCEQ must regulate by law. Examples of "core data" are names, addresses, location information, and type of business.

WHEN MUST I COMPLETE THIS FORM? All tanks meeting the definition of an AST that contain or have contained a petroleum product, unless otherwise exempted or excluded (as described in §334.123-124), are required to be registered with this form. Any change of registration information requires the filing of an amended form. Please see the instructions below for each section of the form. If you have any questions concerning the form, or whether your tanks should be registered, **please contact the Petroleum Storage Tank Registration Team at (512) 239-2160.**

You must submit this form for Central Registry information in two situations only: You are not in our core database (the "Central Registry") yet; or if your core data has changed, and you need to update it with us. In completing this form for Central Registry, the facility where the tank is located is the "regulated entity." The owner of the facility is a "customer," as is the operator of the facility. For example, if the tank is located at a service station, then that service station is the regulated entity, and the owner and operator of the service station are customers.

GENERAL INSTRUCTIONS: You will complete different parts of this form depending on whether you are submitting tank registration information, or Central Registry core data. For the first time submittal of an original form for registration purposes, please complete the entire form so our records will be as current and accurate as possible.

Please mail the completed form to the address provided at the top center of the AST Registration Form.

LINE BY LINE INSTRUCTIONS

ID NUMBERS - These are TCEQ assigned numbers (except for the Tax ID No.) used to identify facilities, owners, and operators.
Customer Reference Number (CN): If this is the first time you have completed this form, you will not have this number yet, so you should leave this space blank. **Do not** enter a permit number, registration number, or license number here. This number always begins with CN. Regardless of how many regulated entities you are associated with, you will have only one customer reference number.
Regulated Entity Reference Number (RN): This unique number identifies the regulated entity and always begins with RN. **Do not** enter a permit number, registration number, or license number here. Enter only one regulated entity reference number here on each respective form.
Facility ID No. and Owner ID No.: These numbers are reported in the upper right hand corner of the form (except for operator ID). The facility ID number must be provided for all filings except for the original registration of a facility. The owner ID number must also be reported if the owner has been assigned an ID previously. The TCEQ will assign operator ID numbers after the initial filing, the operator ID number should be provided with all subsequent filings. The space for Operator ID number is located in top left corner of Section 3.
Federal Tax ID Number: All businesses, except for some small sole proprietors, should have a federal taxpayer identification number. Enter this number here. Use no prefixes, dashes, or hyphens.

1. TANK OWNER INFORMATION - This section contains basic information about the Owner of the ASTs. This entire section (Tank Owner Information) must be filled out whenever the form is submitted to TCEQ for any purpose.

Tank Owner Business or last Name: Enter the legal name of the owner or last name of individual. Include any abbreviations (LLC, Inc., etc.).

Tank Owner First Name: Enter the first name for individuals only.

Owner Mailing Address: Enter a central mailing address for the customer to receive mail from the TCEQ. For example, this address might be the corporate or regional headquarters of a large company.

Country (Outside USA): If the address is outside the United States, enter the country name.

E-Mail Address: This should be a general address that is appropriate for e-mail to your central or regional headquarters.

Owner's Authorized Representative and phone number: Enter information for a contact person at the central or regional headquarters, if applicable.

State Franchise Tax ID Number: Corporations and limited liability companies that operate in Texas are issued a franchise tax identification number. If the owner is a corporation or limited liability company, enter this number here.

DUN Number: Most businesses have a DUNS (Data Universal Numbering System) number issued by Dun and Bradstreet Corp. If the customer has one, enter it here.

Type of Customer: The type of tank owner identifies whether the owner is an individual, corporation, or government entity. Check **one** box.

Location of Records: The location of records indicates if this facility's records are kept onsite or at an alternative location. If records are kept offsite, please indicate the location, contact person, and telephone number.

Fax Number: Enter owner's fax number

Independently Owned and Operated?: Check "No" if the customer is a subsidiary or part of a larger company. Otherwise, check "Yes."

Number of Employees: Enter the number of employees for the customer's entire company, at all locations. This is ***not*** the number of employees at one facility.

2. FACILITY INFORMATION - General information about the AST facility, including facility name, location, and contact person. This entire section must be filled out whenever the form is submitted to TCEQ for any purpose.

Facility Name: This is the name by which you want the regulated entity to be known to the TCEQ.

Physical Location: Enter the physical address of the ASTs. A Post Office or Rural Route Box cannot be accepted for a facility address. Our staff should be able to use this address to find the facility

City: Indicate the city/town or the nearest city/town where the ASTs are located.

ZIP Code: Enter the ZIP code for the area that includes the location of the ASTs.

County: Enter the name of the county in which the ASTs are located.

On Site Contact Person: Enter name, title, and phone number for a contact person at the facility.

E-Mail Address: As with the mailing address, this should be a general address that is appropriate for e-mail to the regulated entity's main office, if applicable.

Fax Number: Enter information for contacting the regulated entity's main office, if applicable.

TYPE OF FACILITY: Indicate activities commonly performed at the location. Check all boxes that apply.

Number of regulated UST/ASTs at this facility: For number of regulated USTs or ASTs at this facility, report the number of active and temporarily out-of-use USTs and the number of active ASTs at this facility.

SIC and NAICS Codes: All regulated entities, with the exception of occupational licensees, should have a Standard Industrial Classification (SIC) code and a North American Industrial Classification System (NAICS) code. These codes may or may not correspond to why the regulated entity is regulated by the TCEQ. When NAICS codes replace SIC codes, you will probably be asked to provide NAICS codes, if you are not providing them now. Enter the SIC code that best describes the main business activity at this regulated entity. Enter the SIC code that best describes the secondary business activity at this regulated entity. For a list of SIC codes on the Web, go to:
www.osha.gov/oshstats/sicser.html

Enter the NAICS code that best describes the main business activities at this regulated entity. Enter the NAICS code that describes the secondary business activity at this regulated entity. For a list of NAICS codes on the Web, go to:
www.census.gov/epcd/www/naicscod.htm

Latitude and Longitude: Enter this information if the regulated entity has a fixed location and the latitude and longitude are known.

3. TANK OPERATOR INFORMATION - This section is the name, mailing address, contact person and certain other basic information about the Operator of the ASTs. The operator is the person in day-to-day control of, and/or having responsibility for the daily operation of the AST system (excluding clerks, cashiers, or other employees of the operator). An operator's employee may, however, be designated as the "operator's authorized representative" on the form if the operator so desires. If tank operator information is identical to tank owner information, check the box at the top of the section marked "mark here if same as owner". If you check this box, completion of Section 3 is not required; however, if the box is not checked, Section 3 must be completed in its entirety.
TCEQ Operator ID: This is a TCEQ assigned number used to identify operators.
Operator's Customer Number: If this is the first time you have completed this form, you will not have this number yet, so you should leave this space blank. **Do not** enter a permit number, registration number, or license number here. This number always begins with CN. Regardless of how many regulated entities you are associated with, you will have only one customer reference number.
Tank Operator Name: Enter the legal name of the customer. Include any abbreviations (LLC, Inc., etc.).
Mailing Address: Enter a central mailing address for the customer to receive mail from the TCEQ. For example, this address might be the corporate or regional headquarters of a large company.
Operator's Authorized Representative: Enter information for a contact person at the central or regional headquarters, if applicable.
Type of tank operator: Identifies whether the operator is an individual, corporation, or government entity. Please check only one.
Date listed person became operator: Is the date (month, day, and year) the operator assumed responsibility for the AST's.

4. REASON FOR FILING - Complete this section to indicate the reason for submitting the form. TCEQ uses this information internally. Please mark all choices that apply.

AST Registration Information - This section is for registration information. Use it to indicate why the form was filed:

- Box 1 - Initial Registration - the first time a AST system is being registered
- Box 2 - Change in Ownership - please provide the effective date of the ownership change
- Box 3 - Amendments - required reporting of changes to current registration information
- Box 4 - Other - reasons other than those listed above

Check the appropriate box, explain "other" answers sufficiently, or use this space to provide additional comments, as necessary. For ownership changes the effective date of the ownership change is required.

5. TANK OWNER/OPERATOR REGISTRATION - In this section, the owner/operator is certifying that they are familiar with the information included on the form and are certifying that the information is true, accurate, and complete. This section must be filled out completely by the UST owner, operator, or their legally authorized representative when filing UST registration information. A signature and date are required or the form will be returned. If an ownership change is involved, only the signature of the current owner or the current owner's legally authorized representative will be accepted.

6. TCEQ PROGRAMS IN WHICH THIS REGULATED ENTITY PARTICIPATES:
Check all that apply. If this regulated entity participates in a TCEQ program not shown on this list, enter that program in one of the blank spaces and check the corresponding box. Check "Unknown" only if you do not know any programs this regulated entity participates in.

TCEQ AST GENERAL INFORMATION

7. DESCRIPTION OF ABOVEGROUND STORAGE TANKS-This section contains descriptive and identifying information for each AST and must be completed with the first submission of the AST Registration form. Please fill out this portion of the form completely to ensure all ASTs are correctly identified. This section also must be used to update or amend the AST registration information. Mark all that apply, unless otherwise indicated.

Tank ID - Indicate individual tanks' ID number, as assigned by the owner/operator
Tank Installation Date - Provide the date the AST was installed at its current location and fuel placed in AST.
Tank Capacity - Provide AST's capacity in gallons. AST must be larger than 1100 gallons.
Tank Status - Enter the current status of the AST. Check the correct box or provide date where specified. When reporting out of use, please provide the date that the AST was emptied.
Product Stored - Enter the current substance stored in the AST.
Material of Construction - Indicate material of construction for tanks.
Containment - Indicate type of containment, if applicable.
Stage I/Stage II Vapor Recovery - Please indicate whether your system has Stage I and/or Stage II vapor recovery equipment and the installation date of the equipment. Applicable requirements may be found in 30 TAC, §115.221-229 and §115.241-249. **If your AST system is not located in a non-attainment county or one of the 95 covered attainment counties, completion of this section is not necessary.** For a complete list of covered attainment counties, please refer to 30 TAC, §115.10.

1. STAGE I - system used to capture vapors from the AST during deliveries. Stage I is required in non-attainment counties and in the 95 covered attainment counties if throughput is greater than 125,000 gallons.
2. STAGE II - system used to capture vapors from vehicle fuel tanks during refueling. Stage II is required only in the 16 non-attainment counties. The counties are: Brazoria, Chambers, Collin, Dallas, Denton, El Paso, Fort Bend, Galveston, Hardin, Harris, Jefferson, Liberty, Montgomery, Orange, Tarrant, and Waller.

If you have any questions concerning the completion of this form or about the information reported on the form, please contact the **Petroleum Storage Tank Registration & Self-Certification Team at (512) 239-2160.**

For TCEQ rules and publications: You can find the official version of TCEQ rules in the Texas Administrative Code on the Secretary of State's Web site (www.sos.state.tx.us/tac/index.shtml).

Other ways to obtain a copy of the rules, or of TCEQ publications, include the following:

- on the Internet, go to the TCEQ's Web site at www.tceq.state.tx.us and click on the link to "Rules, Policy & Legislation";
- fax orders to (512) 239-4488, or order by voice at (512) 239-0028, the TCEQ's publications unit; or
- write to TCEQ Publications, MC 195, P.O. Box 13087, Austin, TX 78711-3087.

MAKE A COPY OF YOUR COMPLETED FORM FOR YOUR RECORDS

TCEQ REGULATORY GUIDANCE
Environmental Assistance Division
RG-475n • Revised August 2017

Aboveground Petroleum Storage Tanks

A guide for owners and operators of ASTs

This is module n of the PST Super Guide, a comprehensive guide to issues relating to petroleum storage tanks (PSTs). This super guide provides an overview to laws and regulations for PSTs and it can be used as an aid in minimizing potential risks. The guide does not replace those laws and regulations, which take precedence over any information in this publication.

Module n explains the rules and procedures for aboveground storage tanks (ASTs).

- You, the owner or operator of a PST, are responsible for ensuring compliance with all applicable laws and regulations.
- If your underground storage tank (UST) system is located in Kinney, Uvalde, Medina, Bexar, Comal, Hays, Travis, or Williamson County, additional requirements related to protecting the Edwards or the Trinity Aquifer may apply (Title 30, Texas Administrative Code [30 TAC], Chapters 213 and 214).
- In addition to the laws and TCEQ rules, local governments, and other state and federal agencies may have rules that apply.

How do I know if my AST is regulated?

TCEQ regulates ASTs with a capacity of more than 1,100 gallons that contain a petroleum product capable of propelling a motor vehicle or airplane (excluding naphtha- or kerosene-type jet fuel). TCEQ rules also state that petroleum product ASTs, not otherwise excluded or exempt, are subject to construction installation notification.

Is my AST excluded?

The following ATSs are excluded from regulation:

- any tank with a capacity of 1,100 gallons or less;
- any tank that contains petroleum products at such dilute concentrations that the mixture is not capable of being used as a fuel for the propulsion of a motor vehicle or aircraft; and any release would not pose any significant threat to human health and safety or the environment; or
- a transformer or other electrical equipment that is used in the transmission of electricity.

Note: Excluded ASTs containing petroleum products are subject to construction notification requirements if located at a retail service station.

Is my AST exempt?

Tanks that meet certain specifications are exempt from registration. These include:

- a farm or residential tank with a capacity of 1,100 gallons or less used for noncommercial purposes;
- a tank used for storing heating oil that is used on the same premises;
- a flow-through process tank;
- a tank associated with the exploration, development, or production of oil, gas, or geothermal resources, or any other activity regulated by the Railroad Commission of Texas;
- a tank that is associated with a petrochemical plant, petroleum refinery, or electric generating facility;
- a septic tank;
- a surface impoundment pit, pond, or lagoon;
- a stormwater or wastewater collection system; or
- a tank located on or above the surface of the floor of an underground area, such as a basement, cellar, mineworking, drift, shaft, or tunnel, if the sole or principal substance in the tank is a hazardous substance.

If you claim an exemption for your tank, you must be prepared to provide appropriate documentation to support the claim at the request of the TCEQ. Documentation must support your exemption claim. Installation records listing tank capacity, facility plans showing tank location or purpose, or recent delivery receipts listing the contents of the tank are examples of possible documentation.

How do I register my tanks?

Complete form TCEQ-00659, available online at <www.tceq.texas.gov/assets/public/permitting/rrr/forms/0659.pdf>. All ASTs located at the same address must be included on the same registration form. If you own or operate tanks at different locations, you must file separate registration forms for each facility.

What if something changes?

You must submit a new AST form TCEQ-00659 to the agency within 30 days of any changes regarding an AST, such as:

- operational status
- condition
- substance stored
- ownership
- location of records
- number of tanks at the facility
- any change in contact information

What if I want to install a new or replacement AST?

If you install a new AST or replace an existing AST:

- At least 30 days before you begin work, submit a construction notification form (TCEQ-00495) to the applicable TCEQ regional office. TCEQ-00495 is available at <www.tceq.texas.gov/assets/public/permitting/rrr/forms/0495.pdf>.
- Between 24 and 72 hours before work begins, contact the appropriate TCEQ office for the region where the activity is to occur to report the time you will begin installation.

To determine which regional office you should notify, please visit <www.tceq.texas.gov/about/directory/region/reglist.html>.

What records do I need to keep?

You must comply with the same general recordkeeping requirements as USTs and legible copies of all original and amended tank registration documents must be kept for the operational life of the AST system. You should keep copies of all records that document compliance with applicable rules. These include, but are not limited to:

- your construction notification;
- application for approval of any proposed AST in the Edwards Aquifer recharge or transition zones;
- registration form;
- tank-manufacturing information;
- receipts of payments; and
- reports, plans, and certifications related to actions taken in response to suspected and confirmed releases.

Do I need a registration certificate?

Yes, you must have your (1) temporary delivery authorization or (2) registration certificate to receive fuel deliveries. For new and replacement tanks:

Temporary Delivery Authorization

- The TCEQ will send you a temporary delivery authorization, upon receipt of your construction notification form (TCEQ-00495).
- Attach the bill of lading for the first fuel delivery to your temporary authorization.
- The temporary delivery authorization expires 90 days after your first delivery.

Registration Certificate

- Submit a registration form (TCEQ-0659) within 30 days of your first fuel delivery.

- The TCEQ will send your registration certificate upon receipt of your registration form.

After the initial 90-day grace period, a common carrier must observe your valid registration certificate to deliver fuel.

What about skid tanks?

Skid tanks must be registered and labeled with the TCEQ-designated ID number. To register your tank:

- Complete a new registration form (TCEQ-00495) each time the tank is moved to a different location or register the tanks at a primary business location.
- Continuously maintain accurate records of the location, status, and type of petroleum product stored to demonstrate compliance at the facility where the tank is registered.

What if my tank leaks?

You must report a suspected or confirmed release to the TCEQ in the following cases:

- A spill or overfill that results in a release to the environment that exceeds 25 gallons or that causes a sheen on nearby surface water.
- A spill or overfill of less than 25 gallons, if the cleanup cannot be accomplished within 24 hours.

Please refer to *Suspected and Confirmed Releases from Petroleum Storage Tanks* (TCEQ publication RG-475h) for more information on this subject.

Do I need financial assurance?

Financial assurance is not required for AST systems.

Do I need operator training?

Operator training is not required for AST systems.

Where to find more information?

State requirements for ASTs appear in 30 TAC 334, Subchapter A available at <texreg.sos.state.tx.us/public/readtac$ext.ViewTAC?tac_view=5&ti=30&pt=1&ch=334&sch=A&rl=Y>, and 30 TAC 334, Subchapter F available at <texreg.sos.state.tx.us/public/readtac$ext.ViewTAC?tac_view=5&ti=30&pt=1&ch=334&sch=F&rl=Y>.

Requirements for tanks in the Edwards Aquifer are in 30 TAC 213 available at <texreg.sos.state.tx.us/public/readtac$ext.ViewTAC?tac_view=4&ti=30&pt=1&ch=213>.

Requirements for tanks over other aquifers are in 30 TAC 214 available at <texreg.sos.state.tx.us/public/readtac$ext.ViewTAC?tac_view=4&ti=30&pt=1&ch=214&rl=Y>.

TCEQ Guidance on PST regulations is available at <www.tceq.texas.gov/agency/pst_cert.html>.

U.S. Environmental Protection Agency requirements, including Spill Prevention, Control, and Countermeasures (SPCC) are in Title 40, Code of Federal Regulations, Part 112 available at <www.ecfr.gov/cgi-bin/text-idx?SID=8ae74a76e3fbf28c91f3358109250f55&mc=true&node=pt40.24.112&rgn=div5>.

Download forms from the TCEQ's website at <www.tceq.texas.gov/search_forms.html>

Download publications at the TCEQ's website at <www.tceq.texas.gov/publications>

For information on cleanup requirements, you may contact the Remediation Division at 512-239-2200.

For confidential environmental compliance assistance for small businesses and local governments, contact Small Business and Local Government Assistance via its hotline at 800-447-2827 or online at <www.TexasEnviroHelp.org>.

PST _ _ RE _ _ NOC

For internal use only

TEXAS COMMISSION ON ENVIRONMENTAL QUALITY
Underground & Aboveground Storage Tank
Construction Notification Form

Facility Name: _______ Facility I D: _______

Address/Location: _______ County: _______

City: _______ Phone: _______

TYPE OF CONSTRUCTION: (INDICATE ALL THAT APPLY)

U S T:

☐ Repair ☐ Removal ☐ Abandonment

☐ Improvement ☐ Return to Service ☐ Stage I

☐ Installation ☐ Replacement (Tank)

A S T:

☐ Installation

Tank Capacity: _______

☐ Stage I

Scheduled date(s) for proposed construction: _______

GENERAL DESCRIPTION OF PROPOSED U S T/A S T ACTIVITY

____

OWNER INFORMATION

Owner Name: _______ Owner I D: _______

Owner's Representative: _______ Phone: _______

Mailing Address (include city/state/zip): Fax: _______

_______ EMAIL: _______

CONTRACTOR INFORMATION

Company: _______

Representative: _______

Mailing Address (include city/state/zip):

____

Phone: _______ Fax: _______

CRP: _______ ILP: _______

CONSULTANT INFORMATION

Company: _______

Representative: _______

Mailing Address (include city/state/zip):

____

Phone: _______ Fax: _______

Submitted by (Print name): _______ Title: _______

Company: _______ Date: _______

Signature: ___________________________________

Mail or Fax completed forms to:

Texas Commission on Environmental Quality
PST Registration & Self-Certification Team (MC-138)
PO Box 13087
Austin, TX 78711-3087
Fax: (512) 239-3398

TCEQ Staff Use Only

Date Received: ____________________

Region: ____________________

Remarks: ____________________

Logged by: ____________________

Individuals are entitled to request and review their personal information that the agency gathers on its forms. They may also have any errors in their information corrected. To review such information, contact us at 512-239-2160.

TCEQ-0495 (06/23/2017)

12. §334.128, Annual Facility Fees for Aboveground Storage Tanks (ASTs) (89)

(a) Fee assessments.

(2) Except as provided in subsection (e) of this section, an annual facility fee of $25 will be assessed by the agency for each aboveground storage tank (AST) subject to the registration provisions of §334.127 of this title (relating to Registration for Aboveground Storage Tanks (ASTs)).

(3) All annual facility fees will be billed to, and must be payable by, the owner of the AST.

(4) Payment of annual facility fees is due no later than 30 days after the date the agency mails a statement of the assessment to the tank owner.

(5) Annual facility fees must be paid by check, certified check, or money order made payable to the Texas Commission on Environmental Quality. Payments must be mailed to the address specified in the billing statement.

(b) Billing schedule.

(1) The agency will establish a schedule for the billing of annual facility fees.

(2) Regardless of the actual billing date, the total amount of annual facility fees billed to and payable by an owner will be based on the total number of regulated ASTs in place on or after the first day of each fiscal year (September 1).

(c) Failure to make payment.

(1) Annual facility fees must be paid at the time and in the manner and amount provided by this section.

(2) The agency will impose interest and penalties on owners who fail to make payment of the annual facility fees assessed under this section when due in accordance with Chapter 12 of this title (relating to Payment of Fees).

(d) Disposition of fees, interest, and penalties. As required by the Texas Water Code (TWC), §26.358(g), the agency will deposit all annual facility fees collected, together with all interest and penalties collected for late payment, in the state treasury to the credit of the storage tank fund.

(e) Exception. An annual facility fee shall cease to be assessed, effective September 1, 2007, and shall not be assessed until such time as reinstated by the commission at an amount determined appropriate by the commission, however, prior owing tank fees are still due as previously described. In addition, at such time as the annual facility fee is reinstated by the commission, it will not be assessed for an AST which is owned by a common carrier railroad, as provided in the TWC, §26.344(g).

13. §334.129 Release Reporting and Corrective Action for Aboveground Storage Tanks (ASTs) (89)

(a) An owner or operator of an aboveground storage tank (AST) must comply with the same release reporting, investigation, and corrective action requirements and procedures applicable to underground storage tanks, as prescribed in Subchapter D of this chapter (relating to Release Reporting and Corrective Action), whenever a suspected or confirmed release of a petroleum product from an AST has occurred.

(b) An owner or operator of an AST which stores other petroleum substances or hazardous substances which are not petroleum products must comply with the release reporting, investigation, and corrective action requirements prescribed in the Texas Water Code, Chapter 26, and Subchapter G.

14. §334.130 Reporting and Recordkeeping for Aboveground Storage Tanks (89) (ASTs).

Commentary: *Please make sure you understand and comply with the specific requirements at minimum identified below. Please note that it is not just registration records but also, installation, payment of annual fees, suspected or actual release data, and any other documentation related to compliance to this section for the "operational Life of the AST" per (b)(2) below.*

(a) Reporting. Owners and operators of aboveground storage tanks (ASTs) must assure that all reporting and filing requirements in this subchapter are met, including the following (as applicable):

(1) installation notification in accordance with §334.126 of this title (relating to Installation Notification for Aboveground Storage Tanks (ASTs));

(2) application for approval of any proposed AST in the Edwards Aquifer recharge or transition zones, in accordance with §334.126(3) of the this title and Chapter 313 of this title (relating to Edwards Aquifer);

(3) registration of ASTs and changes in information, in accordance with §334.127 of this title (relating to Registration for Aboveground Storage Tanks (ASTs));

(4) payment of annual facility fees for ASTs, in accordance with §334.128 of this title (relating to Annual Facility Fees for Aboveground Storage Tanks (ASTs));

(5) all reports, plans, and certifications related to actions taken in response to suspected and confirmed releases of petroleum products, in accordance with §334.129 of this title (relating to Release Reporting and Corrective Action for Aboveground Storage Tanks (ASTs)) and Subchapter D of this chapter (relating to Release Reporting and Corrective Action);

(6) any other reports, filings, notifications, or other submittals required by this subchapter, or otherwise required by the agency or the commission to demonstrate compliance with the provisions of this subchapter.

(b) Recordkeeping.

(1) General record keeping requirements.

(A) Except as provided in subparagraph (B) of this paragraph, owners and operators of ASTs must comply with the same general recordkeeping requirements applicable to underground storage tanks, as prescribed in §334.10(b)(1) of this title (relating to Reporting and Recordkeeping).

(B) Owners and operators of movable or mobile ASTs (e.g., skid tanks) may maintain the records for such tanks in accordance with the provisions of §334.127(f)(2) of this title.

(2) Required records and documents. Owners and operators of ASTs must assure that legible copies of all original and amended tank registration documents, as prescribed in §334.127 of this title, are maintained for the operational life of the AST.

15. §334.132 Other General Provisions for Aboveground Storage Tanks (ASTs). (89)

Commentary: *This section can identify several other issues associated with AST construction and placement in any facility. Of special attention among the agencies you must comply with is the Fire Marshal in your area as they* *may require additional requirements beyond what the TCEQ may require and is not abnormal. Of specific note is that this section informs you that the TCEQ or its agent has the right to inspect your tanks and the surrounding area at any reasonable time, tank samples and require you to do additional monitoring and testing if they deem it appropriate.*

(a) Other regulations. Compliance with the provisions of this subchapter by an owner or operator of an aboveground storage tank (AST) must not relieve such owner or operator from the responsibility of compliance with any other laws and regulations directly and/or indirectly affecting such tanks and the stored petroleum products, including, but not necessarily limited to, all applicable regulations legally promulgated by the United States Environmental Protection Agency, United States Occupational Safety and Health Administration, United States Department of Transportation, United States Nuclear Regulatory Commission, United States Department of Energy, Texas Department of Health, State Board of Insurance (including State Fire Marshal), Railroad Commission of Texas, Texas Department of Agriculture, State Comptroller, Texas Department of Public Safety, Texas Natural Resource Conservation Commission and any other federal, state, and local governmental agencies or entities having appropriate jurisdiction.

(b) Responsibilities of owners and operators. The owners and operators of ASTs subject to the provisions of this subchapter are responsible for ensuring compliance with all applicable provisions of this subchapter. Owners and operators are responsible for any violations or noncompliant activities resulting from the actions or inactions by any person who is employed or otherwise engaged by the owner or operator.

(c) Inspections, monitoring, and testing.

(1)For the purposes of developing or assisting in the development of a regulation, conducting a study, or enforcing this subchapter, an owner or operator of an AST, on the request of the agency must:

(A)furnish information related to the tank, including tank equipment and contents; and permit a designated agent or employee of the agency at all reasonable times to have access to and to copy all records relating to the tank.

(1) For the purposes of developing or assisting in the development of a regulation, conducting a study, or enforcing this subchapter, the agency's designated agent or employee may:

(A) enter at reasonable times an establishment or place in which an AST is located;

(B) inspect and obtain samples of a petroleum substance contained in the tank from any person; and

(C) conduct monitoring or testing of the tank, associated equipment, contents, or surrounding soils, air, surface water, or groundwater.

(3) The agency may direct an owner or operator of an AST to conduct monitoring and testing if the agency finds that there is reasonable cause to believe that a release has occurred in the area in which the tank is located.

PART XIX: ENVIRONMENTAL RELATED TRAINING REQUIREMENTS

Table 10 summarizes the typical environmental training activities you will be required to complete for all staff, supervisors, and managers. This table does not represent all possible training requirements; this guide does not include the required Health, Safety and Department of Transportation (DOT) training. Ensure that you have documented training activities for all compliance requirements and that this documentation is available for agency review on request. Training records can be managed with simple spreadsheets, but computer-based training management programs are available at reasonable costs; these programs will track your training and have the added feature of email notification when an employee requires training, as well as what training is required. Staff and contractor training requirements are based on job responsibilities, and that can impact many of your compliance requirements. Many positions may only require an awareness-level training. Job descriptions should note any specific job requirements for compliance/plan training; these descriptions should also be made available for agency review upon request. Environmental training requirements are very different than Health and Safety requirements, which typically impact every staff person in the facility or contractors that perform work for any duration at the facility. Never hesitate to request any contractor to verify his or her individual training record. You must ensure that all staff and contractors are properly trained. Liabilities are significant and you have a well-established duty to protect your employees and the environment as you conduct business. Please note that this is not legal advice, but lessons learned from many years of experience in compliance support—a significant number of incidents and non-compliance issues arise solely due to lack of training.

Regulatory-mandated training based on permit or plan requirements identified previously are repeated here as part of the overall training summary with only additional reference as needed.

SPCC Training - 40 CFR 112.7(f)

SPCC training must be conducted at least annually with all oil-handling facility staff and typically can be completed in 2-3 hours. This is very dependent on facility size and complexity. Identified below are the specific training requirement; review §112.1(b) for more details related to training specifics.

(f) Personnel, training, and discharge prevention procedures.

(1) At a minimum, train your oil-handling personnel in the operation and maintenance of equipment to prevent discharges; discharge procedure protocols; applicable pollution control laws, rules, and regulations; general facility operations; and, the contents of the facility SPCC Plan.

(2) Designate a person at each applicable facility who is accountable for discharge prevention and who reports to facility management.

(3) Schedule and conduct discharge prevention briefings for your oil-handling personnel at least once a year to assure adequate understanding of the SPCC Plan for that facility. Such briefings must highlight and describe known discharges as described in § 112.1(b) or failures, malfunctioning components, and any recently developed precautionary measures."

FRP Training and Qualified Individual - CFR - Title 40 → Chapter I → Subchapter D → Part 112 → Subpart D → §112.21

Commentary: *FRP training typically takes 8 hours. FRP training is much more complex and time consuming and often involves USCG or GLO staff, either by invitation from you or a request by them. Even though having USCG or GLO staff may seem intimidating it can actually can be a very positive experience. You should have your facility "Qualified Individual" prep and manage the training. From the Model FRP Appendix F to Part 112 "The duties of the designated qualified individual are specified in §112.20(h)(3)(ix). The qualified individual's duties must be described and be consistent with the minimum requirements in §112.20(h)(3)(ix). In addition, the qualified individual must be identified with the Facility Infor-*

mation in section 1.2 of the response plan." There are certification classes for facility staff to get the training to become a "Qualified Individual" required by the FRP but is not technically required as long as the person who is designated meets the requirements identified.

FRP Training (37)

§ 112.21 Facility response training and drills/exercises. (37)

(a) The owner or operator of any facility required to prepare a facility response plan under § 112.20 shall develop and implement a facility response training program and a drill/exercise program that satisfy the requirements of this section. The owner or operator shall describe the programs in the response plan as provided in § 112.20(h)(8).

(b) The facility owner or operator shall develop a facility response training program to train those personnel involved in oil spill response activities. It is recommended that the training program be based on the USCG's Training Elements for Oil Spill Response, as applicable to facility operations. An alternative program can also be acceptable subject to approval by the Regional Administrator.

(1) The owner or operator shall be responsible for the proper instruction of facility personnel in the procedures to respond to discharges of oil and in applicable oil spill response laws, rules, and regulations.

(2) Training shall be functional in nature according to job tasks for both supervisory and non-supervisory operational personnel.

(3) Trainers shall develop specific lesson plans on subject areas relevant to facility personnel involved in oil spill response and cleanup.

(c) The facility owner or operator shall develop a program of facility response drills/exercises, including evaluation procedures. A program that follows the National Preparedness for Response Exercise Program (PREP) (see appendix E to this part, section 13, for availability) will be deemed satisfactory for purposes of this section. An alternative program can also be acceptable subject to approval by the Regional Administrator."

Facility Incident Reporting

Even though formal Incident Command System (ICS) training is not a specific requirement, a facility-specific incident management procedure that all employees are aware of should be in place. This includes who is to be notified and who takes specific actions, such as internal and external reporting requirements when an Environmental Health and Safety (EH&S) incident occurs.

Resource Conservation Recovery Act (RCRA)

Most Used Oil Handlers are typically a Very Small Quantity Generator VSQG (formally a Conditional Exempt Small Quantity Generator (CESQG)). The generation levels that put you a specific Generator Class and trigger associated regulations as well as different training requirements. A reference table from the EPA website is provided below to identify your Generator Class based on monthly generation of hazardous waste and followed by training requirements for each. Also from the same website is the EPA general statement on Personnel Training. Training times vary based on Generator Status (e.g., VSQG training typically takes 2 hours, SQG training typically takes 4 hours, and LQG typically takes 8 hours). You must also add field times that vary based on activity and an individual's responsibilities. Please remember that a person cannot perform any RCRA activity by themselves until they are trained and if a LQG training must take place within 6 months of employment or from the date they assumed a responsibility associated with RCRA management activities.

Personnel Training

"Ensure appropriate personnel complete classroom or on-the-job training to become familiar with proper hazardous waste management and emergency procedures for the wastes handled at the facility." (48)

Table 10 – EPA, RCRA Generator Status, EPA Website

≤100 kg/month, and ≤1 kg/month of acute hazardous waste, and ≤100 kg/month of acute spill residue or soil §260.10	>100 and <1,000 kg/month §260.10	≥1,000 kg/month, or >1 kg/month of acute hazardous waste, or >100 kg/month of acute spill residue or soil §260.10

Very Small Quantity Generator

Commentary: *Formal documented training is not required. However, it is strongly recommended that you still provide some sort of training as suggested in the "Personnel Training" statement from the EPA. Your training can be developed along the SQG requirements identified below.*

§262.14 (48) Conditions for exemption for a very small quantity generator.

(a) Provided that the very small quantity generator meets all the conditions for exemption listed in this section, hazardous waste generated by the very small quantity generator is not subject to the requirements of parts 124, 262 (except §§262.10-262.14) through 268, and 270 of this chapter, and the notification requirements of section 3010 of RCRA and the very small quantity generator may accumulate hazardous waste on site without complying with such requirements. The conditions for exemption are as follows:

(3) If the very small quantity generator accumulates at any time greater than 1 kilogram (2.2 lbs.) of acute hazardous waste or 100 kilograms (220 lbs.) of any residue or contaminated soil, water, or other debris resulting from the cleanup of a spill, into or on any land or water, of any acute hazardous waste listed in §§261.31 or 261.33(e) of this chapter, all quantities of that acute hazardous waste are subject to the following additional conditions for exemption:

(i) Such waste is held on site for no more than 90 days beginning on the date when the accumulated wastes exceed the amounts provided above; and

(ii) The conditions for exemption in §262.17(a) through (g).

Commentary: *This is the section that gives the exemption to training requirements found in the section cited as long as you meet the generator class.*

Small Quantity Generator - §262.16(b)(9)(iii) (48)

Identified below is the training requirement but also included are other requirements that need to be addressed as part of your training at a minimum. Make sure you identify specific individual responsibilities for each requirement.

(9) Emergency procedures. The small quantity generator complies with the following conditions for those areas of the generator facility where hazardous waste is generated and accumulated:

(i) At all times there must be at least one employee either on the premises or on call (i.e., available to respond to an emergency by reaching the facility within a short period of time) with the responsibility for coordinating all emergency response measures specified in paragraph (b)(9)(iv) of this section. This employee is the emergency coordinator.

(ii) The small quantity generator must post the following information next to telephones or in areas directly involved in the generation and accumulation of hazardous waste:

(A) The name and emergency telephone number of the emergency coordinator;

(B) Location of fire extinguishers and spill control material, and, if present, fire alarm; and

(C) The telephone number of the fire department, unless the facility has a direct alarm.

(iii) The small quantity generator must ensure that all employees are thoroughly familiar with proper waste handling and emergency procedures, relevant to their responsibilities during normal facility operations and emergencies;

(iv) The emergency coordinator or his designee must respond to any emergencies that arise. The applicable responses are as follows:

(A) In the event of a fire, call the fire department or attempt to extinguish it using a fire extinguisher;

(B) In the event of a spill, the small quantity generator is responsible for containing the flow of hazardous waste to the extent possible, and as soon as is practicable, cleaning up the hazardous waste and any contaminated materials or soil. Such containment and cleanup can be conducted either by the small quantity generator or by a contractor on behalf of the small quantity generator;

(C) In the event of a fire, explosion, or other release that could threaten human health outside the facility or when the small quantity generator has knowledge that a spill has reached surface water, the small quantity generator must immediately notify the National Response Center (using their 24-hour toll free number 800/424-8802). The report must include the following information:

(1) The name, address, and U.S. EPA identification number of the small quantity generator;

(2) Date, time, and type of incident (e.g., spill or fire);

(3) Quantity and type of hazardous waste involved in the incident;

(4) Extent of injuries, if any; and

(5) Estimated quantity and disposition of recovered materials, if any."

Large Quantity Generator - §262.17(a) (48)

Commentary: *As an LQG increased training requirements compared to the other Generator Classes is just one of the many additional requirements you must insure you are in compliance with. This guide does not cover all RCRA requirements for any of the Generator Classes so please refer to the regulations for more guidance. It also suggested that at least one individual as identified by section (7)(B) below "This program must be directed by a person trained in hazardous waste management procedures, and must include instruction which teaches facility personnel hazardous waste management procedures (including contingency plan implementation) relevant to the positions in which they are employed." Often facilities send at least one person for 3rd party training annually to insure they meet the qualified person requirement who then trains others and by going off-site for training normally allows for getting updated on any new or proposed RCRA regulations. Training must be given to each individual at the facility relevant to the positions. Also note that this includes ALL supervisors that supervise those that do any aspect of RCRA compliance even if they don't directly perform any tasks associated with RCRA, this includes managers where it applies.*

(7) Personnel *training*. (i)

(A) Facility personnel must successfully complete a program of classroom instruction, online training (e.g., computer-based or electronic), or on-the-job training that teaches them to perform their duties in a way that ensures compliance with this part. The large quantity generator must ensure that this program includes all the elements described in the document required under paragraph (a)(7)(iv) of this section.

(B) This program must be directed by a person trained in hazardous waste management procedures, and must include instruction which teaches facility personnel hazardous waste management procedures (including contingency plan implementation) relevant to the positions in which they are employed.

(C) At a minimum, the training program must be designed to ensure that facility personnel are able to respond effectively to emergencies by familiarizing them with emergency procedures, emergency equipment, and emergency systems, including where applicable:

(1) Procedures for using, inspecting, repairing, and replacing facility emergency and monitoring equipment;

(2) Key parameters for automatic waste feed cut-off systems;

(3) Communications or alarm systems;

(4) Response to fires or explosions;

(5) Response to ground-water contamination incidents; and

(6) Shutdown of operations.

(D) For facility employees that receive emergency response training pursuant to Occupational Safety and Health Administration regulations 29 CFR 1910.120(p)(8) and 1910.120(q), the large quantity generator is not required to provide separate emergency response training pursuant to this section, provided that the overall facility training meets all the conditions of exemption in this section.

(ii) Facility personnel must successfully complete the program required in paragraph (a)(7)(i) of this section within six months after the date of their employment or assignment to the facility, or to a new position at the facility, whichever is later. Employees must not work in unsupervised positions until they have completed the training standards of paragraph (a)(7)(i) of this section.

(iii) Facility personnel must take part in an annual review of the initial training required in paragraph (a)(7)(i) of this section.

(iv) The large quantity generator must maintain the following documents and records at the facility:

(A) The job title for each position at the facility related to hazardous waste management, and the name of the employee filling each job;

(B) A written job description for each position listed under paragraph (a)(7)(iv)(A) of this section. This description may be consistent in its degree of specificity with descriptions for other similar positions in the same company location or bargaining unit, but must include the requisite skill, education, or other qualifications, and duties of facility personnel assigned to each position;

(C) A written description of the type and amount of both introductory and continuing training that will be given to each person filling a position listed under paragraph (a)(7)(iv)(A) of this section;

(D) Records that document that the training or job experience, required under paragraphs (a)(7)(i), (ii), and (iii) of this section, has been given to, and completed by, facility personnel.

(v) Training records on current personnel must be kept until closure of the facility. Training records on former employees must be kept for at least three years from the date the employee last worked at the facility. Personnel training records may accompany personnel transferred within the same company."

Clean Air Act (CAA) Air Permit by Rule (PBR), TAC Title 30, Part 1, Chapter 106, §106.8. (8)

Commentary: *Training should focus around PBR operational issues, monitoring, and especially records management as indicated below. Also please note that a CAA - Title V air permits are not addressed in this guide and are a significant facility compliance issue that you should seek professional help for identifying permit requirements, submitting a permit and developing a managing and monitoring program.*

(c) Owners or operators of all other facilities authorized to be constructed and operate under a PBR must retain records as follows:

(1) maintain a copy of each PBR and the applicable general conditions of §106.4 of this title or the general requirements, if any, in effect at the time of the claim under which the facility is operating. The PBR and general requirements claimed should be the version in effect at the time of construction or installation or changes to an existing facility, whichever is most recent. The PBR holder may elect to comply with a more recent version of the applicable PBR and general requirements;

(2) maintain records containing sufficient information to demonstrate compliance with the following:

(A) all applicable general requirements of §106.4 of this title or the general requirements, if any, in effect at the time of the claim; and

(B) all applicable PBR conditions;

(3) keep all required records at the facility site. If however, the facility normally operates unattended, records must be maintained at an office within Texas having day-to-day operational control of the plant site;

(4) make the records available in a reviewable format at the request of personnel from the commission or any air pollution control program having jurisdiction;

(5) beginning April 1, 2002, keep records to support a compliance demonstration for any consecutive 12-month period. Unless specifically required by a PBR, records regarding the quantity of air contaminants emitted by a facility to demonstrate compliance with §106.4 of this title prior to April 1, 2002 are not required under this section; and

(6) for facilities located at sites designated as major in accordance with §122.10(13) of this title (relating to General Definitions) or subject to or potentially subject to any applicable federal requirement, retain all records demonstrating compliance for at least five years. For facilities located at all other sites, all records demonstrating compliance must be retained for at least two years. These record retention requirements supersede any retention conditions of an individual PBR.

The National Pollutant Discharge Elimination System (NPDES)

Regulations at 40 CFR 122.44(d), Texas Water Code, Title 2, Subtitle D, Chapter 26, Subchapter A, Texas Pollutant Discharge Elimination System (TPDES), Clean Water Act (CWA) Storm Water Pollution Prevention Plan (SWPPP), Clean Water Act, Section 402: National Pollutant Discharge Elimination System and Publicly Owned Treatment Works (POTW) permits.

Both the NPDES (TPDES in Texas) and SWPPP requirements in compliance with NPDES requirements have similar training requirements but have significant differences in detail requirements to be trained on depending on the permit granted. Where stormwater reflects surface water management from rainfall or other naturally occurring water intrusion into the facility where NPDES permits may also have industrial generated process waters that when discharged off-site must meet certain requirements.

Water going to a Publically Owned Treatment Works (POTW) if you are classified as an industrial source which is typical for any petroleum related facility in Texas you will be either managed or permitted by the POTW or the TCEQ depending on the POTW in your area may or may not require an NPDES (TPDES in Texas) for discharge besides a site specific POTW permit. As stated in the TCEQ POTW Website: "If you are a categorical industrial user (CIU) discharging process wastewater to a publicly owned treatment works (POTW), you will be regulated by either the local POTW who owns the sanitary sewer system or TCEQ. If you are a significant industrial user not subject to federal categorical pretreatment standards, you may still be regulated by your local POTW." The POTW will ultimately be subjected to NPDES / TPDES discharge which is impacted by the POTW capabilities for treatment prior to discharge of your liquids and including others who also discharge to the POTW.

The following represents training topics that need to be addressed based on your specific TPDES, SWPPP, or POTW permits based on TCEQ and EPA guidelines and regulations as cited. It is important to note that ALL permits require you to identify monitoring, record keeping, and facility staff training requirements. Training in all cases should be done annually and prior to any staff taking action associated with any aspect of any permit.

1. Stormwater or any discharge requirement training is required for all employees who work in areas where industrial activities or material handling activities are exposed to stormwater or where process waters for discharge are responsible for implementing activities necessary to meet the conditions of the permit. These employees include inspectors, operator, maintenance personnel, and all members of your Pollution Prevention Team. The training session or sessions are expected to cover the contents of the facility SWPPP, NPDES (TPDES), POTW control measures implemented to achieve compliance with applicable discharge requirements, spill containment and cleanup procedures, maintenance, monitoring, inspection, planning, reporting, and documentation requirements. EPA and TCEQ requires annual training be conducted for any applicable employees at least annually or if any changes to permit conditions change and is required whenever a new

employee starts who is involved with any operations or documentation associated with any aspect of a permit. You should have a sign-in/sign-out sheet at each training class to document that employees have participated. Keep the sign-in/sign-out sheet with your Plan.

The training program will be reviewed annually by the plan coordinator to determine its effectiveness and to make any necessary changes to the program.

2. This education program will include background on the components and goals of the plan and hands-on training in spill prevention and response, good housekeeping, proper material handling, disposal and control of waste, container filling and transfer, and proper storage, washing, and inspection procedures. All new employees will be trained within one week of their start date at not work at any job that has any impact on the plan unaccompanied prior to training.
3. At a minimum, employee training must include:
   a. proper material handling practices for specific chemicals, fluids, and any other materials used or commonly encountered at the facility
   b. spill prevention measures
   c. the location of materials and equipment necessary for spill clean-up
   d. spill cleanup techniques
   e. proper spill reporting procedures
   f. familiarization with good housekeeping measures, Best Management Practices (BMPs), and the overall goals of the plan
   g. When developing the schedule for employee training sessions, you must consider pollutant potential, employee turnover rate, and other factors, as needed."
4. Complete and up-to-date records shall be maintained as needed to demonstrate compliance with §115.142 and §115.143 of this title (relating to Control Requirements; and Alternate Control Requirements) which are sufficient to demonstrate the characteristics of wastewater streams and the qualification for any exemptions claimed under §115.147 of this title (relating to Exemptions).
5. Records shall be maintained of the results of any inspection or monitoring conducted in accordance with §115.144 of this title (relating to Inspection and Monitoring Requirements). Records shall be sufficient to demonstrate proper functioning of applicable control equipment to design specifications to ensure compliance with §115.142 and §115.143 of this title.

6. Records shall be maintained of the results of any testing conducted in accordance with §115.145 of this title (relating to Approved Test Methods).
7. Record Retention: All records shall be maintained at the plant for at least three years or in some permits 5 years or more and be made available upon request to representatives of the executive director, EPA, or any local air pollution control agency having jurisdiction in the area.
8. All plans are ever-changing documents. In order to remain in compliance with the Clean Water Act, POTW permitting program, a plan which details the sites' stormwater management initiatives, or industrial discharge must be developed and maintained throughout the facility. As the facility progresses and goes through changes, plan(s) are required to be revised to reflect those changes.
9. You also must offer education to employees at the facility who are not directly responsible for implementing or maintaining activities identified in the SWPPP, and who do not participate in the employee training program. At a minimum, educate these employees of the SWP3's basic goal and how to contact the facility's Pollution Prevention Team about stormwater related issues.

Used Oil and Waste Sampling

The primary focus for sampling training is normally on requirements identified in your Material Analysis Plan (MAP) as identified in section XV. Material Analysis Plan (MAP) of this guide. However other sampling related to any waste materials (hazardous, non-hazardous, and universal) managed at the site need to have a qualified person sample for characterization, profiling, operational parameters etc.

> ***Commentary:*** *Of significant importance to a Used Oil Handler is the sampling requirements to validate the "Rebuttable Presumption" criteria either for on-site or off-site materials. A program should be developed on-site that is accompanied by standard operating procedures (SOP) identifying specific requirements associated with material types, volumes, containers to be sampled, analysis requirements, chain of custody, documentation requirements, and especially safety requirements. Additional training associated with more challenging aspects of sampling such as sample plans development, interpretation of analysis, and waste characterization normally require the assistance*

of a qualified professional or as often takes place you would send one person off-site to receive formal training and then train others at the facility as needed. What is critical to consistency, and regulatory documentation requirements is that all procedures are written and updated at least every two years or as needed. Any sampling training needs to be documented for each individual and knowledge re-verified every year and re-training as required along with additional documentation. Of note is that a significant amount of training programs exist on line and available on disk but still require some field training to insure skills are complete.

HAZWOPER Training: "Hazardous Waste Operations and Emergency Response" Standard (HAZWOPER), 29 CFR 1910.120 (183)

Commentary: *As a Used Oil Handler, it is often a debated issue in the industry as to the HAZWOPER training requirement. From the Occupational Safety and Health Administration (OSHA) which is the federal governing body over HAZWOPER requirements identifies in Part 1910, Occupational Safety and Health Standards, Subpart H, Subpart Title Hazardous Materials, Standard 1910.120, Title Hazardous waste operations and emergency response. Section* 1910.120(a)(1)(v) *states "Emergency response operations for releases of, or substantial threats of releases of, hazardous substances without regard to the location of the hazard." Even with most understanding focused on the HAZWOPER need by clean-up response groups, first responders, and Hazardous Waste Facilities the standard also identifies facilities that release a hazardous material. Used oil in Texas by itself is not classified as a hazardous material by the US DOT however if the used oil is mixed with something that causes it to be classified as a hazardous material by the US DOT (e.g., it is mixed with gasoline such that the flash point is below 141 degrees F) is a DOT hazardous material but when it is released to the ground it is considered a waste and is often a hazardous waste based on the typical constituents when analysis shows it meets hazardous waste criteria. With material that can be released into the environment as a* *hazardous material and or a hazardous waste it appears HAZWOPER training requirements exist.*

Texas and Federal waste regulations only exempt used oil if it is destined for recycle which is a presumption as stated by the TCEQ Used Oil Handlers Guide "Used oil that does not meet the standards for recycling is regulated as hazardous waste under 30 TAC Chapter 335." Therefor if you release Used Oil into the environment, even on-site, it could be a hazardous waste qualifying your facility under OSHA above to have staff on-site that are HAZWOPER trained to a minimum "*First responder awareness level.*" per 1910.120(q)(6)(i) **or** *First responder operations level* per 1910.120(q)(6)(ii) **identified below** which typically requires 8 hours of training to be certified and annual refresher training. In most interpretations, based on personal discussions, is that they have contractors who are HAZWOPER trained at a higher level that respond as needed. An argument, in this writer's view, that if the regulations clearly identify a HAZWOPER requirement for the facility that a facility would need facility staff trained at all levels.

General HAZWOPER training requirements and responder levels with minimum training times are identified below: *(Note: All citations below are directly from regulations.)*

1910.120(e)(1)(i)

All employees working on site (such as but not limited to equipment operators, general laborers and others) exposed to hazardous substances, health hazards, or safety hazards and their supervisors and management responsible for the site shall receive training meeting the requirements of this paragraph before they are permitted to engage in hazardous waste operations that could expose them to hazardous substances, safety, or health hazards, and they shall receive review training as specified in this paragraph.

1910.120(e)(1)(ii)

Employees shall not be permitted to participate in or supervise field activities until they have been trained to a level required by their job function and responsibility.

1910.120(p)(8)(iii)(A)

Training for emergency response employees shall be completed before they are called upon to perform in real emergencies. Such training shall include the elements of the emergency response plan, standard operating procedures the employer has established for the job, the personal protective equipment to be worn and procedures for handling emergency incidents.

Exception #1: an employer need not train all employees to the degree specified if the employer divides the work force in a manner such that a sufficient number of employees who have responsibility to control emergencies have the training specified, and all other employees, who may first respond to an emergency incident, have sufficient awareness training to recognize that an emergency response situation exists and that they are instructed in that case to summon the fully trained employees and not attempt control activities for which they are not trained.

Exception #2: An employer need not train all employees to the degree specified if arrangements have been made in advance for an outside fully-trained emergency response team to respond in a reasonable period and all employees, who may come to the incident first, have sufficient awareness training to recognize that an emergency response situation exists and they have been instructed to call the designated outside fully-trained emergency response team for assistance.

1910.120(q)(6)(i) *First responder awareness level.*

Competency training usually 8 hrs.

First responder awareness level. First responders at the awareness level are individuals who are likely to witness or discover a hazardous substance release and who have been trained to initiate an emergency response sequence by notifying the proper authorities of the release. They would take no further action beyond notifying the authorities of the release. First responders at the awareness level shall have **sufficient training or have had sufficient experience to objectively demonstrate competency** in the following areas:

1910.120(q)(6)(ii) First responder operations level.- 8 hrs.

First responder operations level. First responders at the operations level are individuals who respond to releases or potential releases of hazardous substances as part of the initial response to the site for the purpose of protecting nearby persons, property, or the environment from the effects of the release. They are trained to respond in a defensive fashion without actually trying to stop the release. Their function is to contain the release from a safe distance, keep it from spreading, and prevent exposures. First responders at the operational level shall have received at least **eight hours of training** or have had sufficient experience to objectively demonstrate competency in the following areas in addition to those listed for the awareness level and the employer shall so certify:

191020(e)(3)(ii) 24 hrs. Training + 1 Day Field Training

Workers on site only occasionally for a specific limited task (such as, but not limited to, ground water monitoring, land surveying, or geophysical surveying) and who are unlikely to be exposed over permissible exposure limits and published exposure limits **shall receive a minimum of 24 hours of instruction off the site**, and the minimum of one day actual field experience under the direct supervision of a trained, experienced supervisor.

1910.120(e)(3)(iii) 24 hrs. + 1 Day Field Training

Workers regularly on site who work in areas which have been monitored and fully characterized indicating that exposures are under permissible exposure limits and published exposure limits where respirators are not necessary, and the characterization indicates that there are no health hazards or the possibility of an emergency developing, **shall receive a minimum of 24 hours of instruction off the site, and the minimum of one day actual field experience** under the direct supervision of a trained, experienced supervisor.

1910.120(e)(3)(iv) 24 hrs. + 16 hrs. + 2 Day Field Training

Workers with 24 hours of training who are covered by paragraphs (e)(3)(ii) and (e)(3)(iii) of this section, and who become general site workers or **who are required to wear respirators, shall have the additional 16 hours and two days of training necessary to total the training specified in paragraph (e)(3)(i).**

1910.120(e)(3)(i) 40 hrs. Training + 3 Days Field Training

General site workers (such as equipment operators, general laborers and supervisory personnel) engaged in hazardous substance removal or other activities which expose or **potentially expose workers to hazardous substances and health hazards shall receive a minimum of 40 hours of instruction off the site, and a minimum of three days actu-**

al field experience under the direct supervision of a trained experienced supervisor.

1910.120(e)(4) Management and Supervision Training 40 hrs. + 3 Days Field Training or as applicable 24 hrs. + 1 Day Field Training

Management and supervisor training. On-site management and supervisors directly responsible for, or who supervise employees engaged in, hazardous waste operations **shall receive 40 hours initial training, and three days of supervised field experience (the training may be reduced to 24 hours and one day if the only area of their responsibility is employees covered by paragraphs (e)(3)(ii) and (e)(3)(iii)) and at least eight additional hours of specialized training at the time of job assignment on such topics as, but not limited to, the employer's safety and health program and the associated employee training program, personal protective equipment program, spill containment program, and health hazard monitoring procedure and techniques.**

1910.120(q)(6)(v) On scene incident commander 24 hrs. + Meet Competency In The Listed Areas

On scene incident commander. Incident commanders, who will assume control of the incident scene beyond the first responder awareness level, shall receive at least 24 hours of training equal to the first responder operations level and in addition have competency in the following areas and the employer shall so certify:

1910.120(q)(6)(v)(A)
Know and be able to implement the employer's incident command system.

1910.120(q)(6)(v)(B)
Know how to implement the employer's emergency response plan.

1910.120(q)(6)(v)(C)
Know and understand the hazards and risks associated with employees working in chemical protective clothing.

1910.120(q)(6)(v)(D)
Know how to implement the local emergency response plan.

1910.120(q)(6)(v)(E)

Know of the state emergency response plan and of the Federal Regional Response Team.

1910.120(q)(6)(v)(F)

Know and understand the importance of decontamination procedures.

1910.120(e)(5)

Qualifications for trainers. Trainers shall be qualified to instruct employees about the subject matter that is being presented in training. Such trainers shall have satisfactorily completed a training program for teaching the subjects they are expected to teach, or they shall have the academic credentials and instructional experience necessary for teaching the subjects. Instructors shall demonstrate competent instructional skills and knowledge of the applicable subject matter.

1910.120(e)(6)

Training certification. Employees and supervisors that have received and successfully completed the training and field experience specified in paragraphs (e)(1) through (e)(4) of this section shall be certified by their instructor or the head instructor and trained supervisor as having completed the necessary training. A written certificate shall be given to each person so certified. Any person who has not been so certified or who does not meet the requirements of paragraph (e)(9) of this section shall be prohibited from engaging in hazardous waste operations.

1910.120(e)(7)

Emergency response. Employees who are engaged in responding to hazardous emergency situations at hazardous waste clean-up sites that may expose them to hazardous substances shall be trained in how to respond to such expected emergencies.

1910.120(e)(8)

Refresher training. Employees specified in paragraph (e)(1) of this section, and managers and supervisors specified in paragraph (e)(4) of this section, shall receive eight hours of refresher training annually on the items specified in paragraph (e)(2) and/or (e)(4) of this section, any critique of incidents that have occurred in the past year that can serve as training examples of related work, and other relevant topics.

1910.120(e)(9)

Equivalent training. Employers who can show by documentation or certification that an employee's work experience and/or training has resulted in training equivalent to that training required in paragraphs (e)(1) through (e)(4) of this section shall not be required to provide the initial training requirements of those paragraphs to such employees and shall provide a copy of the certification or documentation to the employee upon request. However, certified employees or employees with equivalent training new to a site shall receive appropriate, site specific training before site entry and have appropriate supervised field experience at the new site. Equivalent training includes any academic training or the training that existing employees might have already received from actual hazardous waste site experience.

Environmental Records Retention:

> ***Commentary:*** *It is strongly recommended that you create a record retention policy training session that needs to be a formal SOP and company specific. You also assign at least one person as the company Records Coordinator who oversees the policy and typically in a small facility would also do the training of all staff. Typically the training can be accomplished in a one 8 hr. session. Below are typical areas you would address in the training.*

1. Record Retention Training Content

a. **Purpose of Record Retention Policy**

   1. The purpose of this policy is to clarify the requirements for the retention and disposition of all company records.
   2. It is the responsibility of each company to ensure effective record retention management so that records are not arbitrarily destroyed and that legal record-keeping requirements are met.

b. **What is a Record?**

   1. The official records of the company consist of recorded information, created or received by all company employees and contractors in the course of performing

daily operations on behalf of the company and held as evidence of its organization, function, policies, decisions, procedures, operations and internal or external transactions.

c. **Confidential Records**

1. Confidential Records contain any information that is received or created, which includes protected health information or any personal financial information. This includes, but is not limited to, name, address, date of birth, social security number, bank account numbers, financial information, medical information, etc.
2. These records will be protected from disclosure and have special disposition requirements.
3. **<u>These records will be stored at the company office or other designated secure location.</u>**

d. **Electronic Records**

1. The following emails, not listed in the records table, are to be stored and disposed of accordingly:
    a. Emails related to compliance issues will be saved in an accessible format for one year and permanently saved to a disc, specifically emails containing documents or information submitted to a governing agency for consideration, approval or other action.
    b. Emails regarding an employee grievance relating to personnel policies or working conditions are to be considered confidential, will be stored for a minimum of two years and follow confidential disposition requirements.
    c. Emails regarding a specific project must be saved until project completion and then must follow all applicable retention requirements.

e. **Original Facility Records**

1. Any facility records required to be copied, due to the determination by the Director of Compliance or facility manager requirement to maintain a copy at another office, will be copied at the originating facility without exception unless specific approval is given by Facility Management.
2. In no case shall facility staff or EH&S staff remove or authorize another person to remove an original document or single copy from a facility.

f. Record **Categories:**

Identify all environmental and safety documents including, but not limited to, reports, permits, plans, manifests, shipping documents related to an accident or spill, agency communications, logs and E&S training records.

g. **Record Retention Times:**

Retention times need to be record / document specific and insure all compliance related records are addressed and a detailed reference is provided as part of any SOP as well as in any training materials.

h. **Categorizing Your Records:**

Develop a numbering system with associated names that allow for adding records to a specific group without impacting the numbering of all categories.

i. **Tracking Your Records:**

How and who will verify records are being kept correctly and in a timely manner in each department. What is going to be the frequency of any audits on the record management process and who will be doing it,

j. **Storing your Records:**

Specific locations for each record need to be identified and where records are sent to for long term retention. Be sure to identify any record that must remain available for agency review at the site and specifically how the binder or file is labeled for immediate access if required,

Table 11 – Typical Environmental Related Training Activities

Course Description	Typical Training Hours	Regulatory Reference / Comment
Environmental and Response Related Training		
SPCC Plan Review	2-3	40 CFR 112.7(f) - Plan Specific
FRP Plan	8	40 CFR §112.21 - Plan Specific
Facility Incident Report-ing	2	Even though formal Incident Command System (ICS) training is not a specific re-quirement you should have a Facility specific incident management procedure.
RCRA - Very Small Quantity Generator (VSQG)	2	Not required -§262.14 It is recommended that you do have an "Awareness" level courses modeled after the SQG requirements.
RCRA - Small Quantity Generator (SQG)	4	§262.16(b)(9)(iii)
RCRA - Large Quantity Generator (LQG)	8	§262.17(a)(7)
Clean Air Act (CAA) Air Permit By Rule	4	TAC Title 30 Part 1 Chapter 106
Clean Water Act (CWA) Storm Water Pollution Prevention Plan (SWPPP), National Pol-lutant Discharge Elimina-tion System (NPDES)	8	The National Pollutant Discharge Elimina-tion System (NPDES) regulations at 40 CFR 122.44(d), Texas Water Code, Title 2, Subtitle D, Chapter 26, Subchapter A, Texas Pollutant Discharge Elimination Sys-tem (TPDES)
Used Oil and Waste Sam-pling	8	Sampling Techniques focused on your MAP compliance and Waste analysis
ER/First Responder	8	OSHA 29 CFR 1910.120

HAZWOPER Supervisor	16	OSHA 29 CFR 1910.120
HAZWOPER 40 hour	40	OSHA 29 CFR 1910.120
HAZWOPER 24 hour	24	OSHA 29 CFR 1910.120
HAZWOPER 8 hour Refresher	8	OSHA 29 CFR 1910.120
Environmental Records Retention	2	Even though not a specific training course requirement record retention is part of your overall environmental compliance requirements associated with all Plans and Permits and all staff associated with any documentation in support to compliance issues should be aware of requirements and facility specific management requirements.

PART XX: ENFORCEMENT

Commentary: *Enforcement at first thought by most people is that it is a very bad thing and yes you need to avoid an enforcement action but knowing what is enforcement's role within the TCEQ is also important. With that understanding should be a general knowledge of their functions and objectives as well as processes you may be subjected to. You should also know the protections you have under the Texas Environmental, Health, and Safety Audit Privilege Act (Audit Act) as you plan your auditing process that should be part of your normal business practices no matter how big your business is. This section utilizes various sources from the TCEQ and the EPA primarily the EPA Practical Guide to Environmental Management for Small Businesses (2002),* ***(184)*** *Texas Vernon's Civil Statutes* ***(126)*** *Title 71. Health-Public Chapter 1 Health Boards and Laws Art. 4447cc. Environmental, Health, and Safety Audit Privilege Act.,* ***(121)*** *TCEQ "A Guide to the Texas Environmental, Health, and Safety Audit Privilege Act",* ***(21)*** *TCEQ Annual Enforcement Report Fiscal Year 2016, 2015 TCEQ Presentation Compliance and Enforcement: Investigation Overview.* ***(94)*** *As in other Guide sections large redacted portions of the writings are utilized with additional italicized comments added. As always this is not to be implied and not expressed as legal advice in any way and you should always seek assistance by a qualified environmental professional and / or legal advice when involved with any non-voluntary enforcement action. But, as always you should also not hesitate to seek assistance with any questions.*

To avoid enforcement the most obvious reason is to avoid penalties associated with non-compliance-fines for violations can be in the thousands-a severe blow to a small *or large* business (not to mention the black mark of being labeled a violator).

Commentary: *Perhaps the above statement from the EPA Guide is one of the most significant statements regarding impact to your business as once you develop a reputation as a violator it can spiral due to oversight increase because the agencies are there to protect the environment and public as well as your employees. Also, your non-compliance can also impact your customers and what customer would want to continue to do business with you if there was a possibility that they would be swept up in any enforcement action caused by you, even if they are innocent it still drains there resources. In addition to business actions you could have public and private individuals take actions that impact your business in so many ways as identified further by the EPA guide below. How does this happen? Certain agencies at the federal, state, and local levels are there to see that businesses follow environmental regulations. TCEQ and the EPA in order to make this determination, they can send inspectors to any business to see firsthand if it is playing by the rules. Of course, when they find problems, your business will be required to fix them. The agency will send an official letter with the list of the problems or violations. They will spell out what they expect you to do about each violation and a due date that it must be completed by. Depending on the violations, your business can be required to pay monetary fines. The regulator may even close down your business temporarily until problems are fixed. It is possible that the person at your company who has completely failed to take notice of or severely disregarded the regulations may be criminally prosecuted and may even serve jail time. Please note the previous sentence as it is a very real issue and despite what people may tell you my experience, based on 40 years of doing compliance support, is that it DOES HAPPEN and yes even to a Used Oil Handler and nobody wants that impact on them or their family or your business. Often, your business will be re-inspected to make sure the problems have really been fixed. If just dealing with the inspection wasn't enough, your problems can make the news. Ironically the same newspaper that couldn't be bothered mentioning the food drive your business sponsored, will be all too willing to plaster the story on the front page about your business being fined or forced to temporarily close.*

The regulatory agencies keep records of all violations and they are public documents. Anyone can ask to see them-they are allowed to under the "Freedom of Information Act." People are expressing more interest these days in knowing what the companies in their community are doing. The EPA's website is experiencing 40 million hits per month! Real estate search engines can find environmental problems in areas surrounding a property. In this modern day of computer access by everyone and activities by special interest groups it is critical to understand that what happens to your business from a compliance perspective is available to everyone, good and bad. Not only can negative image hurt sales in your local market, it can also hinder any expansion activities. This is because some expansions require an environmental permit before you can proceed. When a company is applying for a new permit, there is a public comment period and often a public hearing. If your company has a bad reputation for environmental harm, members of your community are likely to oppose the permit, causing a delay. If the public is really opposed, the permit may not be issued at all, jeopardizing your expansion plans. This negative image identified above is also a very important reason why to insure you have a positive reputation with your neighbors by being in compliance and responding to concerns expressed by anyone not just the agency can have a real impact on your business now and future plans.

Compliance with environmental regulations will help retain the value of your property. Any negative environmental impact can decrease your property value. Before loaning money, banks usually require buyers to pay for a professional review of previous property uses to see if the property may have been contaminated in some way. So if you ever need to change locations or use your property for collateral, it will become important to be able to show that your activities have not caused contamination. Good documentation that your company is in compliance with environmental regulations can protect you from having to pay for past contamination to your site. Likewise compliance with regulations can lower the cost of liability insurance, because insurance companies will have less concern about the future costs of a clean-up or the risk of harm to the health of your employees and community.

Compliance Perspective

First, not understanding what regulations apply to your business is not an acceptable excuse for failing to meet the requirements. *We all have probably heard and used the expression from Thomas Jefferson "Ignorance of the law is no excuse"* ignorance of the rules does not protect you from enforcement, even criminal enforcement.

Secondly, although an inspection will generate a handy list of things that you need to do, dropping everything to meet the imposed deadlines can cost more than would have been invested to get into compliance in the first place. You will not have the luxury of time to look for the least expensive and disruptive way to work within the rules Keep in mind that if your business seeks assistance with understanding how regulations apply, you may be given some leeway in correcting problems that are found as part of that effort.

Third, finding out what rules apply to your business and getting started on addressing them by putting an environmental management program in place is well worth the effort. You will be able to talk knowledgably about what you have done to meet the requirements and defend your methods as offering equal protection.

Fourth, everyone at your company will be more prepared for an inspection and more organized to follow up on any issues that result. To put it simply, an environmental management program focusing on regulations is a way for your business to emphasize effectiveness, efficiency, and continuous improvement instead of crisis management.

TCEQ Enforcement Activity Review

Based on the TCEQ Annual Enforcement Report Fiscal Year 2016 below are:

1. Milestones and operation issues associated with the Enforcement group.
   a. Fiscal Year (FY) 2016 was a year of significant accomplishments for the TCEQ. Enforcement efforts resulted in 1,404 administrative orders issued with over $8.9 million to be paid as penalties
   b. Compliance Assistance to the regulated community was provided to more than 66,000 Texas small businesses and local governments keeping them informed about changes and trends in environmental regulations to help them comply.
   c. TCEQ provided compliance assistance and technical information to nearly 7,900 Texans through agency-sponsored seminars and workshops.
   d. The agency conducted over 102,000 routine investigations, including 4,873 complaint investigations.

e. Nearly 18,000 Notices of Violation (NOVs) were issued regarding investigations conducted.
f. The voluntary environmental self-audit program resulted in the receipt, evaluation, and response to 2,724 notices of intent to audit and 1,191 disclosures of violation.

2. Those who comply with the conditions of the Texas Environmental, Health, and Safety Audit Privilege Act may qualify for immunity from penalties if swift compliance is achieved.
   a. TCEQ also partners with the Texas Environmental Enforcement Task Force to pursue criminal enforcement. The agency participated in seven search warrants and finalized 19 cases with convictions against 22 individuals and five corporations. The finalized cases included 28 felony counts and 22 misdemeanor counts. These cases resulted in a total of $1,771,600 in criminal fines, over 53 years of community supervision, 178 months of confinement, and over $87,760,000 in restitution.

Commentary: *As the above statement indicates, criminal prosecution does happen.*

   b. Tier II Chemical Reporting Program: Tier II Program initiated compliance actions to correct report deficiencies for 10,155 facilities and resolved 7,400 of those deficiencies.
   c. Results of Enforcement Actions: Enforcement Orders resulted in more than 100 million pounds of pollutants eliminated, reduced, or the routes of exposure reduced, and an estimated cost of more than $71 million that respondents will spend to achieve compliance.
   d. The agency pursues strong and vigorous enforcement, used within an overall strategy for achieving maximum compliance.
   e. The TCEQ uses a broad range of tools to enable and require businesses and governments, large and small, to comply with environmental rules. These tools include:

1) on-site and off-site facility assistance,
2) small business and local government assistance,

3) general education events,
4) regulatory workshops,
5) publications and guidance documents,
6) telephone hotlines, and
7) the agency's expansive Internet site,
8) as well as traditional investigations and enforcement.

Investigations and Enforcement Policy

Strong enforcement also means ensuring that enforcement measures are applied in an equitable manner. Equitable treatment requires that violators not come out ahead economically to the disadvantage of those entities that spend substantial amounts of money to comply with the law. The TCEQ continues to implement the provisions of the enforcement penalty policy which was last revised in April 2014. The cornerstone of this policy emphasizes compliance to protect our citizens from harm, coupled with swift, sure and firm enforcement for those who do not comply. The TCEQ has the authority to levy penalties in 27 environmental program areas. As of September 1, 2011, penalties can be as much as $25,000 per day, per violation for most administrative cases and $25,000 per day, per violation for most civil judicial cases.

1. Application of a minimum penalty of $2,500 and maximum penalty of $25,000 per violation.
2. Application of a strict penalty of $10,000 per day of violation (with an exception), rather than defining a maximum.
3. Providing different maximum penalty amounts based upon the particular violation cited.
4. Some statutes provide for a particular penalty amount total per violation, rather than per Day.
5. When determining the penalty amount, the TCEQ considers the nature, circumstances, extent, duration, and gravity of the violation; the severity of impacts to human health and the environment; compliance history; culpability; good faith efforts to comply; economic benefit; deterrence; and other factors as justice may require. It is important to note that the TCEQ's enforcement activities tend to originate, not in Austin at the Commission's headquarters, but in Texas communities.

Scientific, technical, and environmental specialists in the agency's 16 regional offices are responsible for the vast majority of both investigations conducted and complaints investigated.

6. In FY 2016, the TCEQ conducted over 102,000 investigations of which 61,550 were on-site investigations.
7. The agency investigated 4,873 complaints in FY 2016
8. The number of complaint investigations has continued to increase over the last two years.
9. Investigation time line is typically 1-14 days from contact, 1-14 days for investigation then 60-90 days for findings determined and letter to be issued.
10. Advance notification of routine compliance investigations ensures that proper personnel and records will be available at the time of the investigation. The agency provided advance notice for 14,493 on-site investigations and conducted 44,379 unannounced on-site investigations. Please note that a single notice may be given for an investigation that includes multiple activities that contribute to the total investigation count. Investigations that do not receive advance notification include those that are complaint related, enforcement follow-up investigations, or otherwise warrant an unannounced investigation.
11. The Criminal Element: Texas Environmental Enforcement Task Force (the "Task Force") Texas is a leader and national model in the investigation and prosecution of environmental crime. Created in 1991, the Task Force has included participants from more than 25 local, state, and federal agencies. The TCEQ and particularly the Environmental Crimes Unit play a key investigative role and head the task force. Members of the Task Force include the TCEQ, the Governor's Office, the Texas Parks and Wildlife Department, the OAG, the Texas Railroad Commission, and the General Land Office. These agencies coordinate activities with the U.S. Environmental Protection Agency's Criminal Investigation Division, the U.S. Attorney's Office, the Federal Bureau of Investigations, and local district and county attorneys. Numerous other federal, state, and local agencies also participate in Task Force investigations. The TCEQ Environmental Crimes Unit with assistance from other members of the Task Force also conducts environmental crimes enforcement training for local law enforcement officers. The agency participated in seven search warrants and finalized 19 cases with convic-

tions against 22 individuals and 5 corporations during FY 2016 (see Table 3-1). The finalized cases included 28 felony counts and 22 misdemeanor counts. These cases resulted in a total of $1,771,600 in criminal fines, over 53 years of community supervision, 178 months of confinement, and over $87,760,000 in restitution.

12. The most common tool used in Texas to encourage compliance with environmental laws is the "Notice of Violation," known as the "NOV." An NOV is sent to an entity when it is determined that an environmental violation may have occurred. The NOV is a multi-purpose, flexible tool that is used thousands of times each year. In many cases it is enough to encourage compliance, thereby halting possible damage to the environment. In other cases it can be the first step in a process that ultimately can result in administrative enforcement, civil enforcement, and/or possible criminal charges for violation of environmental laws.

Texas Environmental, Health, and Safety Audit Privilege Act ("Audit Act")

The TCEQ's traditional enforcement efforts have been enhanced by voluntary environmental self-audits conducted at facilities under the Texas Environmental, Health, and Safety Audit Privilege Act ("Audit Act"). This legislation encourages businesses and governments subject to environmental regulation to perform comprehensive assessments of compliance with environmental laws, regulations, and permits for their own facilities. Organizations who participate in the Audit Act are required to notify the TCEQ of their intent to self-audit and then fully disclose and resolve violations resulting from the audit. The TCEQ ensures that all violations disclosed under this program are corrected, and, provided that certain conditions of the Audit Act are complied with, the participants in this program may not be subject to civil and administrative penalties. *The Audit Act statute detail can be found at Vernon's Civil Statutes Title 71. Health – Public Chapter 1. Health Boards and Laws Art. 4447cc. Environmental, Health, and Safety Audit Privilege Act available from* http://www.statutes.legis.state.tx.us/Docs/CV/htm/CV.71.1.htm

This is a guide for those who plan to use the provisions of the Texas Environmental, Health, and Safety Audit Privilege Act (Audit Act). Under the Audit Act, certain documents and information gathered as part of an environmental self-audit are privileged from disclosure. The Audit Act also provides certain immunities from administrative or civil penalties for violations voluntarily disclosed and corrected within a reasonable amount of time.

> ***Commentary:*** *An important part of the process is the submission to the Texas Commission on Environmental Quality of a letter indicating intent to initiate a self-audit and a letter disclosing violations discovered. Content and submission requirements are identified in the reference statute above.*

Please note that, although the Audit Act is applicable to issues within the jurisdiction of other state agencies, or even litigation between private parties, this guide focuses exclusively on the Act as it relates to the TCEQ's jurisdiction.

Administrative Orders

The TCEQ issued 1,404 administrative orders in FY 2016 (see Figure 2-1). The number of orders issued by the TCEQ has remained relatively high over the years due to improvements in efficiencies related to processing enforcement cases.

The 1,404 administrative orders issued in FY 2016 required payments of $8,985,063 in penalties and an additional $3,206,058 was required to be paid for Supplemental Environmental Projects (SEPs). In addition, there were two administrative orders for which stipulated penalties were collected during FY 2016. These orders contain provisions that require payment of stipulated penalty amounts for specified future violations for the duration of the order. During FY 2016, equal amounts were paid in penalties and SEPs ($276,500 each for a total of $553,000) for these two orders.

Texas Tier II Chemical Reporting Program and Texas LEPC

On September 1, 2015, the Texas Tier II Chemical Reporting Program (Tier II Program) was transferred from the Texas Department of State Health Services to the TCEQ pursuant to House Bill 942, enacted by the 84th Texas Legislature and signed into law by the Governor on June 16, 2015.

On March 1, 2016 the Tier II Program finished its first annual reporting period at the TCEQ with 78,302 chemical reports received from the regulated community. As of August 2016, the Tier II Program staff had handled 4,467 phone calls and had completed audits on 78,273, or 99%, of the chemical inventory reports received during this reporting period. The Tier II Program also collected a total of $1.159 million in fees for its first year at TCEQ. The Tier II Program has initiated compliance actions to correct report deficiencies for 10,155 facilities that either submitted partial or incorrect information or did not make the correct fee payments. By August 2016, deficiencies at over 7,400 facilities had been resolved.

As a part of legislative implementation, the TCEQ also created the Texas Local Emergency Planning Committee (LEPC) Grant Program, which began accepting applications from Texas LEPCs on July 22, 2016. The Grant Program will seek to award approximately $4.42 million to Texas LEPCs during its first year in FY 2017, while future grant years will provide roughly $210,000 annually. These grant funds will assist LEPCs in fulfilling their responsibilities under the Emergency Planning and Community Right-to-Know Act.

PART XXI: TIER II REPORTING (EPCRA)

An additional state and federal reporting requirement that most Used Oil Handlers are subject to is Tier II Chemical Reporting. The Tier II program is part of The Emergency Planning & Community Right-to-Know Act EPCRA, 42 U.S.C. 11011 et seq. (1986), **(17)**. Also known as the Superfund Amendments & Reauthorization Act (SARA), and also identified as Title III Texas Law: Texas Community Right-to-Know Acts (TCRAs) Health & Safety Code. **(122)**. The information included in this section is based on various Texas and EPA developed references accessible from their websites; reference these websites for more detail on applicability and filing either Tier II or TRI reports. Typically Used Oil Handlers as addressed in this Guide are not subject to TRI reporting, but please ensure your applicability.

TIER II REPORTING OVERVIEW

The three Texas Community Right-to-Know Acts (TCRAs), Chapters 505-507 of the Health and Safety Code, effective September 1, 1993, are: (122)

1. **Manufacturing Facility Community Right-to-Know Act:** A community right-to-know law that applies to facilities listed under Standard Industrial Classification (SIC) Codes 2011-3999.
2. **Public Employer Community Right-to-Know Act:** A community right-to-know law that applies to Public Employers facilities (i.e., city, county, and state agencies; public schools, colleges, and universities; tax base-supported hospitals; river authorities; volunteer emergency service organizations; and agencies created by state law) listed under SIC Codes 8011-9999. This law is also applicable to water supply and water treatment facilities in SIC Code 4941 that are operated by local government agencies.
3. **Non-Manufacturing Facilities Community Right-to-Know Act:** A community

right-to-know law that applies to all facilities NOT listed under SIC Codes 2011-3999 and that are NOT Public Employers listed under SIC Codes 8011-9999.

Identified requirements for reporting are provided below and can also be found on the TCEQ website. (76)

Which companies might need to submit reports (if the company exceeds the threshold limits)? Tier II reporting companies are "Any company using chemicals that could present a physical or health hazard." Hazardous Chemicals are identified per OSHA's Hazard Communication Standard 29 CFR 1910.1200(c) – Hazardous Chemical: Any chemical which is classified as a physical or health hazard, a simple asphyxiant, combustible dust, pyrophoric gas, or hazard not otherwise classified.

Toxics Release Inventory (TRI) reporting companies are "Companies with 10 employees or more, and hazardous waste treatment, chemical distributors, metal mining, coal mining, petroleum bulk stations (wholesale), and electric utilities. Industry sector must be covered by the NAICS code list found in Table 1 of the December 2012 EPA publication 260-R-10-001, Toxic Chemical Release Inventory Reporting Forms and Instruction, and manufacture, process, or otherwise use more than the triggering amount for that classification of a TRI chemical.

EPCRA and Texas Community Right-To-Know Acts (TCRAs) Overview (110)

EPCRA was passed in 1986 in response to concerns about environmental and safety hazards posed by the storage and handling of toxic chemicals. These concerns were triggered by the disaster in Bhopal, India, in which more than 3,800 people died and adverse health effects in greater than 170,000 survivors resulted from the accidental release of methyl isocyanate.

The Community Right-to-Know Program has been established under both federal and state laws. As a result of these laws, all facilities that store significant quantities of hazardous chemicals must share this information with state and local emergency responders and planners. Facilities in Texas share this information by filing annual hazardous chemical inventories called Texas Tier II Forms with the state, with Local Emergency Planning Committees (LEPCs), and with local fire departments.

The Texas Tier II Reports contain facility identification information and detailed chemical data about hazardous chemicals stored at the facility. Emergency response personnel, such as fire fighters, can use information contained in Texas Tier II Reports to plan response strategies in the event that an emergency situation arises. Private citizens in the community may request and receive copies of the Texas Tier II Reports, as well as custom reports generated from Texas Tier II data.

The TCEQ serves as the state repository for community right-to-know information, provides outreach for compliance on both the federal and state laws, supports Local Emergency Planning Committees (LEPCs) in community right-to-know endeavors, and administers an enforcement program. In 1999, the agency (the Department of Health Services at the time) received and processed over 50,000 Texas Tier II Reports, which are required to be maintained for 30 years. Annual inventory filing fees, which are collected under the TCRAs, provide the funding for both the community and worker right-to-know programs.

The aim of the Texas Tier II Chemical Reporting Program at the TCEQ is to protect the public health and environment by providing current and accurate information about hazardous chemicals and their health effects and by ensuring that the regulated community complies with the requirements of the state and federal community right-to-know laws (Community Right-to-Know for Citizens, Community Planners and Emergency Responders).

Requirements for the Texas Community Right-to-Know Acts (TCRAs, Health & Safety Code, Chapters 505, 506, & 507) (122)

The TCEQ is the state repository for EPCRA-required Emergency Planning Letters (EPLs), which are one-time notifications to the state from facilities that have certain extremely hazardous chemicals in specified amounts.

Tier II Reporting (110)

All Texas facility operators who submit the Texas Tier II Report MUST submit a report in an electronic format that was prepared using the Tier2 Submit software program.

All Texas Tier II Reports received by TCEQ MUST be accompanied by a filing fee. A limited number of facility reports may be consolidated under a single filing fee, if certain criteria are met.

All facilities that submit the Texas Tier II Report to TCEQ MUST submit copies of the report to the local fire department having jurisdiction over the reporting facility and to the Local Emergency Planning Committee (LEPC) for the county or area of the reporting facility.

The following items are reportable:

1. Hazardous Chemical 10,000 pounds or more present at any one time (e.g., diesel fuel is reportable at 10,000 pounds or more)
2. EPA's Extremely Hazardous Substance (EHS) List: 40 CFR Part 355, Appendices A & B
3. Extremely Hazardous Substance (EHS)
4. Threshold Planning Quantity (TPQ) 500 pounds or TPQ, whichever is less (e.g., chlorine gas is reportable at 100 pounds (TPQ))

The following items are not reportable:

1. Hazardous waste
2. Tobacco
3. Wood or wood products
4. Articles (items that do not result in exposure)
5. Food, drugs, cosmetics, or alcoholic beverages in retail
6. Hazardous consumer products in transit sealed and not in a facility longer than 5 days
7. Radioactive waste
8. Items that do not meet threshold amount

When to report:

1. Annual reporting period January 1 – March 1, for previous calendar year Initial Report
2. New chemical onsite, or new facility = 90 days (ammonium nitrate storage facility = 72 hours)
3. Update report when a significant change in information on previously reported chemical (i.e., a change in chemical weight range = 90 days or ammonium nitrate storage facility = 72 hours)

What to report about your facility:

1. Facility information
2. Facility name
3. Facility address
4. Facility lat/long
5. Manned/unmanned
6. North American Industry Classification
7. System (NAICS) Code
8. TCEQ RN/CN (optional)
9. TXT2 Number

What to report about your chemicals:

1. Chemical and location
2. Chemical name/CAS
3. Chemical physical properties/health effects
4. Chemical amount range
5. Extremely hazardous substances (EHS)
6. Specific chemical location within facility (site plan optional)
7. Confidential location

What to report about who is reporting:

1. Contact information
    a. Tier II information contact
    b. Emergency contact (two phone numbers; one phone number must be a 24-hour number)
    c. Owner/operator contact
    d. Emergency coordinator (if an EHS is present)

Where to file your report:

1. TCEQ – Electronic Reporting
2. State Emergency Response Commission (SERC)
4. Local Emergency Planning Committee (LEPC) – list available at www.TexasTier2.org
5. Local fire department – list available at www.TexasTier2.org

Tier II Chemical Reporting

Program contacts

a. TCEQ Critical Infrastructure Division 800-452-2791 (in Texas only) 512-239-5060
b. General questions may be submitted to Tier2Help@tceq.texas.gov
c. Report submissions can be emailed to STEERS at T2Sdatafiles@tceq.texas.gov (www.TexasTier2.org (108))

Tier II Report Filing Fees

Filing fees are collected under the authority provided in the TCRAs (17) and provide the funding for the agency's community and worker right-to-know programs. These funds may also be used by the TCEQ to provide grants to LEPCs. For fee determination calculation, go to https://www.tceq.texas.gov/permitting/tier2/man-fee-calc.

Who Is Using the Tier II Information? (17)

Under state and federal open records laws, any person may submit a request to receive or view a copy of any Tier II Report or other document submitted to the Program under EPCRA or the TCRAs. Any information in the requested documents that is determined to be confidential must be withheld from release to the public. Requests for open records must be submitted in writing and must specify the scope of the records search (ranging from very specific requests for a single facility's current documents to very broad requests based on geographic areas, industry classifications, reporting years, and so forth.). Tier II Report information is also used by emergency response personnel (fire, police, and emergency medical services) in responding to accidents and other community emergencies and by Local Emergency Planning Committees who use this information for developing community emergency response plans.

What Is the State Emergency Response Commission (SERC)?

The State Emergency Response Commission (SERC) is a multi-agency work group charged with performing certain state emergency planning, community right-to-know, and response functions relating to hazardous materials. Every state is required to establish a SERC under the federal Emergency Planning and Community Right-to-Know Act

(EPCRA, also known as SARA, Title III). In Texas, the SERC is a standing element of the Texas Emergency Management Council, a group of more than 20 state agencies, the American Red Cross, and the Salvation Army, that was originally created under the 1975 Texas Disaster Act. The Texas SERC is headquartered at the Texas Department of Public Safety, Division of Emergency Management. As a member agency of the SERC, the TCEQ serves as the state repository for the Tier II Chemical inventory reports.

Tier II Program Transfer

Effective Sept. 1, 2015, Tier II Chemical Reporting Program moved to the TCEQ from the Department of State Health Services. House Bill 942 was passed during the 84th Texas Legislature, Regular Session, and signed into law by the Governor. The bill requires the transfer of the duties authorized by Health and Safety Codes 505, 506, and 507 (Community Right-to-Know Acts or Tier II Program) from DSHS to the TCEQ.

Comparison of Tier II, TRI, and 112(r) Requirements

For more information, contact the TCEQ's Small Business Assistance program at 1-800-447-2827.

Question	Tier II	Toxics Release Inventory (TRI)	Section 112(r)
Who administers this program?	Texas Commission on Environmental Quality (TCEQ) 800/452-2791 or 512/239-5060 Tier II Chemical Reporting Email: Tier2Help@tceq.texas.gov	Texas Commission on Environmental Quality (TCEQ) 512/239-4874 or toxic@tceq.texas.gov and U.S. Environmental Protection Agency (EPA) 800/424-9346 (option 3) or 800/553-7672 (TDD) Toxic Release Inventory Program Overview Exit	U.S. Environmental Protection Agency (EPA) 800/424-9346 or 800/553-7672 (TDD) Risk Management Program Overview Exit Risk Management Program Frequently Asked Questions Exit
What is a general description of the program?	Emergency Planning (the EP in EPCRA): reporting significant quantities of hazardous chemicals maintained at fixed facilities for emergency planning purposes.	Community Right-to-Know (the CRA in EPCRA): reporting on air emissions, waste disposal, and wastewater discharges.	Risk Management for Accident Prevention: responding to emergencies and planning to prevent accidents.
Which companies might need to submit reports (if the company **exceeds the threshold limits**)?	Any company using chemicals that could present a physical or health hazard.	Companies with 10 employees or more, and hazardous waste treatment, chemical distributors, metal mining, coal mining, petroleum bulk stations (wholesale), and electric utilities. Industry sector must be covered by the NAICS code list found in Table 1 of the	Businesses such as: farm cooperatives, chemical manufacturers, chemical distributors, metal product manufacturers, food processors, drinking water systems, and utilities.

Question	Tier II	Toxics Release Inventory (TRI)	Section 112(r)
		December 2012 EPA publication 260-R-10-001, *Toxic Chemical Release Inventory Reporting Forms and Instruction*, and ; manufacture, process, or otherwise use more than the triggering amount for that classification of a TRI chemical.	
What reporting is required?	• One-time Emergency Planning Letters (EPLs) for facilities that have certain hazardous chemicals in specified amounts • Texas Tier II Chemical Inventory Reports • As of August 1, 2006, all EPLs and Tier II Chemical Inventory Reports are required to be submitted in electronic format. TCEQ provides software called Tier2Submit for reporting.	Form A: a shorter report for companies that meet the qualifications for reporting but have lower chemical on-site amounts and off-site transfers. Form R: an inventory of on-site releases and off-site waste transfers of substances (not products) listed under TRI. The EPA requires facilities to use TRI-MEweb to submit TRI reporting forms (except for trade secret information, which facilities will still complete on paper).	Risk Management Plan should include: • hazard assessment • prevention program • emergency response program Submit electronically through RMP*eSubmit
When is the report or plan due?	March 1st of every year for the previous calendar year.	July 1st of every year for the previous calendar year.	June 20, 1999, initially; update every 5 years.
Is there a list of reportable chemicals with thresholds?	Yes; however, there are hazardous chemicals that are not on the threshold list, which the company may need to report.	Yes, on the EPA's TRI-Listed Chemicals site.	Yes, in 40 CFR §68.130

Question	Tier II	Toxics Release Inventory (TRI)	Section 112(r)
How many reportable chemicals are on this threshold list?	There are 356 chemicals on the Extremely Hazardous Substances list; however, 650,000 reportable hazardous substances are **not on the list.**	There are 650 toxic chemicals and chemical categories.	There are 77 acutely toxic substances and 63 flammable gases and volatile liquids.
What are the thresholds?	Each limit represents the **peak storage quantity** of that chemical over a year. • 10,000 lb for 650,000 hazardous substances • 500 lb or the Threshold Planning Quantity (TQP), whichever is less, for 356 chemicals listed on the Extremely Hazardous Substances list	The TRI program classifies chemicals as normal TRI chemicals or as persistent, bioaccumulative toxins (PBT) chemicals. If the company manufactures or processes more than 25,000 lb cumulative total in a year of any one normal TRI chemical, or otherwise uses (such as cleaning machine parts) more than 10,000 lb of any TRI chemical, then the company must report any on-site releases or off-site waste transfers of those chemicals (no minimum for releases or transfers). Lower manufacturing, processing, and otherwise use thresholds for triggering reporting of PBT chemicals are found in section B.4 of the December 2012 EPA publication 260-R-10-001, *Toxic Chemical Release Inventory Reporting Forms and Instruction.*	Each chemical has a different threshold quantity. If the peak storage quantity of any chemical exceeds the limit on the 112(r) list at any point in the year, the company must report to EPA.

PART XXII: WASTE MANAGEMENT IN TEXAS

Waste management in Texas requires compliance with federal and state requirements. Texas has accepted the federal rules under the Resource Conservation Recover Act (RCRA) found in 40 CFR and Texas requirements found in Title 30 of the Texas Administrative Code (TAC).

This guide is not intended to provide all regulations or guidance associated with waste management in Texas, but is intended to provide a general overview of the issues. The redacted information is primarily taken from the documents listed below, as well as the TCEQ websites on the issues with italicized comments. It is strongly recommended that you review the following primary reference documents indicated below for reference on the many complex issues regarding waste management in Texas. A reference table regarding generator requirements is provided from the EPA document referenced; please note that any reference to a Conditionally Exempt Small Quantity Generator (CESQG) is now identified as a Very Small Quantity Generator (VSQG). The changes in the federal waste rules that redesignated CESQG to VSQG among other improvements are identified by the EPA website.

CESQG to VSQG 2017 Change

This final rule became effective on May 30, 2017 and allows a hazardous waste generator to avoid increased burden of a higher generator status when generating episodic waste provided the episodic waste is properly managed. It also allows a VSQG to send its hazardous waste to a large quantity generator under control of the same person.

Primary Reference Documents

It is important to reference federal and state websites as well as the following documents that present a good overview of the generator requirements:

1. RCRA Orientation Manual (2014), available from https://www.epa.gov/sites/production/files/2015-07/documents/rom.pdf (45)
2. Guidelines for the Classification and Coding of Industrial and Hazardous Wastes Texas Commission on Environmental Quality Waste Permits Division RG-022 (revised November, 2014). available from https://www.tceq.texas.gov/assets/public/comm_exec/pubs/rg/rg-022.pdf (100)
3. EPA RCRA Training Module, Introduction to United States Environmental Protection Agency Generators (40 CFR Part 262), available from https://www.epa.gov/sites/production/files/2014-12/documents/gen05.pdf (48)

Resource Conservation and Recovery Act (RCRA) Overview (24)

The Resource Conservation and Recovery Act (RCRA) sets forth an approach for handling the volumes of waste generated in the United States each year. Based on the authority granted by RCRA Subtitle C, EPA developed regulations for the cradle-to-grave management of hazardous waste. Persons who produce hazardous waste, called hazardous waste generators, are the first link in this cradle-to-grave system.

> ***Commentary:*** *Cradle-to-grave is the time period from the initial generation of hazardous waste to its ultimate disposal. It is important to recognize that when you generate and dispose of a hazardous waste, your company (and possibly you personally as well as other staff members) are responsible for any impacts of the material forever, even if you send it to a permitted facility. Even though someone may indemnify you from any costs associated with how your hazardous waste is managed by them, you are still subject to RCRA liability. Your cradle-to-grave liability is why you must insure your hazardous waste is managed effectively.*

The RCRA regulations establish basic hazardous waste management standards for generators. The generator regulations ensure that hazardous waste is appropriately identified and handled safely to protect human health and the environment, while minimizing interference with daily business operations. A solid foundation in the generator regulations is critical to a thorough understanding of the regulations governing the management of hazardous waste from the moment it is produced, or the point of generation, through final disposition.

The RCRA regulations establish a comprehensive hazardous waste management system under the authority of RCRA Subtitle C. RCRA regulates hazardous waste from its point of generation through its point of final disposal. Hazardous waste generators are the first link in the cradle-to-grave hazardous waste management system. Pursuant to the authority granted by RCRA §3002(a), EPA has developed generator standards that address on-site accumulation of hazardous waste, cradle-to-grave tracking (manifest system), labeling, and recordkeeping and reporting requirements. These standards are found in 40 CFR Part 262. Recognizing that generators produce waste in different quantities, Congress established three categories of generators in the statute: large quantity generators (LQGs), small quantity generators (SQGs), and conditionally exempt small quantity generators (CESQGs) *(now Very Small Quantity Generators (VSQG)).* The extent of regulation to which hazardous waste generators are subject depends on the volume of hazardous waste each generator produces.

Hazardous and Solid Waste Determination

40 CFR §262.11 **(162)** requires generators to determine if their wastes are hazardous. The hazardous waste determination requirement applies to all generators of solid waste. Section 262.11 sets forth the following three-step hazardous waste determination process that any person who generates a solid waste, as defined in §261.2, must follow. First, the generator must determine if the solid waste is excluded from RCRA regulation in §261.4. Second, if the waste is not excluded, the generator must determine if it is listed in Subpart D of Part 261. Third, for purposes of compliance with the land disposal restrictions or if the waste is not listed in Subpart D of Part 261, the generator must identify all relevant hazardous waste characteristics in Subpart C of Part 261. The modules entitled Hazardous Waste Identification, Definition of Solid Waste and Hazardous Waste Recycling, and Solid and Hazardous Waste Exclusions together present the process of hazardous waste determination in detail. It is the generator's responsibility to determine if the wastes are defined as hazardous and therefore subject to regulation under Subtitle C. As discussed in Sections 2.3 and 2.4 of this module, once a generator determines a material meets the definition of a hazardous waste, the generator must determine the extent of regulation to which he or she is subject. Generators are not required to perform analytical testing to identify their wastes. They are, however, required to accurately characterize

their wastes. This means they may make a hazardous waste determination by testing *or* applying their knowledge (generator knowledge may be process knowledge and/or analytical testing but must be documented including sampling).) of the waste's chemical and physical properties as specified in §262.11 (e.g., knowledge of the process, inputs, reactions, or operating status for the day).

It is important to note that waste determination starts with the federal rules; if the waste is not a RCRA hazardous waste, you will still need to determine Texas waste classification that will be discussed later in this section. RCRA and its impact on your generator status does turn on hazardous waste, but RCRA also identifies solid waste including municipal solid waste and industrial waste as other discarded materials. This includes solid, semisolid, liquid, or contained gaseous materials resulting from industrial, commercial, mining, agricultural, and community activities (e.g., boiler slags). The definition of solid waste is not limited to wastes that are physically solid. Many solid wastes are liquid, while others are semisolid or gaseous. The term solid waste, as defined by the Statute, is very broad, including not only the traditional nonhazardous solid wastes, such as municipal garbage and industrial wastes, but also hazardous wastes. Hazardous waste, a subset of solid waste, is regulated under RCRA Subtitle C. For purposes of regulating hazardous wastes, EPA established by regulation a separate definition of solid waste.

Federal regulations identify municipal solid waste as a subset of solid waste; it is defined as durable goods (e.g., appliances, tires, batteries), nondurable goods (e.g., newspapers, books, magazines), containers and packaging, food wastes, yard trimmings, and miscellaneous organic wastes from residential, commercial, and industrial non-process sources.

Federal regulations identify industrial waste as another subset of solid waste it is defined as solid waste generated by manufacturing or industrial processes that is not a hazardous waste regulated under Subtitle C of RCRA. Such waste may include, but is not limited to, waste resulting from the following manufacturing processes: electric power generation; fertilizer or agricultural chemicals; food and related products or by-products; inorganic chemicals; iron and steel manufacturing; leather and leather products; nonferrous metals manufacturing or foundries; organic chemicals; plastics and resins manufacturing; pulp and paper industry; rubber and miscellaneous plastic products; stone, glass, clay, and concrete products; textile manufacturing; transportation equipment; and water treatment. Industrial waste does not include mining waste or oil and gas production waste. Note that both municipal and industrial waste may have different definitions and management practices for those materials generated in Texas.

GENERATOR ISSUES

Section 260.10 **(162)** defines a generator as "any person, by site, whose act or process produces hazardous waste identified or listed in Part 261 or whose act first causes a hazardous waste to become subject to regulation." This definition contains three important terms you need to understand to apply the generator regulations. The first term, *by site*, refers to where a hazardous waste is generated. The regulations do not explicitly define the term *site*. EPA tracks hazardous waste generation on a site-specific basis or by "individual generation site." To do this, EPA issues unique identification numbers to identify generators by site. Activities occurring under the control of an owner or operator on a single piece of property should be evaluated collectively for hazardous waste generation. One qualifier to the previous statement is that if you would acquire a piece of property to expand your facility it may very well require you to have two EPA numbers if that property has ever been issued one and then would need to track each site generation separately. For example, if Company A operates three operational areas or laboratories on a single piece of property, all three laboratories may share one EPA ID number, and the waste from all three laboratories may be evaluated together. If, however, Company A operates three laboratories or facilities at three different locations that are not considered to be on contiguous property, each laboratory is viewed as a separate potential generator and is required to obtain an individual EPA ID number. The second term is *person*. Person is defined in §260.10 as "an individual, trust, firm, joint stock company, federal agency, corporation (including a government corporation), partnership, association, state, municipality, commission, political subdivision of a state, or any interstate body." The definition of person encompasses any entity involved with a process that generates hazardous waste. The third key component of the generator definition is the phrase "act or process." Because a generator is defined as the person whose act or process first causes a hazardous waste to become subject to regulation, sometimes the generator of a waste may not necessarily be the person who actually produced the waste. For example, if a cleaning service removes residues from a product storage tank excluded in §261.4(c), the person removing the residues is the first person to cause the waste to become subject to regulation, not the owner of the tank (i.e., the person who produced the waste).

Co-Generator (45)

Co-generators in the above residue removal example, the person removing the waste from the unit is not the owner or operator of the unit, but he or she may be considered a generator. The owner or operator of the unit may also be considered a generator since the act of operating the unit led to the generation of the hazardous waste. In other words, both the remover of the waste and the owner or operator of the tank are considered to be co-generators. In cases where one or more persons meet the definition of generator, all persons are jointly and severally liable for compliance with the generator regulations. The parties may through a mutual decision have one party assume the duties of generator, but in the event that a violation occurs, all persons meeting the definition of generator could be held liable for the improper management of the waste (45 FR 72024, 72026; October 30, 1980).

Generator Status (45)

Congress amended the definition of generator in the Hazardous and Solid Waste Amendments of 1984 (HSWA), requiring EPA to regulate more stringently persons who generate between 100 and 1,000 kg of hazardous waste in a calendar month. On March 24, 1986, EPA issued final regulations establishing a third class of generators (persons generating 100-1,000 kg per month), and narrowed the scope of the conditional exemption to include only persons generating 100 kg or less of hazardous waste in a calendar month (51 FR 10146). Generators who produce 1 kg or less of an acute hazardous waste were also granted the conditional exemption.

Generators now fall into one of three general groups according to the amount of waste generated in a calendar month. The three classes of generators are described in Table 1. Since the regulations become increasingly stringent as the volume of waste generated grows, accurate counting of the waste is critical. This complex counting issue will be addressed in the next section of the module.

1. Large Quantity Generator **(45)**: "LQGs produce 1,000 kg or more of hazardous waste per calendar month, or more than 1 kg of acutely hazardous waste (i.e., waste codes denoted with the hazard code "H" and all P-listed wastes). Wastes from these generators are subject to full regulation in Part 262.
2. Small Quantity Generator **(45)**: "SQGs produce between 100 and 1,000 kg of hazardous waste per calendar month and are subject to modified regulations found in

Part 262. Generally, SQGs must comply with some but not all of the regulations that apply to LQGs. Simplified requirements are specified for SQGs in place of some LQG requirements.

3. Very Small Quantity Generator **(45)**: "VSQG produce 100 kg or less of hazardous waste per calendar month. In addition, generators who produce 1 kg or less of acutely hazardous waste, or 100 kg or less of contaminated soil, waste, or debris resulting from the cleanup of an acute hazardous waste spill, VSQGs are exempt from Parts 262 through 270 if they comply with the requirements in §261.5.

Episodic Generators (45)

Generators may periodically exceed or fall below their normal generation limits in any given calendar month. If the amount of waste generated in a given calendar month places the generator in a different category, he or she is responsible for complying with all applicable requirements of that category for all waste generated during that calendar month. For example, if a generator produces 300 kg of hazardous waste in March, that waste must be managed in accordance with the SQG regulations; if the same generator produces 1,500 kg of hazardous waste in April, that waste must be managed in accordance with the LQG regulations (51 FR 10146, 10153; March 24, 1986).

Compliance Issues

Regarding other waste management issues, you should be aware of a list below is provided but complete definitions are found in the reference documents identified above. The listing below is not intended to be all the associated requirements with your waste management activities. Your level and method of compliance is impacted by your generator status.

1. COUNTING WASTE – *Identify what waste counts toward your generator status.*
2. ACCUMULATION STANDARDS – *Identify how long, types of storage and other requirements associated with your hazardous waste accumulation.*
3. TREATMENT WITHOUT A PERMIT – *Types of waste treatment you can perform without a permit.*
4. PRE-TRANSPORT REQUIREMENTS – *Activities associated with transportation of hazardous waste that includes, requirement for an* ***EPA*** *Identification*

Number, Uniform Hazardous Waste Manifest (EPA FORM 8700-22), Designated Facility identification, and permissible "On-Site" Transportation.

5. REPORTING AND RECORDKEEPING – *Reporting and recordkeeping requirements associated with your generator status.*
6. EXPORT AND IMPORT REQUIREMENTS - *Contains several provisions that apply to exporters and importers of hazardous waste. Even though unlikely you should be aware of the associated issues if you consider this option.*
7. WASTE MINIMIZATION - One of the mandates of RCRA is to reduce or eliminate the generation of hazardous waste as expeditiously as possible (RCRA §1003(b)). Hazardous waste generators, when preparing a manifest, are required to certify that they have taken steps to minimize the amount of hazardous waste that they generate. LQGs must certify that they have "a program in place" to reduce the volume and toxicity of the hazardous waste they generate; SQGs must certify that they have made a good faith effort to minimize their waste generation. EPA outlined six basic elements that should be included in a waste minimization program in the May 28, 1993, Federal Register (58 FR 31114). *Beyond the regulatory requirement it is good business to do all that you can to minimize any waste generated no matter hazardous or not as besides minimizing compliance burden and overall cost for disposal it is the responsible thing to do as an environmentally responsible company.*

Table 12 – EPA Generator Compliance Requirement Summary (Note: CESQG is now VSQG, EPA Website)

5. GENERATOR SUMMARY CHART

	CESQG	SQG	LQG
Quantity Limits	≤ ≤100 kg/month ≤≤ 1 kg/month of acute hazardous waste ≤≤ 100 kg/month of acute spill residue or soil §§261.5(a) and (e)	between 100-1,000 kg/month §262.34(d)	≥1000 kg/month > 1 kg/month of acute hazardous waste > 100 kg/month of acute spill residue or soil Part 262 and §261.5(e)
EPA ID Number	Not required §261.5	Required §262.12	Required §262.12
On-Site Accumulation Quantity	≤≤ 1,000 kg ≤ ≤1 kg acute ≤ ≤100 kg acute spill residue §§261.5(f)(2) and (g)(2)	≤ 6000 kg §262.34(d)(1)	No Limit
Accumulation Time Limits	None §261.5	≤180 days or ≤270 days (if > 200 miles) §§262.34(d) and (e)	≤90 days §262.34(a)
Storage Requirements	None §261.5	Basic requirements with technical standards for tanks or containers §§262.34(d)(2) and (3)	Full compliance for management of tanks, containers, drip pads, or containment buildings §262.34(a)
Off-site Management of Waste	State approved or RCRA permitted/interim status facility §§261.5(f)(3) and (g)(3)	RCRA permitted/interim status facility §262.20(b)	RCRA permitted/interim status facility §262.20(b)
Manifest	Not required §261.5	Required §262.20	Required §262.20
Biennial Report	Not required §261.5	Not required §262.44	Required §262.41
Personnel Training	Not required §261.5	Basic training required §262.34(d)(5)(iii)	Required §262.34(a)(4)
Contingency Plan	Not required 261.5	Basic plan §262.34(d)(5)(i)	Full plan required §262.34(a)(4)
Emergency Procedures	Not required §261.5	Required §262.34(d)(5)(iv)	Required §262.34(a)(4)
DOT Transport Requirements	Yes (if required by DOT)	Yes §§262.30-262.33	Yes §§262.30-262.33

TEXAS WASTE MANAGEMENT ISSUES

Even though Texas has adopted the federal waste regulations, it has various state-specific requirements you need to be aware of as a generator and as a person offering waste for transport. Again, refer to the guidance identified above for more detail. Below are only a few of the state-specific issues to point out the types of differences present.

1. Solid Waste Classification Process: The solid waste classification process, as defined by the Resource Conservation and Recovery Act (RCRA) and associated state laws/rules, is not easy. In order to classify your waste as "hazardous, nonhazardous, industrial class 1, 2, or 3, special waste, or excluded" you need all the information you can possibly gather about the waste you are classifying. Below is a chart from the TCEQ guidance that shows the classification of Texas waste materials. As noted, you need to identify your waste from a RCRA perspective rather than as a nonhazardous waste; Texas has Industrial Class 1, Class 2, and Class 3 and Nonhazardous that require specific waste code application, disposal, permitting, and reporting requirements.

Flowchart 7 – TCEQ Texas Hazardous and Nonhazardous Wastes

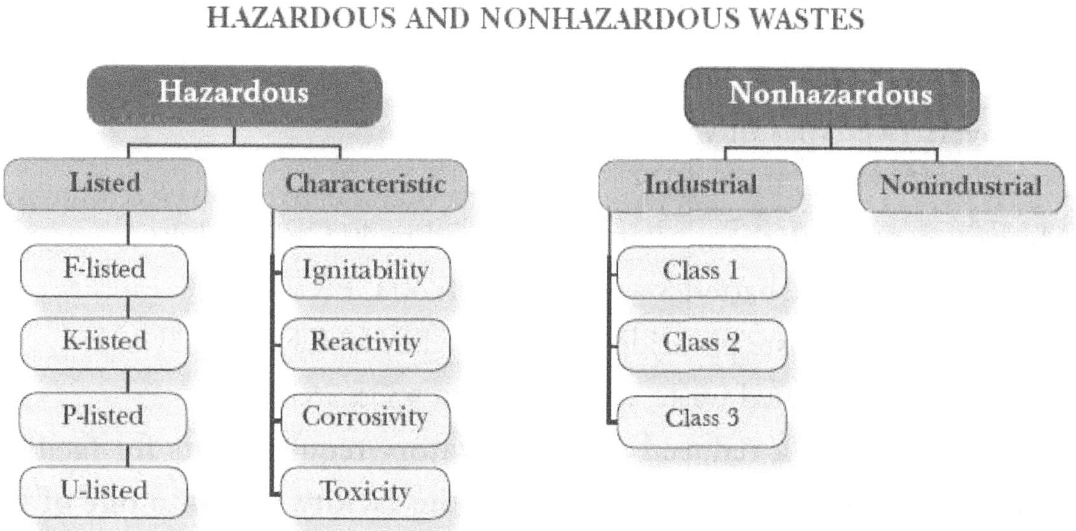

Waste code required for transportation documents and recordkeeping and reporting are referred to as Texas waste code (30 TAC Section 335.503). This 8-digit code identifies a waste stream. The first 4 digits are the sequence number, the next 3 digits are the form code, and the last digit is the waste's classification (sequence number + form code + classification code = waste code). (Some of the "digits" referred to here actually may be letters or a combination of letters and numbers.)

Texas Regulated Generators (30 TAC Chapter 335 Subchapters A and C): If you generate the following amounts of waste, you are a regulated generator and must follow regulations in Chapter 335: If you generate less than the amounts shown below, you are considered a Conditionally Exempt Small-Quantity Generator and are not subject to regulations requiring notification, manifesting, and fees.

Table 13 – TCEQ Generator Amounts Identifying Regulated Status

Waste Type	Monthly Amount
Class 1	100 kg (220 lbs) or more
hazardous	100 kg (220 lbs) or more
acutely hazardous	1 kg (2.2 lbs) or more

Universal Waste (30 TAC Section 335.261 and 40 CFR Part 273)

This rule covers five types of waste:

1. Lamps as described in 40 CFR §273.5, and §335.261(b)(16)(F);
2. Mercury-containing thermostats as described in 40 CFR 273.4;
3. All hazardous waste batteries as described in 40 CFR 273.2;
4. Some hazardous waste pesticides as described in 40 CFR 273.3; and
5. Paint and paint-related waste as described in §335.262(b).

The rule establishes a reduced set of regulatory requirements for facilities managing universal waste, depending on whether the facility falls into one of four categories:

1. small-quantity handler of universal waste (SQHUW),
2. large-quantity handler of universal waste (LQHUW),
3. transporter of universal waste, or
4. final destination facilities.

In addition, the rules establish a petitioning procedure whereby additional wastes may be added to the universal waste rule.

PART XXIII: CLEAN WATER ACT (CWA)

The Clean Water Act (CWA) can have a significant impact on your Used Oil management in Texas. Your facility operational permits typically will include some form of water management permit that is either controlled or impacted by the requirements under the CWA.

Typical permits and requirements are a Texas NPDES (TNPDES), Storm Water Pollution Prevention Plan (SWPPP), Public Owned Treatment Works (POTW), and associated retention (evaporation) pond effluent management.

This guide is not intended to provide all the requirements for each of the activities above and it is strongly suggested that you refer to the following primary references used for the content below that include italicized comments. Additionally, a listing of typical term definitions are provided as reference.

Primary CWA References

1. Federal EPA website, located at https://www.epa.gov/laws-regulations/summary-clean-water-act (13)
2. Federal, EPA, Copeland C., 2016, EPA - Clean Water Act: A Summary of the Law, available from https://fas.org/sgp/crs/misc/RL30030.pdf (11)
3. Federal, EPA, Introduction to the Clean Water Act, available from https://cfpub.epa.gov/watertrain/pdf/modules/introtocwa.pdf (26)
4. Federal, 33 U.S.C., Chapter 26, Sections 1251-1387 (131)
5. Federal, United States Department of Agriculture, National Resources Conservation Services, Clean Water Act, available from https://www.nrcs.usda.gov/Internet/FSE_DOCUMENTS/stelprdb1187162.pdf (130)
6. Federal, EPA, 2009, Developing Your Stormwater Pollution Prevention Plan: A Guide for Industrial Operators, available from

https://www3.epa.gov/npdes/pubs/industrial_swppp_guide.pdf (15)

7. Texas, TCEQ, General Permit to Discharge Under the Texas Pollutant Discharge Elimination System (TPDES), available from https://www.tceq.texas.gov/assets/public/permitting/stormwater/txr050000.pdf (82)
8. Texas, Texas Water Code (TWC) §26.001 (127)
9. Texas, Title 30 Texas Administrative Code (TAC) Chapter 305 (124)
10. Texas, GENERAL PERMIT TO DISPOSE OF WASTEWATER under provisions of Chapter 26 of the Texas Water Code and 30 Texas Administrative Code Chapter 205 Wastewater generated from industrial or water treatment facilities located in the state of Texas may be disposed of by evaporation from surface impoundments adjacent to water in the state only according to wastewater limitations, monitoring requirements, and other conditions set forth in this general permit, as well as the rules of the Texas Commission on Environmental Quality (TCEQ), the laws of the State of Texas, and other orders of the Commission of the TCEQ (Commission). Available from https://www.tceq.texas.gov/assets/public/permitting/wastewater/wqg100000_draft_permit.pdf (99)
11. Texas, Evaporation Pond General Permit, TPDES General Permit No. WQG100000, effective March 28, 2014 – March 28, 2019; available from https://www.tceq.texas.gov/assets/public/permitting/wastewater/general/wqg100000_draft.pdf (97)
12. Texas, TCEQ, 2014, Fact Sheet and Executive Director's Preliminary Decision General Permit WQG100000 (authorizes wastewater generated by industrial or water treatment facilities to be disposed of by evaporation from surface impoundments adjacent to water in the state) available from https://www.tceq.texas.gov/assets/public/permitting/wastewater/general/wqg100000_factsheet.pdf (98)

Clean Water Act (CWA) Overview (13)

The principal law governing pollution of the nation's surface waters is the Federal Water Pollution Control Act, or Clean Water Act. Originally enacted in 1948, it was totally revised by amendments in 1972 that gave the act its current dimensions. The 1972 legislation spelled out ambitious programs for water quality improvement that have since been expanded and are still being implemented by industries and municipalities. The act has been termed a technology-forcing statute because of the rigorous demands placed on those who are regulated by it to achieve higher and higher levels of pollution abatement under deadlines specified in the law. Early on, emphasis was on controlling discharges of conventional pollutants (e.g., suspended solids or bacteria that are biodegradable and occur naturally in the aquatic environment), while control of toxic pollutant discharges has been a key focus of water quality programs more recently. Prior to 1987, programs were primarily directed at point source pollution, that is, wastes discharged by industrial and municipal facilities from discrete sources such as pipes and outfalls. Amendments to the law in that year authorized measures to address nonpoint source pollution (runoff from farm lands, forests, construction sites, and urban areas), which is estimated to represent more than 50% of the nation's remaining water pollution problems. The act also prohibits discharge of oil and hazardous substances into U.S. waters. Under this act, federal jurisdiction is broad, particularly regarding establishment of national standards or effluent limitations. Certain responsibilities are delegated to the states, and the act embodies a philosophy of federal-state partnership in which the federal government sets the agenda and standards for pollution abatement, while states carry out day-to-day activities of implementation and enforcement. To achieve its objectives, the act is based on the concept that all discharges into the nation's waters are unlawful, unless specifically authorized by a permit, which is the act's principal enforcement tool. The law has civil, criminal, and administrative enforcement provisions and also permits citizen suit enforcement. Financial assistance for constructing municipal sewage treatment plants and certain other types of water quality improvements projects is authorized under Title VI. It authorizes grants to capitalize State Water Pollution Control Revolving Funds, or loan programs. States contribute matching funds, and under the revolving loan fund concept, monies used for wastewater treatment construction are repaid to states, to be available for future construction in other communities.

The Federal Water Pollution Control Act of 1948 was the first comprehensive statement of federal interest in clean water programs, and it specifically provided state and local governments with technical assistance funds to address water pollution problems, including research. Water pollution was viewed as primarily a state and local problem, hence, there were no federally required goals, objectives, limits, or even guidelines. When it came to enforcement, federal involvement was strictly limited to matters involving interstate waters and only with the consent of the state in which the pollution originated. During the latter half of the 1950s and well into the 1960s, water pollution control programs were shaped by four laws that amended the 1948 statute. They dealt largely with federal assistance to municipal dischargers and with federal enforcement programs for all dischargers. During this period, the federal role and federal jurisdiction were gradually extended to include navigable intrastate, as well as interstate, waters. Water quality standards became a feature of the law in 1965, requiring states to set standards for interstate waters that would be used to determine actual pollution levels and control requirements. By the late 1960s, there was a widespread perception that existing enforcement procedures were too time-consuming and that the water quality standards approach was flawed because of difficulties in linking a particular discharger to violations of stream quality standards. Additionally, there was mounting frustration over the slow pace of pollution cleanup efforts and a suspicion that control technologies were being developed but not applied to the problems. These perceptions and frustrations, along with increased public interest in environmental protection, set the stage for the 1972 amendments. The 1972 statute did not continue the basic components of previous laws as much as it set up new ones. It set optimistic and ambitious goals, required all municipal and industrial wastewater to be treated before being discharged into waterways, increased federal assistance for municipal treatment plant construction, strengthened and streamlined enforcement, and expanded the federal role while retaining the responsibility of states for day-to-day implementation of the law. The 1972 legislation declared as its objective the restoration and maintenance of the chemical, physical, and biological integrity of the nation's waters. Two goals also were established: zero discharge of pollutants by 1985 and, as an interim goal and where possible, water quality that is both "fishable" and "swimmable" by mid-1983. While those dates have passed, the goals remain, and efforts to attain them continue."

The Clean Water Act (CWA) Today

CWA today consists of two parts, broadly speaking, one being the Title II and Title VI provisions, which authorize federal financial assistance for municipal sewage treatment plant construction. The other is regulatory requirements, found throughout the act, that apply to industrial and municipal dischargers. The act has been termed a technology-forcing statute because of the rigorous demands placed on those who are regulated by it to achieve higher and higher levels of pollution abatement. Industries were given until July 1, 1977, to install "best practicable control technology" (BPT) to clean up waste discharges. Municipal wastewater treatment plants were required to meet an equivalent goal, termed "secondary treatment," by that date. (Municipalities unable to achieve secondary treatment by that date were allowed to apply for case-by-case extensions up to July 1, 1988. According to EPA, 86% of all cities met the 1988 deadline; the remainder were put under administrative or court-ordered schedules requiring compliance as soon as possible.

However, many cities continue to make investments in building or upgrading facilities needed to achieve secondary treatment, and funding needs remain high; see discussion below.) Cities that discharge wastes into marine waters were eligible for case-by-case waivers of the secondary treatment requirement, where sufficient showing could be made that natural factors provide significant elimination of traditional forms of pollution and that both balanced populations of fish, shellfish, and wildlife and water quality standards would be protected. The primary focus of BPT was on controlling discharges of conventional pollutants, such as suspended solids, biochemical oxygen demanding materials, fecal coliform and bacteria, and pH. These pollutants are substances that are biodegradable (i.e., bacteria can break them down), occur naturally in the aquatic environment, and deplete the dissolved oxygen concentration in water, which is necessary for fish and other aquatic life. The act also mandated greater pollutant cleanup than BPT by no later than March 31, 1989, generally requiring that industry use the "best available technology" (BAT) that is economically achievable. BAT level controls generally focus on toxic substances. Compliance extensions of as long as two years are available for industrial sources utilizing innovative or alternative technology. Failure to meet statutory deadlines could lead to enforcement action (see below).

The CWA utilizes both water quality standards and technology-based effluent limitations to protect water quality. Technology-based effluent limitations are specific numerical limitations established by EPA and placed on certain pollutants from certain sources. They are applied to industrial and municipal sources through numerical effluent limitations in discharge permits issued by states or EPA (see discussion of "Permits, Regulations, and Enforcement," below). Water quality standards are standards for the overall quality of water. They consist of the designated beneficial use or uses of a waterbody (recreation, water supply, industrial, or other), plus a numerical or narrative statement identifying maximum concentrations of various pollutants that would not interfere with the designated use. The act requires each state to establish water quality standards for all bodies of water in the state. These standards serve as the backup to federally set technology-based requirements by indicating where additional pollutant controls are needed to achieve the overall goals of the act. In waters where industrial and municipal sources have achieved technology-based effluent limitations, yet water quality standards have not been met, dischargers may be required to meet additional pollution control requirements. For each of these waters, the act requires states to set a total maximum daily load (TMDL) of pollutants at a level that ensures that applicable water quality standards can be attained and maintained. A TMDL is both a planning process for attaining water quality standards and a quantitative assessment of pollution problems, sources, and pollutant reductions needed to restore and protect a river, stream, or lake. Based on state reports, EPA estimates that more than 40,000 U.S. waters are impaired and require preparation of TMDLs. Control of toxic pollutant discharges has been a key focus of water quality programs. In addition to the BPT and BAT national standards, states are required to implement control strategies for waters expected to remain polluted by toxic chemicals even after industrial dischargers have installed the best available cleanup technologies required under the law. Development of management programs for these post-BAT pollutant problems was a prominent element in the 1987 amendments and is a key continuing aspect of CWA implementation. Prior to the 1987 amendments, programs in the Clean Water Act were primarily directed at point source pollution, wastes discharged from discrete and identifiable industrial and municipal sources, such as pipes and other outfalls. In contrast, except for general planning activities, little attention had been given to nonpoint source pollution (runoff of stormwater or snowmelt from agricultural lands, forests, construction sites, and urban areas), despite estimates that it represents more than 50% of the nation's remaining water pollution problems. As it travels across land surface towards rivers and streams, rainfall and snowmelt runoff picks up pollutants, including sediments, toxic materials, and conventional wastes (e.g., nutrients) that can degrade water quality. The 1987 amendments authorized measures to address such pollution by directing states to develop and implement nonpoint pollution management programs (Section 319 of the act). States were encouraged to pursue groundwater protection activities as part of their overall nonpoint pollution control efforts. Federal financial assistance was authorized to support demonstration projects and actual control activities. These grants may cover up to

60% of program implementation costs.

The CWA provides for special regulation of the discharge of oil or hazardous substances, because of the potentially catastrophic effects of such events on public health and welfare. Section 311 prohibits the discharge of oil or hazardous substances into U.S. waters. It also requires higher standards of care in the management and movement of oil, including a requirement for spill prevention plans; it enables the government to recover the costs of cleaning up oil and hazardous substance discharges; and it provides for penalties for such discharges. In 1990, Congress enacted the Oil Pollution Act, which partially amended Section 311 and established a comprehensive system for the cleanup of oil spills, adding a mechanism to impose liability for such spills. While the act imposes great technological demands, it also recognizes the need for comprehensive research on water quality problems. This is provided throughout the statute, on topics including pollution in the Great Lakes and Chesapeake Bay, in-place toxic pollutants in harbors and navigable waterways, and water pollution resulting from mine drainage. The act also authorizes support to train personnel who operate and maintain wastewater treatment facilities.

Federal and State Responsibilities Under this act, federal jurisdiction is broad, particularly regarding establishment of national standards or effluent limitations. The EPA issues regulations containing the BPT and BAT effluent standards applicable to categories of industrial sources (such as iron and steel manufacturing, organic chemical manufacturing, petroleum refining, and others). Certain responsibilities can be assumed by qualified states, in lieu of EPA, and this act, like other environmental laws, embodies a philosophy of federal-state partnership in which the federal government sets the agenda and standards for pollution abatement, while states carry out day-to-day activities of implementation and enforcement. Responsibilities under the act that may be carried out by qualified states include authority to issue discharge permits to industries and municipalities and to enforce permits; 46 states have been authorized to administer this permit program. EPA issues discharge permits in the remaining states—Idaho, Massachusetts, New Hampshire, New Mexico—the District of Columbia, and most U.S. territories. In addition, as noted above, all states are responsible for establishing water quality standards.

Permits, Regulations, and Enforcement

To achieve its objectives, the CWA embodies the concept that all discharges into the nation's waters are unlawful, unless specifically authorized by a permit. Thus, more than

65,000 conventional industrial and municipal dischargers must obtain permits from EPA (or qualified states) under the act's National Pollutant Discharge Elimination System (NPDES) program (authorized in Section 402 of the act). NPDES permits also are required for more than 150,000 industrial and municipal sources of stormwater discharges. An NPDES permit requires the discharger (source) to attain technology-based effluent limits (BPT or BAT for industry, secondary treatment for municipalities, or more stringent where needed for water quality protection). Permits specify the effluent limitations a discharger must meet, and the deadline for compliance. Sources also are required to maintain records and to carry out effluent monitoring activities. Permits are issued for up to five years and must be renewed thereafter to allow continued discharge. The NPDES permit incorporates numerical effluent limitations issued by EPA.

The initial BPT limitations focused on regulating discharges of conventional pollutants, such as bacteria and oxygen-consuming materials. The more stringent BAT limitations emphasize controlling toxic pollutants—heavy metals, pesticides, and other organic chemicals. In addition to these limitations applicable to categories of industry, EPA has issued water quality criteria for more than 115 pollutants, including 65 named classes or categories of toxic chemicals, or "priority pollutants." These criteria recommend ambient, or overall, concentration levels for the pollutants and provide guidance to states for establishing water quality standards that will achieve the goals of the act.

A separate type of permit is required to dispose of dredged or fill material in the nation's waters, including wetlands. Authorized by Section 404 of the act, this permit program is administered by the U.S. Army Corps of Engineers, subject to and using EPA's environmental guidance. Some types of activities are exempt from permit requirements, including certain farming, ranching, and forestry practices which do not alter the use or character of the land; some construction and maintenance; and activities already regulated by states under other provisions of the act. EPA may delegate certain Section 404 permitting responsibility to qualified states and has done so twice (Michigan and New Jersey). For some time, the act's wetlands permit program has been one of the most controversial parts of the law. Some who wish to undertake development projects in wetlands maintain that federal regulation intrudes on and impedes private land-use decisions, while environmentalists seek more protection for remaining wetlands and limits on activities that are authorized to take place in wetlands. Nonpoint sources of pollution, which EPA and states believe are responsible for the majority of water quality impairments in the nation, are not subject to CWA permits or other regulatory requirements under federal law. They are covered by state programs for the management of runoff, under Section 319 of the act. Other EPA regulations under the CWA include guidelines on using and disposing of sewage sludge and guidelines for discharging pollutants from land-based sources into the ocean. (A related law, the Ocean Dumping Act, 33 U.S.C. §§1401-45, regulates the intentional disposal of wastes into ocean waters.) EPA also provides guidance on technologies that will achieve BPT, BAT, and other effluent limitations. The NPDES permit, containing effluent limitations on what may be discharged by a source, is the act's principal enforcement tool. EPA may issue a compliance order or bring a civil suit in U.S. district court against persons who violate the terms of a permit. The penalty for such a violation can be as much as $25,000 per day. Stiffer penalties are authorized for criminal violations of the act—for negligent or knowing violations—of as much as $50,000 per day, three years' imprisonment, or both. A fine of as much as $250,000, 15 years in prison, or both, is authorized for "knowing endangerment"—violations that knowingly place another person in imminent danger of death or serious bodily injury.

Finally, EPA is authorized to assess civil penalties administratively for certain well-documented violations of the law. These civil and criminal enforcement provisions are contained in Section 309 of the act. EPA, working with the Army Corps of Engineers, also has responsibility for enforcing against entities who fail to obtain or comply with a Section 404 permit. While the CWA addresses federal enforcement, the majority of actions taken to enforce the law are undertaken by states, both because states issue the majority of permits to dischargers and because the federal government lacks the resources for day-to-day monitoring and enforcement. Like most other federal environmental laws, CWA enforcement is shared by EPA and states, with states having primary responsibility.

However, EPA has oversight of state enforcement and retains the right to bring a direct action where it believes that a state has failed to take timely and appropriate action or where a state or local agency requests EPA involvement. Finally, the federal government acts to enforce against criminal violations of the federal law. In addition, individuals may bring a citizen suit in U.S. district court against persons who violate a prescribed effluent standard or limitation or permit requirement. Citizens also may bring suit against the Administrator of EPA for failure to carry out a nondiscretionary duty under the act."

CWA primary used oil facility impacts focus on your NPDES (TPDES in Texas) and associated SWPPP. The following additional explanation of both issues for your reference:

NPDES (TPDES)

The Clean Water Act prohibits anybody from discharging "pollutants" through a "point source" into a "water of the United States" unless they have an NPDES permit. The permit will contain limits on what you can discharge, monitoring and reporting requirements, and other provisions to ensure that the discharge does not hurt water quality or people's health. In essence, the permit translates general requirements of the Clean Water Act into specific provisions tailored to the operations of each person discharging pollutants.

The term point source is also defined very broadly in the Clean Water Act because it has been through 25 years of litigation. It means any discernible, confined and discrete conveyance, such as a pipe, ditch, channel, tunnel, conduit, discrete fissure, or container.

> ***Commentary:*** *Point source understanding as it relates to your specific facility is critical to your compliance so if you have any questions seek agency or other professional advice.*

It also includes vessels or other floating craft from which pollutants are or may be discharged.

It depends on where you send your pollutants. If you discharge from a point source into the waters of the United States, you need an NPDES permit. If you discharge pollutants into a municipal sanitary sewer system, you do not need an NPDES permit, but you should ask the municipality about their permit requirements. If you discharge pollutants into a municipal storm sewer system, you may need a permit depending on what you discharge. You should ask the NPDES permitting authority. NPDES permits are issued by Texas that has obtained EPA approval.

A National Pollutant Discharge Elimination System (NPDES) individual permit is written to reflect site-specific conditions of a single discharger (or in rare instances to multiple co-permittees) based on information submitted by that discharger in a permit application and is unique to that discharger whereas an NPDES general permit is written to cover multiple dischargers with similar operations and types of discharges based on the permit writer's professional knowledge of those types of activities and discharges. Individual permits are issued directly to an individual discharger whereas a general permit is issued to no one in particular with multiple dischargers obtaining coverage under that general permit after it is issued, consistent with the permit eligibility and authorization provisions. As such, dischargers covered under general permits know their applicable requirements before obtaining coverage under that permit. Furthermore, obtaining coverage under a general permit is typically quicker than an individual permit with coverage under a general permit often occurring immediately (depending on how the permit is written) or after a short waiting period. Coverage under an individual permit may take six months or longer.

National Pollutant Discharge Elimination System (NPDES) general permits do not require that Operators "apply" for coverage; rather, general permits typically rely on the submission of a document called a Notice of Intent (NOI). An NOI differs from an individual permit application in that it is submitted by Operators after the general permit is issued by the permitting authority. An NOI for a general permit is a notice to the NPDES permitting authority (TCEQ) of an Operator's intent to be covered under a general permit, and typically contains basic information about the Operator and the planned discharge for which coverage is being requested.

Yes. The NPDES administrative procedures require that the public be notified and allowed to comment on NPDES permit applications. When EPA authorizes a state to issue NPDES permits, EPA requires that the state provide the public with this same access. *(As indicated, public availability to participate in the CWA permitting process and obtain information on your facility is another reason to insure you are doing compliance correctly.)* If any member of the general public finds that a facility is violating its NPDES permit, that member can independently start a legal action, unless EPA or the state regulatory agency has taken an enforcement action.

There are various methods used to monitor NPDES permit conditions. The permit will require the facility to sample its discharges and notify EPA and the state regulatory agency of these results. In addition, the permit will require the facility to notify EPA and the state regulatory agency when the facility determines it is not in compliance with the requirements of a permit. EPA and state regulatory agencies also will send inspectors to companies in order to determine if they are in compliance with the conditions imposed under their permits.

2. CWA Storm Water Pollution Prevention Plan (SWPPP):
   a. In 1987, the Water Quality Act (WQA) added provisions to the CWA that allowed the EPA to govern storm water discharges from industrial activities. EPA published the final notice for Phase I of the Multi-Sector General Storm Water Permit program (Federal Register Volume 60 No. 189, September 20, 1995, page 50804) in 1995 which included provisions for the development of a Storm Water Pollution Prevention Plan (SWPPP) by each industrial facility discharging storm water, including automobile salvage yards. Development, implementation, and maintenance of the SWPPP.

SWPPP

The SWPPP will provide tools to reduce pollutants contained in storm water discharges and comply with the requirements of the General Storm Water Permit issued by the State of Texas. The primary goals of the SWPPP will be to:

1) Identify potential sources of pollutants that affect storm water discharges from the site;
2) Describe the practices that will be implemented to prevent or control the release of pollutants in storm water discharges; and
3) Create an implementation schedule to ensure that the practices described in this SWPPP are in fact implemented and to evaluate the plan's effectiveness in reducing the pollutant levels in storm water discharges
4) This SWPPP includes all of the following:
   a) Identification of the SWPPP coordinator with a description of this person's duties;
   b) Identification of the SWPPP implementation team members;
   c) Description of the facility including information regarding the facility's location and activities as well as a site description, three maps, and a summary of the storm water drainage system;
   d) Identification of potential storm water contaminants;
   e) Description of storm water management controls and various Best Manage-

ment Practices (BMPs) necessary to reduce pollutants in storm water discharge;

f) Description of the facility monitoring plan; and a
g) Description of the implementation schedule and provisions for amendment of the plan.

PART XXIV: CLEAN AIR ACT (CAA)

The Clean Air Act (CAA) can have various impacts on a Use oil Handler (UOH) in Texas from Title V permits that are the most significant for UOH that require significant support by environmental specialists to file and maintain may require a full time environmental specialists (who would also presumably be managing many other environmental, health and safety issues) on site but it is rare for a typical small Texas UOH unless you are a processor, re-refiner or have a used oil filter treatment process that requires a Title V. What is common is a Permit by Rule (PBR) permit associated typical oil storage and oil filter management associated with heater use. Also included in many facilities are boiler and or other heater permits that are also covered under PBR. Requirements for any CAA permit in Texas is significant and you should seek professional help when any permit is required. Below is a listing of Primary References utilized for this section and it is recommended that you review applicable documents as your determine your CAA facility compliance requirements. Also includes are a brief CAA overview, limited set of terms you should be familiar with and sections from the TCEQ regulations and Websites associated with various aspects of CAA management in Texas.

Primary CAA References

1. Texas - Texas Commission on Environmental Quality Storage Tank and Change of Service Air Permits by Rule (PBR) Checklist Title 30 Texas Administrative Code § 106.478 available from https://www.tceq.texas.gov/assets/public/permitting/air/Forms/PermitsByRule/Checklists/10144.pdf (107)
2. Texas - TCEQ Website - Air PBR 106.183: Boilers, Heaters, and Other Combustion Devices, available from https://www.tceq.texas.gov/permitting/air/permitbyrule/subchapter-g/big_boilers.html (92)

3. Texas – TCEQ - The State of Texas Environmental Electronic Reporting System (STEERS) can be found at https://www3.tceq.texas.gov/steers/.(105)
4. Texas – Health and Safety Code Title 5, Sanitation and Environmental Quality, Subtitle C Air Quality, Chapter 382, Clean Air Act, Subchapter A, General Provisions (123)
5. Federal - EPA - The Plain English Guide to the Clean Air Act, available from https://www.epa.gov/sites/production/files/2015-08/documents/peg.pdf (9)
6. Texas – TCEQ - Permit by Rule Applicability Checklist (Title 30 Texas Administrative Code § 106.4) available from https://www.tceq.texas.gov/assets/public/permitting/air/Forms/PermitsByRule/Checklists/10149.pdf (107)
7. Federal - 42 U.S.C. United States Code, 2010 Edition Title 42 – The Public Health and Welfare Chapter 85, Air Pollution Prevention and Control, available from https://www.gpo.gov/fdsys/pkg/USCODE-2010-title42/html/USCODE-2010-title42-chap85.htm (128)
8. Federal, EPA CAA Website, available from https://www.epa.gov/clean-air-act-overview/clean-air-act-text#what (8)
9. Federal, EPA Website, Clean Air Act Title V – Permits, available from https://www.epa.gov/clean-air-act-overview/clean-air-act-title-v-permits (10)
10. Texas, TCEQ – Website, Types of New Source Review Authorizations, available from https://www.tceq.texas.gov/permitting/air/guidance/authorize.html (111)
11. Texas – TCEQ - Fact Sheet - Air Quality Permitting, available from http://tceq.net/permitting/air/guidance/newsourcereview/permit_fact5_09.html/at_download/file (93)
12. Texas - Boilers, Heaters, and Other Combustion Devices Air Permits by Rule (PBR) Checklist Title 30 Texas Administrative Code § 106.183, available from https://www.tceq.texas.gov/permitting/air/permitbyrule/subchapter-g/big_boilers.html (92)

Clean Air Act (CAA) Overview (9)

The Clean Air Act (CAA) is the comprehensive federal law that regulates air emissions from stationary and mobile sources. Among other things, this law authorizes EPA to establish National Ambient Air Quality Standards (NAAQS) to protect public health and public welfare and to regulate emissions of hazardous air pollutants.

One of the goals of the Act was to set and achieve NAAQS in every state by 1975 in order to address the public health and welfare risks posed by certain widespread air pollutants. The setting of these pollutant standards was coupled with directing the states to develop state implementation plans (SIPs), applicable to appropriate industrial sources in the state, in order to achieve these standards. The Act was amended in 1977 and 1990 primarily to set new goals (dates) for achieving attainment of NAAQS since many areas of the country had failed to meet the deadlines.

Section 112 of the Clean Air Act addresses emissions of hazardous air pollutants. Prior to 1990, CAA established a risk-based program under which only a few standards were developed. The 1990 Clean Air Act Amendments revised Section 112 to first require issuance of technology-based standards for major sources and certain area sources. "Major sources" are defined as a stationary source or group of stationary sources that emit or have the potential to emit 10 tons per year or more of a hazardous air pollutant or 25 tons per year or more of a combination of hazardous air pollutants. An "area source" is any stationary source that is not a major source.

For major sources, Section 112 requires that EPA establish emission standards that require the maximum degree of reduction in emissions of hazardous air pollutants. These emission standards are commonly referred to as "maximum achievable control technology" or "MACT" standards. EPA is required to review those standards periodically to determine whether any residual risk exists for that source category and, if necessary, revise the standards to address such risk.

EPA's Role Under the Clean Air Act

EPA sets limits on certain air pollutants, including setting limits on how much can be in the air anywhere in the United States. This helps to ensure basic health and environmental protection from air pollution for all Americans. The Clean Air Act also gives EPA the authority to limit emissions of air pollutants coming from sources like chemical plants, utilities, and steel mills. Individual states or tribes may have stronger air pollution laws, but they may not have weaker pollution limits than those set by EPA. EPA must approve state, tribal, and local agency plans for reducing air pollution. If a plan does not meet the necessary requirements, EPA can issue sanctions against the state and, if necessary, take over enforcing the Clean Air Act in that area. EPA assists state, tribal, and local agencies by providing research, expert studies, engineering designs, and funding to support clean air progress. Since 1970, Congress and the EPA have provided several billion dollars to the states, local agencies, and tribal nations to accomplish this.

1. EPA's mission is to protect human health and the environment. To achieve this mission, EPA implements a variety of programs under the Clean Air Act that focus on:
    a. reducing outdoor, or ambient, concentrations of air pollutants that cause smog, haze, acid rain, and other problems;
    b. reducing emissions of toxic air pollutants that are known to, or are suspected of, causing cancer or other serious health effects; and
    c. phasing out production and use of chemicals that destroy stratospheric ozone.
2. These pollutants come from stationary sources (like chemical plants, gas stations, and power plants) and mobile sources (like cars, trucks, and planes).

State and Local Governments' Role

It makes sense for state and local air pollution agencies to take the lead in carrying out the Clean Air Act. They are able to develop solutions for pollution problems that require special understanding of local industries, geography, housing, and travel patterns, as well as other factors. State, local, and tribal governments also monitor air quality, inspect facilities under their jurisdictions and enforce Clean Air Act regulations. States have to develop State Implementation Plans (SIPs) that outline how each state will control air pollution under the Clean Air Act. A SIP is a collection of the regulations, programs and policies that a state will use to clean up polluted areas. The states must involve the public and industries through hearings and opportunities to comment on the development of each state plan.

Selected example terms to be familiar with regarding the CAA from federal and Texas regulation and guidance. Where terms are materially different both federal and state are listed.

This listing is not intended to represent all terms with which you may need to be familiar.

1. Texas - De Minimis Facilities/Sources - Facilities/sources that meet the condi-

tions of 30 TAC § 116.119(a) are considered to be de minimis and do not have to obtain any registration or authorization prior to construction.

2. Texas - Flexible Permits - A flexible permit allows an owner/operator more flexibility in managing the operations by staying under an overall emissions cap or individual emission limitation. The owner/operator is allowed to structure the flexible permit to best serve their needs.
3. Federal - Federal Clean Air Act - The Federal Clean Air Act (FCAA) is the legal foundation for the national air pollution control program. The FCAA requires each state to produce and regularly update a SIP. The FCAA also requires that SIPs include a description of control strategies, or measures to deal with pollution, for areas that fail to achieve national ambient air quality standards (NAAQS). Finally, this Act grants powers of enforcement to the EPA.
4. Texas - Federal Source - A facility, group of facilities, or other source that is subject to the permitting requirements of Title IV or V of the federal Clean Air Act Amendments of 1990 (Pub.L. No. 101-549)
5. Federal - Major Source - The term "major source" means any stationary source (or any group of stationary sources located within a contiguous area and under common control) that is either of the following: (A) A major source as defined in section 7412 of this title. (B) A major stationary source as defined in section 7602 of this title or part D of subchapter I of this chapter.
6. Texas - New Source Review (NSR) Permits - Owners/operators with facilities that do not qualify for PBRs or storage facility can submit a NSR permit application (30 TAC Chapter 116). The pre-construction permitting requirements consist of an administrative review and a technical review. Administrative review will take less than 30 days for a complete application. Once administratively complete, the applicant publishes notices in a local newspaper and post signs around the proposed facility location. Publication starts a 30-day comment period. If a public hearing request is received within the period, the applicant may be required to undergo a second public notice which by statute is a 30- day notice period. The technical review primarily relates to source identification and air emission quantification, analysis of the off-property health impacts of those emissions, determination of best available control technology, and applicability of any source category or emission based state and federal regulations.

7. Texas - Nonattainment Permits - If the facility is located in a non-attainment area, designated by U.S. Environmental Protection Agency, additional permitting requirements may apply. Non-attainment permit review is required if the facility has emissions above the major source threshold for the specific county designated as non-attainment. Nonattainment permitting requires the installation of lowest achievable control technology emission controls, the acquisition of emission reductions to offset the proposed emissions increases, and a mandatory second public notice.
8. Texas - Permits by Rule - Facilities that will not make a significant contribution of air contaminants to the atmosphere if operated or constructed with certain restrictions.
9. Texas - Permit by Rule Authorizations - If the operation emits less than 250 tons per year (tpy) of carbon monoxide (CO) or nitrogen oxides (NOx); or less than 25 tpy of total particulate matter (PM10), sulfur dioxide (SO2), total volatile organic compounds (VOC); or 25 tpy of any other air contaminant except carbon dioxide, water, nitrogen, methane, ethane, hydrogen, and oxygen then the operation may qualify for permit-by-rule (PBR) requirements, contained in 30 TAC Chapter 106. There are 120 individual PBRs which may be claimed. Note that the facility must meet all the established PBR requirements to claim a PBR.
10. Texas - Permits to Construct - Describes what type of permit is needed before construction can begin, and provides links to those permits, which provide forms and any guidance available.
11. Texas - Portable Permit Relocation and Change of Locations - Portable permits contain "relocation conditions" which typically contain control requirements and future site authorization instructions for seeking approval from the appropriate regional office and distance limitations.
12. Texas - Prevention of Significant Deterioration (PSD) Permits - If the facility is a major stationary source, per Title 40 Code of Federal Regulations § 52.21(b)(1)(i), a PSD permit will be required. The PSD review will require additional modeling to determine if the new emissions will have a negative impact on the National Ambient Air Quality Standards of the surrounding air quality. The PSD review has a mandatory second public notice.

13. Texas - Standard Permits - Standard permits are authorized under 30 TAC Chapter 116, Subchapter F and are available for the following project types: Animal carcass incinerators, boilers, concrete batch plants, concrete batch plants with enhanced controls, electric generating units, municipal solid waste landfills, oil and gas facilities, pollution control projects, sawmills, temporary and permanent rock and concrete crushers, and temporary and permanent hot mix asphalt plants. Owners/operators with facilities that meet the established standard permit criteria may qualify for a standard permit.
14. Texas - State Implementation Plan (SIP) is the state's comprehensive plan to clean the air and meet federal air quality standards.
15. Texas - Title V Federal Operating Permits - If the facility is major, per 30 TAC Chapter 122, the owner/operator of the facility must submit an abbreviated Title V permit application before start-up. The TCEQ shall inform the applicant in writing of the deadline for submitting the remaining information. Note, that submittal of a Title V permit application will not affect start-up of the facility.

Texas PBR Regulations and Reporting

The State of Texas Environmental Electronic Reporting System (STEERS) is a mandatory system in Texas for CAA reporting but is also available for Texas PBR permit filing. The reporting electronic system which is available at
https://www3.tceq.texas.gov/steers/STEERS

skips the Air Permits Initial Review Team (APIRT), which could allow for a quicker permit approval. STEERS, or ePermits, is a system where notifications, certifications, and registrations can be submitted electronically.

Storage Tank Permit: This PBR (§ 106.478) requires registration for storage tanks with a capacity of 25,000 gallons or greater and located in a designated ozone nonattainment area with the commission's Office of Air in Austin before construction begins. The registration shall include a list of all tanks, calculated emissions for each compound in tons per year for each tank, and a Table 7 for each different tank design. The facility may be registered by completing Form PI-7, "Registration for Permits by Rule," or Form PI-7-CERT, "Registration and Certification for Permits by Rule." This checklist should accompany the registration form.

Boilers, heaters, drying or curing ovens, furnaces, or other combustion units that meet the conditions of Title 30, Texas Administrative Code (30 TAC), Section 106.183, are permitted by rule. To qualify for this permit by rule (PBR), your facility must also meet the general conditions stated in 30 TAC Section 106.4. This PBR does not include stationary internal combustion engines or turbines. To find out whether your facility meets the general conditions stated in Section 106.4, use our PBR Applicability Checklist, form TCEQ-10149. To find out whether it meets the specific conditions for this PBR, complete our PBR 106.183 Checklist, TCEQ Form -10114, If your facility cannot comply with the conditions of this PBR, see Types of New Source Review Authorizations for other options. *Important following requirement* - This PBR does not require you to register or pay a fee. Under 30 TAC **Section 106.8**, your records must identify the rules under which your facility is operating and demonstrate compliance. This PBR requires you to maintain at least these records for a rolling 2-year period:

a. §106.183 – Boilers, Heaters, and Other Combustion Devices

Boilers, heaters, drying or curing ovens, furnaces, or other combustion units, but not including stationary internal combustion engines or turbines are permitted by rule, provided that the following conditions are met.

(1) The only emissions shall be products of combustion of the fuel.

(2) The maximum heat input shall be 40 million British thermal unit (Btu) per hour with the fuel being:

- (A) sweet natural gas;
- (B) liquid petroleum gas;
- (C) fuel gas containing no more than 0.1 grain of total sulfur compounds, calculated as sulfur, per dry standard cubic foot; or
- (D) combinations of the fuels in subparagraphs (A) - (C) of this paragraph.

(3) Distillate fuel oil shall be fired as a backup fuel only. Firing shall be limited to 720 hours per year. The fuel oil shall contain less than 0.3% sulfur by weight and shall not be blended with waste oils or solvents.

(4) All gas fired heaters and boilers with a heat input greater than ten million Btu per hour (higher heating value) shall be designed such that the emissions of nitrogen oxides shall not exceed 0.1 pounds per million Btu heat input.

(5) Records of hours of fuel oil firing and fuel oil purchases shall be maintained on-site on a two-year rolling retention period and made available upon request to

the commission or any local air pollution control agency having jurisdiction.

1. TAC Title 30, Part 1, Chapter 106, Subchapter A, §106.4 – Requirements for Permitting by Rule per Subchapter A.

(a) To qualify for a permit by rule, the following general requirements must be met.

(1) Total actual emissions authorized under permit by rule from the facility shall not exceed the following limits, as applicable:

(A) 250 tons per year (tpy) of carbon monoxide (CO) or nitrogen oxides (NOX);

(B) 25 tpy of volatile organic compounds (VOC), sulfur dioxide (SO2), or inhalable particulate matter (PM);

(C) 15 tpy of particulate matter with diameters of 10 microns or less (PM10);

(D) 10 tpy of particulate matter with diameters of 2.5 microns or less (PM2.5); or

(E) 25 tpy of any other air contaminant except:

(i) water, nitrogen, ethane, hydrogen, and oxygen; and

(ii) not withstanding any provision in any specific permit by rule to the contrary, greenhouse gases as defined in §101.1 of this title (relating to Definitions).

(2) Any facility or group of facilities, which constitutes a new major stationary source, as defined in §116.12 of this title (relating to Nonattainment and Prevention of Significant Deterioration Review Definitions), or any modification which constitutes a major modification, as defined in §116.12 of this title, under the new source review requirements of the Federal Clean Air Act (FCAA), Part D (Nonattainment) as amended by the FCAA Amendments of 1990, and regulations promulgated thereunder, must meet the permitting requirements of Chapter 116, Subchapter B of this title (relating to New Source Review Permits) and cannot qualify for a permit by rule under this chapter. Persons claiming a permit by rule under this chapter should see the requirements of §116.150 of this title (relating to New Major Source or Major Modification in Ozone Nonattainment Areas) to ensure that any applicable netting requirements have been satisfied.

(3) Any facility or group of facilities, which constitutes a new major stationary

source, as defined in 40 Code of Federal Regulations (CFR) §52.21, or any change which constitutes a major modification, as defined in 40 CFR §52.21, under the new source review requirements of the FCAA, Part C (Prevention of Significant Deterioration) as amended by the FCAA Amendments of 1990, and regulations promulgated thereunder because of emissions of air contaminants other than greenhouse gases, must meet the permitting requirements of Chapter 116, Subchapter B of this title and cannot qualify for a permit by rule under this chapter. Notwithstanding any provision in any specific permit by rule to the contrary, a new major stationary source or major modification which is subject to Chapter 116, Subchapter B, Division 6 of this title due solely to emissions of greenhouse gases may use a permit by rule under this chapter for air contaminants that are not greenhouse gases. However, facilities or projects which require a prevention of significant deterioration permit due to emissions of greenhouse gases may not commence construction or operation until the prevention of significant deterioration permit is issued.

(4) Unless at least one facility at an account has been subject to public notification and comment as required in Chapter 116, Subchapter B or Subchapter D of this title (relating to New Source Review Permits or Permit Renewals), total actual emissions from all facilities permitted by rule at an account shall not exceed 250 tpy of CO or NOX ; or 25 tpy of VOC or SO2 or PM; or 15 tpy of PM10 ; or 10 tpy of PM2.5 ; or 25 tpy of any other air contaminant except water, nitrogen, ethane, hydrogen, oxygen, and GHGs (as specified in §106.2 of this title (relating to Applicability)).

(5) Construction or modification of a facility commenced on or after the effective date of a revision of this section or the effective date of a revision to a specific permit by rule in this chapter must meet the revised requirements to qualify for a permit by rule.

(6) A facility shall comply with all applicable provisions of the FCAA, §111 (Federal New Source Performance Standards) and §112 (Hazardous Air Pollutants), and the new source review requirements of the FCAA, Part C and Part D and regulations promulgated thereunder.

(7) There are no permits under the same commission account number that contain a condition or conditions precluding the use of a permit by rule under this

chapter.

(8) The proposed facility or group of facilities shall obtain allowances for NOX if they are subject to Chapter 101, Subchapter H, Division 3 of this title (relating to Mass Emissions Cap and Trade Program).

(b) No person shall circumvent by artificial limitations the requirements of §116.110 of this title (relating to Applicability).

(c) The emissions from the facility shall comply with all rules and regulations of the commission and with the intent of the Texas Clean Air Act (TCAA), including protection of health and property of the public, and all emissions control equipment shall be maintained in good condition and operated properly during operation of the facility.

(d) Facilities permitted by rule under this chapter are not exempted from any permits or registrations required by local air pollution control agencies. Any such requirements must be in accordance with Texas Health and Safety Code, §382.113 and any other applicable law.

PART: XXV: THE COMPREHENSIVE ENVIRONMENTAL RESPONSE, COMPENSATION AND LIABILITY ACT (CERCLA) (73)

The Comprehensive Environmental Response, Compensation and Liability Act (CERCLA) is a very expansive subject and this Guide will not to cover all issues associated with the Act. What will be provided is the typical impacts associated with Used Oil Handlers (UOH) associated All Available Inquiry (AAI) associated with any property transfer a UOH may be involved in and potential liability protections CERCLA does provide when certain investigation activities are done related to site contamination and associated environmental impacts short and long term. Also provided will be due – diligence related to typical property transactions that you could expect to be exposed to regarding a purchase or sale of a facility. A brief overview of CERCAL will be provided along with selected documents you should review as well as a brief listing of terms related to associated activities. Comments are provided as italicized statements within the presented reference citations.

CERCLA Overview

From the opening CERCLA (Act) text the following statement summarizes the intent of the "COMPREHENSIVE ENVIRONMENTAL RESPONSE, COMPENSATION, AND LIABILITY ACT OF 1980 (SUPERFUND). An Act to provide for liability, compensation, cleanup, and emergency response for hazardous substances released into the environment and the cleanup of inactive hazardous waste disposal sites."

From the EPA Website: (73) "The Comprehensive Environmental Response, Compensation, and Liability Act (CERCLA), commonly known as Superfund, was enacted by Congress on December 11, 1980. This law created a tax on the chemical and petroleum industries and provided broad Federal authority to respond directly to releases or threatened releases of hazardous substances that may endanger public health or the environment. Over five years, $1.6 billion was collected and the tax went to a trust fund for cleaning up abandoned or uncontrolled hazardous waste sites. The Comprehensive Environmental Response, Compensation, and Liability Act of 1980 (CERCLA):

- Established prohibitions and requirements concerning closed and abandoned hazardous waste sites;
- Provided for liability of persons responsible for releases of hazardous waste at these sites; and
- Established a trust fund to provide for cleanup when no responsible party could be identified.

The law authorizes two kinds of response actions:

1. Short-term removals, where actions may be taken to address releases or threatened releases requiring prompt response.
2. Long-term remedial response actions, that permanently and significantly reduce the dangers associated with releases or threats of releases of hazardous substances that are serious, but not immediately life threatening. These actions can be conducted only at sites listed on EPA's National Priorities List.

CERCLA also enabled the revision of the National Contingency Plan (NCP). The NCP provided the guidelines and procedures needed to respond to releases and threatened releases of hazardous substances, pollutants, or contaminants. The NCP also established the National Priorities List.

CERCLA was amended by the Superfund Amendments and Reauthorization Act on October 17, 1986.

Most have probably heard of the Superfund and how it impacts all activities associated with clean-up of designated superfund sites. Part of the management of those site or potential sites is the determination of liabilities which this Guide will provide information on. As previously stated this Guide does not provide any implied or expressed legal advice and as always if you are involved in any CERCAL actions or investigations you need to seek qualified advice from those familiar with CERCAL that may include consultants and legal. As always you should manage your compliance to prevent any involvement but unfortunately no matter how you conduct your business as soon as oil leaves your property for disposition how that person / company manages the material can impact your potential liability as well.

As a Used Oil Handler it is important to have an understanding of how your actions regarding how you handle your used oil collected, stored, and its disposition can significantly impact your liability that is both financial and potential criminal liability.

The CERCLA liability related to Used Oil is a difficult subject but based on the EPA, Memorandum (1987) Scope of the CERCLA Petroleum Exclusion Under Sections 1O1(14) and 104(a)(2) identifies used oil as part of the petroleum exclusion as derived from crude oil, however if the used oil is a waste when discarded (includes spilling) per RCRA or is sent to a Superfund site and contributes to the contamination because it has hazardous substances contained in it not naturally part of the crude oil because of use, as used oil does, it would be applicable to CERCLA. In essence the petroleum would be exempt but the additional hazardous substances could make it a hazardous waste triggering CERCLA liabilities. Additionally, the memorandum also identifies that under the same application used oil generators, transporters, processors, storage facilities. And filter handlers could be liable under CERCLA. It should also be noted that even though someone can indemnify you from costs associated with CERCLA financial liabilities it does not mean you could not be impacted due to something you or your representatives did.

The above statement does not preclude mandated clean-up, remediation, or site environmental management that have significant liabilities under many other state and federal requirements that include but not limited to the CWA, CAA, and TCEQ regulations that are not controlled by CERCAL requirements.

Potential Responsible Party (PRP) (3)

Liability (civil and criminal) extends from a release management as stated at the EPA CERCLA website where it identifies criminal liability under CERCLA can be created by "A person in charge of a facility or vessel "Knowingly" Fails to notify the appropriate agency of the U.S. Government immediately as soon as he/she became aware of the release into the environment of a hazardous substance in an amount equal to or greater than a reportable quantity without a federal permit " but also any activity associated with compliance with any CERCLA requirement once involved in the process from initial investigation to completion of all requirements. Those that may be involved in any CERCAL action are identified as a "potentially responsible party" (PRP). The following from an article by Cornell Law School, Comprehensive Environmental Response, Compensation and Liability Act (CERCLA), provides a brief overview of the issue.

Scope of Liability

CERCLA casts a wide net in bringing in responsible parties.

> ***Commentary:*** *This initial comment is critical to your understanding as initial actions include taking the step of identifying everyone who ever sent something to a site now involved in any Superfund action. But the net goes far beyond just the UOH and a specific facility actions but many of those connected to ownership and management at all levels, but also to financial institutions.*

Through the broad definition of hazardous substance (§101**(14)** - liability is extended to anything that presents an "imminent and substantial danger"), the loose interpretation of "release" (there is no time constraint on this) and the wide scope of liability (each party involved may be responsible for the entire clean up) the statute seeks to fund the cleanup of these sites. Although there have been challenges to the apparently retroactive nature of this (because people are being held liable for actions that predate the statute), the courts have held that as the waste is continuing to cause problems, and the statute is not a punishment but rather a reimbursement obligation, the statute is not retroactive, and thus not unconstitutional. (US v. Monsanto, 858 F. 2d 160 (4th Cir. 1988)). (Note: additional challenges under the commerce clause have been rejected as well. See US v. Olin (107 F.3d 1506 (11th Cir. 1997)).

Potentially Responsible Parties (PRP) (30)

CERCLA §107(a) lays out the responsible parties and in doing so establishes expansive liability.

Current *(land)* Owners / Operator. Although this excludes property acquired through foreclosure or a security interest, this will still include current owners even if they made no contribution to the hazardous release. In NY v. Shore, Shore, a landowner who had purchased a contaminated site with the intuition of cleaning it up and developing it, was held liable for cleanup costs as he was aware of the condition of the land when he purchased it (and that dumping was still on-going at time of purchase).

Past Owner / Operator at time the pollution occurred. Prior owners / operators will be held liable as well for any release of toxic substance that occurred during their possession / control of the land. Note that even a party who does not / did not own a facility may be liable as an "operator" if they had substantial control over the activities that lead to the release of hazardous substances. (See US v. Bestfoods (524 U.S. 51, 1998)

Arrangers (person who arranged for disposal of a hazardous substance at a site). Under the "Useful products doctrine" if you are not arranging for disposal, then you are out of the system. Thus, a producer of a chemical that is sold for use (not disposed of) is not an arranger. See Burlington Northern note below.

Transporters (person who transported a hazardous substance to the site). Transporter liability is premised on the transporter playing some role i the selection of the site where the hazardous substances are taken. (Tippins, Inc v. USX Corp., 37 F.3d 87 (3d. Cir. 1994). Additionally, even if the transporter moves the substance from one contaminated area of a parcel to another, uncontaminated area, they will still be held liable. (Kaiser Aluminum & Chemical Corp. v. Catellus Development Corp. (976 F.2d 1338 (9th Cir. 1992).

DEFENSES

Although CERCLA imposes Strict Liability *(see definitions below)* as well as Joint and Several Liabilities, creating major liability for involved parties, there are limited defenses available to avoid CERCLA liability.

Mere knowledge is not sufficient. In Burlington Northern (Docket 07-1601, 2009), the Supreme Court found that Shell, who supplied chemical products to Brown & Bryant Inc (a chemical distribution business) was not an arranger and thus not responsible for Brown and Bryant's spills. The court noted that Shell took numerous steps to reduce the likelihood of spills and that Shell's mere knowledge that spills and leaks contented to occur was insufficient grounds for including Shell as an "arranger" of disposal.

107(b) Defenses (30) provides potential, but difficult to use defenses: an act of God, an act of War, and a third-party defense:

1. **Act of God / Act of War:** Courts have held these defenses do not apply to events that could be anticipated, such as heavy rains, or earthquakes; rather these defenses would be limited to an exceedingly rare and unforeseen set of circumstances.
2. **Third-party defense:** This defense will only apply if (1) the PRP can show they exercised due care and (2) there was no contractual relationship with the egregious third party. Thus, this defense will not apply if the dumper was an employee, agent, parent company, subsidiary, or anyone in a contract with the PRP. This defense could apply in the case of a midnight dumper - unbeknownst to you, someone secretly dumps toxic waste on your land, and you had no reason to suspect this might be happening.
3. **Innocent Landowners:** In response to criticism, amendments to CERCLA known as SARA include a defense for innocent landowners who "undertook all appropriate inquiries" before purchasing the land, and "had no actual or constructive knowledge" of the hazardous substance.

Commentary: *This area of "all appropriate inquires" is one that will be discussed in more detail below as it relates to facility purchase and sales and assumption of operational control at any site. Analytical Methods is usually accomplished utilizing a Phase I and if triggered by a Phase I findings a Phase II Environmental Site Assessment typically utilizing American Society for Testing and Materials (ASTM), ASTM E1527-13, Standard Practice for Environmental Site Assessments: Phase I Environmental Site Assessment Process performed by a qualified Environmental Professional. Technically you do not have to use the ASTM procedure identified above as long as the report meets the §312.31 of the final rule identified below. Real world application of the process is that a lending institution, buyer or seller would require the ASTM procedure when any land sale or purchase, operations agreement typically associated with asset purchase is involved but does not require an asset purchase. No matter how small the transaction is or what the area has been used for historically or any information made available* *through the due-diligence process a Phase I and Phase II if triggered by Phase I should be done to protect your liability. It is not just to protect the buyer but is also the mechanism that can provide a defense to the seller if something comes up later by establishing a baseline for a change of ownership and or operational control. Of course nothing is a complete defense based on other facts that may come up later but without the Phase I at a minimum you are potentially at more risk without it. Again I am not offering a legal opinion but it is based on experience with parties on both sides.*

De minims settlements: Should the EPA determine a PRP contributed only a minimal amount of hazardous waste (as compared to other wastes at the site), and this waste did not contribute significantly to the cleanup, they may offer a settlement to that PRP. This is designed to protect small businesses who were not major contributors to the mess.

Separate Harm: Likewise, if a PRP can show their mess was contained and removed (and thus contributed in no way to the hazardous release), they may limit their liability.

§107(q) Contiguous Properties: This protects neighbors to polluted lands - a person who take reasonable care and who is unaware of hazardous waste leaking onto their land will not be liable for the clean-up costs.

4. **Service Station Dealers:** An issue associated with UOH is the CERCLA application to many of the Generators you interact with and defined per CERCLA Section 101(37)(A) defines "service station dealer" (3) as any person (i) who owns or operates a motor vehicle service station, filling station, or garage, or similar retail establishment engaged in the business of selling, repairing, or servicing motor vehicles, where a significant percentage of the gross revenue of the establishment is derived from the fueling, repairing, or servicing motor vehicles, and (ii) who accepts for collection, accumulation, and delivery to an oil recycling facility, recycled oil that (I) has been removed from the engine of a light duty motor vehicle or appliance, and (II) is presented, by such owner, to such person for collection, accumulation, and delivery to an oil recycling facility.

Commentary: *SSDs cannot accept just any oil and cannot mix it with a hazardous substance but if used oil is managed under the identified conditions they would be exempt from CERCLA liability.*

EPA, 2001, RCRA, Superfund & EPCRA Call Center Training Module, Used Oil identifies more detail associated with the SSD liability exemption.

"Since 1986, §114(c) of the Comprehensive Environmental Response, Compensation, and Liability Act (CERCLA) has offered a recycled used oil exemption from CERCLA liability to service station dealers (SSD), as defined in CERCLA §101**(37)**. Because of three conditions set forth in the statute, no person could be eligible for the exemption until the present recycled used oil management standards in Part 279 were created. The first condition requires SSDs to be in compliance with management standards promulgated pursuant to §3014 of RCRA. Since no standards had been promulgated until September 10, 1992, the exemption had been previously unusable. The second condition does not allow SSDs to mix used oil with any hazardous substance. The third condition requires the SSDs to accept DIY generated used oil for recycling. The exemption is limited to generator and transporter liability under CERCLA §§107(a)(3) and (4), so the SSD would be responsible for conducting any corrective action measures if a release of used oil occurs from any container, aboveground tank, or underground storage tank on-site."

All Appropriate Inquiry (AAI) (5)

Now that we have identified that a UOH can be a PRP what you should really focus on besides operations in a compliant manner never to contaminate your facility. But prevention of your possible liabilities starts with knowledge of the possible known or suspected Recognized Environmental Conditions (REC), historical recognized environmental conditions, and de minimis conditions that are associated with the property to be purchased or land below the area you are going to have operating control or as indicated above if you are going to sell the facility or transfer operational control.

In order to limit or protect you completely from any CERCLA liability you must show that you met the requirements for All Appropriate Inquiry (AAI) prior to purchase of any property or taking over operational control. It is important that you do not take possession or begin control until AAI is complete as you could be a PRP because of your involvement at the site.

From the suggested documents to review below identified as EPA, Memorandum 2003, Interim Guidance Regarding Criteria Landowners Must Meet in Order to Quality for Bona Fide Prospective Purchaser, Contiguous Property Owner, or Innocent Landowner Limitations on CERCLA Liability ("Common Elements") contains several issues you should be aware of and I have provided several of the points identified as to protecting your innocent landowner conditions. Of note in the statements below from the AAI section is the statement as indicated above "Purchasers of property wishing to avail themselves of a landowner liability protection cannot perform all appropriate inquiry after purchasing contaminated property." Also, as indicated you could still purchase property with known contamination and still avail yourself of the protections but typically it is extremely problematic for a small facility owner or operator based on my experience without your own qualified environmental staff to interact with the property owners or previous operators as they clean-up or monitor existing contamination. Real world problems also include how you address any release you may have that may impact the already existing contamination and do you get drawn into their problems when if the property was not contaminated prior you could clean up your own release and move on. Practically, based on my experience, if a lending institution is involved they would never let you do the deal anyway even if you thought it could be managed.

A party claiming to be a bona fide prospective purchaser, contiguous property owner, or section 101(35)(A)(i) innocent landowner bears the burden of proving that it meets the conditions of the applicable landowner liability protection. Ultimately, courts will determine whether landowners in specific cases have met the conditions of the landowner liability protections and may provide interpretations of the statutory conditions. EPA offers some general guidance below regarding the common elements. This guidance is intended to be used by Agency personnel in exercising enforcement discretion. Evaluating whether a party meets these conditions will require careful, fact-specific analysis.

Threshold Criteria: To qualify as a bona fide prospective purchaser, contiguous property owner, or innocent landowner, a person must perform "all appropriate inquiry" before acquiring the property. Bona fide prospective purchasers and contiguous property owners must, in addition, demonstrate that they are not potentially liable or "affiliated" with any other person that is potentially liable for response costs at the property.

To meet the statutory criteria of a bona fide prospective purchaser, contiguous property owner, or innocent landowner, a person must perform "all appropriate inquiry" into the previous ownership and uses of property before acquisition of the property. CERCLA §§ 101(40)(B), 107(q)(1)(A)(viii), 101(35)(A)(i),(B)(i). Purchasers of property wishing to avail themselves of a landowner liability protection cannot perform all appropriate inquiry after purchasing contaminated property. As discussed above, bona fide prospective purchasers may acquire property with knowledge of contamination, after performing all appropriate inquiry, and maintain their protection from liability. In contrast, knowledge, or reason to know, of contamination prior to purchase defeats the contiguous property owner liability protection and the innocent landowner liability protection. The Brownfields Amendments specify the all appropriate inquiry standard to be applied. The Brownfields Amendments state that purchasers of property before May 31, 1997 shall take into account such things as commonly known information about the property, the value of the property if clean, the ability of the defendant to detect contamination, and other similar criteria. CERCLA § 101(35)(B)(iv)(I). For property purchased on or after May 31, 1997, the procedures of the American Society for Testing and Materials ("ASTM"), including the document known as Standard E1527 - 97, entitled "Standard Practice for Environmental Site Assessments: Phase 1 Environmental Site Assessment Process," are to be used. CERCLA § 101(35)(B)(iv)(II). The Brownfields Amendments require EPA, not later than January 2004, to promulgate a regulation containing standards and practices for all appropriate inquiry and set out criteria that must be addressed in EPA's regulation. CERCLA § 101(35)(B)(ii), (iii). The all appropriate inquiry standard will thus be the subject of future EPA regulation and guidance.

Affiliation: To meet the statutory criteria of a bona fide prospective purchaser or contiguous property owner, a party must not be potentially liable or affiliated with any other person who is potentially liable for response costs. Neither the bona fide prospective purchaser/contiguous property owner provisions nor the legislative history define the phrase "affiliated with," but on its face the phrase has a broad definition, covering direct and indirect familial relationships, as well as many contractual, corporate, and financial relationships. It appears that Congress intended the affiliation language to prevent a potentially responsible party from contracting away its CERCLA liability through a transaction to a family member or related corporate entity. EPA recognizes that the potential breadth of the term "affiliation" could be taken to an extreme, and in exercising its enforcement discretion, EPA intends to be guided by Congress' intent of preventing transactions structured to avoid liability.

The innocent landowner provision does not contain this "affiliation" language. In order to meet the statutory criteria of the innocent landowner liability protection, however, a person must establish by a preponderance of the evidence that the act or omission that caused the release or threat of release of hazardous substances and the resulting damages were caused by a third party with whom the person does not have an employment, agency, or contractual relationship. Contractual relationship is defined in section 101(35)(A)."

Continuing Obligations: Several of the conditions a landowner must meet in order to achieve and maintain a landowner liability protection are continuing obligations. This section discusses those continuing obligations: (1) complying with land use restrictions and institutional controls; (2) taking reasonable steps with respect to hazardous substance releases; (3) providing full cooperation, assistance, and access to persons that are authorized to conduct response actions or natural resource restoration; (4) complying with information requests and administrative subpoenas; and (5) providing legally required notices. *Please refer to the referenced document for more discussion on this issue for more detail on your continuing obligations.*

Conducting All Appropriate Inquiry

The following section addresses specifics of the AAI. The section content is based on the EPA, Fact Sheet (5) All Appropriate Inquiries Rule: Reporting Requirements and Suggestions on Report Content document that addresses the AAI elements. Even though you are not required to utilize the identified ASTM procedure, you must have a written report generated by a qualified individual. 4(a) identifies the requirements of the "written report" and following are the elements of a Phase I conducted using the ASTM procedure that meet the "written report" requirement. If the Phase I indicates conditions indicative of releases or threatened releases where further investigation would need to be done that typically includes site sampling and or other engineering determinations to prevent threatened releases such as the existence of insufficient containment. ASTM E1903 - 11 Standard Practice for Environmental Site Assessments: Phase II Environmental Site Assessment Process can be used for a Phase II assessment but it is not required. If the Phase I identifies a need for further investigation and you do not complete this, you would not be completing AAI.

> ***Commentary:*** *A critical issue associated with the need for an AAI activity is that the consultant must be a Qualified Environmental Professional. Included as a suggested reference is the EPA Fact Sheet, All Appropriate Inquiries*

Rule: Definition of Environmental Professional, also defined in §312.10. From a practical application, a qualified environmental professional can be held liable for his or her actions associated with the AAI, as well as to the company for which services are performed. In addition, the consultant must act as an independent third-party decision maker regarding the identified hazardous conditions. Even though they will typically work with you during the process and you are allowed to refute identified findings, they must make the final decision as to their work product. In some cases, you pick the Phase I assessor; in others, another involved party may select the assessor. If for some reason you are not satisfied with the selection of the assessor, you will need to object prior to any draft report as it becomes part of the site record forever when final. Attempting to refute a Final Phase I at a later date will be problematic.

1. What are the documentation requirements for ALL APPROPRIATE INQUIRIES? The final rule requires that the results of an all appropriate inquiries investigation be documented in a written report. The specific reporting requirements for all appropriate inquiries are provided in 40 CFR §312.21 (Results of Inquiry by an Environmental Professional) and §312.31 of the final rule and include:

a. An opinion as to whether the inquiry has identified conditions indicative of releases or threatened releases of hazardous substances on, at, in, or to the subject property.
b. An identification of data gaps (as defined in §312.10) in the information collected for the inquiry that affect the ability of the environmental professional to identify conditions indicative of releases or threatened releases of hazardous substances on, at, in, or to the subject property, as well as comments regarding the significance of these data gaps.
c. Qualifications and signature of the environmental professional(s). The environmental professional must place the following statements in the document and sign the document: "[I, We] declare that, to the best of [my, our] professional knowledge and belief, [I, we] meet the definition of Environmental Professional as defined in §312.10 of this part." "[I, We] have the specific qualifications based on

education, training, and experience to assess a property of the nature, history, and setting of the subject property. [I, We] have developed and performed the all appropriate inquiries in conformance with the standards and practices set forth in 40 CFR Part 312."

d. In compliance with §32.31(b), the environmental professional must include in the final report an opinion regarding additional appropriate investigation, if the environmental professional has such an opinion.

2. **Is there a required format for reporting results OF ALL APPROPRIATE INQUIRIES? (7)**

The final rule requires no specific format, length, or structure of the written report. However, EPA offers the following suggestions regarding the potential content of a written report. The following suggestions generally are consistent with recommendations published in ASTM E1527-13, Standard Practice for Environmental Site Assessments: Phase I Environmental Site Assessment Process. The ASTM E 1527-13 standard is consistent with the requirements of the final rule and may be used to comply with the provisions of the rule. The following are suggestions regarding format and content of an all appropriate inquiries written report. Please note that the suggestions below do not represent regulatory requirements. Prospective landowners and environmental professionals may design their own format for a written report, as long as the report contains the four documentation requirements listed above (and as noted below in full text from Fact Sheet guidance).

Introduction: An introduction could include descriptions of: the purpose and objectives of the assessment; scope of services provided; methodology used to complete the inquiry; any significant assumptions made; limitations and exceptions; any modifications or deviations from the final rule requirements or from the ASTM E 1527-13 process; special terms and conditions; and information obtained from the landowner or user. The environmental professional and the person(s) who conducted the site reconnaissance and interviews may be identified.

Site Description: This section may describe the property location; site and vicinity characteristics; structures, roads, site improvements, and utilities; current and historic use(s) of the property; site topography, geology, and surface/ground water resources; and current and historic use(s) of adjacent properties.

User Provided Information: The report may describe any information provided by the prospective landowner, or user, to the environmental professional. This information may include: title records; information of recorded environmental cleanup liens; recorded activity and use limitations (e.g., engineering controls, land use restrictions, institutional controls); specialized knowledge or experience held by the user related to the property or nearby properties; commonly known or reasonably ascertainable information; and relationship of the purchase price to the fair market value of the property, if it were not contaminated.

Records Review: The written report may include a section that summarizes the information found during the records review. This section may describe records that were reviewed to complete the inquiry including: physical setting sources (e.g., topographic maps); historical use sources (e.g., aerial photographs, fire insurance maps, street directories, newspaper archives); federal, state, tribal, and local records or databases of government records; and other environmental record sources (e.g., prior investigation reports, tank/transformer inventories, spill records, permits, etc.).

Site Reconnaissance: The written report may include a section dedicated to describing the methodology used to conduct the visual inspection of the subject and adjoining properties. The description may include: when and who performed the reconnaissance; physical imitations (e.g., snow-covered ground, limited access, safety concerns, etc.); general site setting; exterior observations; and interior observations. Additional information on evidence of staining, spills, odors, stressed vegetation, corrosion, pools of liquids, discolored water, ground surface alterations, and other conditions that might suggest a release or threatened release of hazardous substances also may be provided.

Interviews: A summary of the interviews conducted could include a description of when and with whom the interviews were conducted (e.g., current property owner and occupants, site manager, attorneys, financial manager, local/state/federal government officials, past site owners, and two occupants) and the method used to conduct the interviews (e.g., in person, written, telephone). If property is abandoned, this section may describe which neighboring property owners were interviewed and if applicable, which past owners and occupants were interviewed.

Findings: A findings section could describe the results of the assessment including the identified known or suspected recognized environmental conditions, historical recognized environmental conditions, and de minimis conditions. This section also may include findings related to, but not limited to: current and historic site usage; adjoining and nearby properties; hazardous substances and petroleum products; non-hazardous, solid, and hazardous waste management; water pollution; pits, ponds, and lagoons; drains and sumps; waste water; wells; septic systems; spills or releases; air emissions; storage tanks and drums; soil and groundwater contamination, polychlorinated biphenyls (PCB) contaminants, or other contaminants.

Opinion of the Environmental Professional: In compliance with the all appropriate inquiries final rule at §312.21(c)(1), the written report must include the environmental professional's opinion(s) as to whether the inquiry identified conditions indicative of releases or threatened releases of hazardous substances on, at, in, or to the subject property. The opinion likely will be based on conditions identified during the inquiries (and potentially noted in a findings section), and include a discussion of the logic, reasoning, and rationale used by the environmental professional in developing the opinion. The environmental professional also must include in the final report an opinion regarding additional appropriate investigation to detect the presence of contamination at the property, if the environmental professional has such an opinion.

Data Gaps: As required in §312.21(c)(2) of the final rule, the report should document and discuss significant data gaps that affect the ability of the environmental professional to identify conditions indicative of releases or threatened releases.

Conclusions: A conclusions section may be included that summarizes all identified conditions indicative of releases or threatened releases of hazardous substances (or recognized environmental conditions) (*see definition below*) connected with the property. The final rule does not require that any specific statements be made regarding these conditions, however, ASTM E 1527-13 requires that the report include a statement substantially similar to one of the following written statements:

"We have performed a Phase I Environmental Site Assessment in conformance with the scope and limitations of ASTM Practice E 1527 of [insert address or legal description], the property. Any exceptions to, or deletions from, this practice are described in Section [ ] of this report. This assessment has revealed no evidence of recognized environmental conditions in connection with the property," or

"We have performed a Phase I Environmental Site Assessment in conformance with the scope and limitations of ASTM Practice E 1527 of [insert address or legal description], the property. Any exceptions to, or deletions from, this practice are described in Section [ ] of this report. This assessment has revealed no evidence of recognized environmental conditions in connection with the property except for the following: (list)."

Additional Services: If applicable, it may be useful to include a description of any additional services performed as part of the assessment that are beyond the scope of the final rule, and were contracted for between the user and the environmental professional. Additional services could include, but are not limited to: non-scope considerations (e.g., lead-based paint, mold, radon, asbestos, regulatory compliance assessment, indoor air quality, etc.); broader scope of assessment; liability or risk evaluations; Phase II sampling and analysis; health and safety; evaluation of remediation techniques; etc.

References: A reference section may be included that lists the published sources relied upon to complete the assessment.

Signature(s) and Qualifications of the Environmental Professional(s): Include the statements and environmental professional(s) signature required by §312.21(d), as discussed above in "What are the Documentation Requirements for All Appropriate Inquiries?"

Appendices: Appendices could include: regulatory records documentation; environmental database report; site map/plan; vicinity maps; site photographs; historical source documentation (building department records, local street records, chain of title documents, property tax records, zoning/land use records, aerial photos, fire insurance maps, USGS topographical maps); interview documentation; and qualifications of the environmental professional(s).

3. Phase II and Phase III further investigation based on Phase I findings. The following from James Witkin (2002) Environmental Aspects of Real Estate and Commercial Transactions: From Brownfields to Green Buildings describes descriptions for Phase II and Phase III activities. Based on my typical industry prac-

tices, normally it would be very unusual for the buyer to pay for anything beyond the Phase I and Phase II assessment and the seller would determine if they wanted to follow-up with the Phase III if required either to pursue the sale or required by regulation based on the Phase II findings.

Phase II Environmental Site Assessment: Phase II is an "intrusive" investigation which collects original samples of soil, groundwater or building materials to analyze for quantitative values of various contaminants.[10] This investigation is normally undertaken when a Phase I ESA determines a likelihood of site contamination. The most frequent substances tested are petroleum hydrocarbons, heavy metals, pesticides, solvents, asbestos and mold.

Phase III Environmental Site Assessment: Phase III is an investigation involving remediation of a site. Phase III investigations aim to delineate the physical extent of contamination based on recommendations made in Phase II assessments. Phase III investigations may involve intensive testing, sampling, and monitoring, "fate and transport" studies and other modeling, and the **design** of feasibility studies for remediation and remedial plans. This study normally involves assessment of alternative cleanup methods, costs and logistics. The associated reportage details the steps taken to perform site cleanup and the follow-up monitoring for residual contaminants.

4. The overview of AIA and utilization of the Phase I or equivalent process issues provided above is only a very limited overview of all the issues associated with CERCLA (Superfund) activities and does not represent all the complex actions and possible fact patterns that would trigger CERCLA liability by any PRP. It is critical to understand is that a UOH needs to consider impacts related to CERCLA as business activities are pursued especially associated to facility operational control and or land purchase but as always with an understanding of additional compliance liabilities can be created when you allow used oil to get into the environment.

PART XXVI: DUE DILIGENCE

Due diligence is another complex subject as it relates to all aspects of any business sale or purchase and is utilized to provide a process to evaluate any environmental risk associated with a potential purchase. It is related to AAI discussed above as it is part of any due diligence activity. This Guide will address the environmental related due diligence approach related to a typical UOH transaction that is usually not as complex as most petroleum related facility purchases. In addition to personal experience with this issue include content references from the EPA, 2016, Prepared Workbook Process for Risk Evaluation, Property Analysis and Reuse Decisions **(44)** and a presentation from a Texas law firm of Thompson & Knight **(129)** "Conducting Environmental Due Diligence and Audits". I have also included a typical Environmental Due Diligence request I have developed over the years and used often based on many sites that I have been the Environmental Professional responsible for the collection, assessment and review of related documents either related to support to the buyer or seller. Even though in many situations any due diligence request is generated by lawyers and almost always presented to the seller by legal representatives and it is prepared by someone who has an environmental background that may be a lawyer as well. In all situations a confidentiality agreement is executed by the lawyers involved when the purchase process begins to protect everyone and information related to the inquiry. From a compliance and business management perspective you should never enter into a due diligence process without a confidentiality agreement executed by both parties. Remember none of the information in this Guide is implied or expressed as legal advice and you are not to rely on any information presented.

EPA Prepared Workbook Process for Risk Evaluation, Property Analysis and Reuse Decisions Discussion of Due Diligence

The environmental due diligence process typically begins with an all appropriate inquires investigation or a Phase I Environmental Site Assessment (Phase I ESA). There is often confusion about the distinction between the terms "due diligence" and "all appropriate inquiries." The most common way to conduct a Phase I ESA, and the generally accepted business practice for doing so, is the ASTM E1527- 13 Standard Practice. ASTM International (ASTM) — an international standards organization that develops and publishes voluntary consensus technical standards — developed voluntary consensus standards for conducting Phase I and II ESAs. Based on the results of an AAI investigation or Phase I ESA, a Phase II Environmental Site Assessment (Phase II ESA) may be necessary to better understand the type and extent of any potential environmental contamination at the property.

While there is substantial overlap between the two, due diligence conducted for a property is generally broader in scope than what is required for an all appropriate inquiries investigation (including, for example, a Phase II ESA, building structural analysis). From a risk management perspective, these additional inquiries and activities can provide information necessary to more fully assess both the source and magnitude of potential risks. It is for this reason that this Workbook uses the broader term "due diligence" when describing such information gathering efforts. The term "all appropriate inquiries" is used when specifically referring to the statutory requirements of 40 CFR 312.

An important component of the due diligence process is commonly referred to as "environmental due diligence" and is conducted to identify and/or address:

1. Presence and management of hazardous substances and petroleum products on the property
2. Conditions indicative of releases or threatened releases of hazardous substances at, in, from, or to a property
3. Inspections of environmental conditions at the property, including conditions of buildings and environmental media (e.g., soil, ground water, surface water, sediment) located on the property
4. Historical, current, and potential future cleanup action on the property
5. Regulatory status of the property • Current and historical ownership and property access issues
6. Potential risks and liabilities associated with the presence of hazardous substances and petroleum products on the property and potential cleanup action needed

CONDUCTING ENVIRONMENTAL DUE DILIGENCE

The following information is provided as an overview of issues typically associated with you or an environmental professional's environmental due diligence and auditing activities with added italicized comments as appropriate.

1. It is imperative that all efforts are conducted to understand and manage risk and avoid or at a minimum limit possible liabilities.
   a. Determination as to liability for conditions on the property as outlined in the CERCLA discussion which is one of the reasons AAI is a focus. *(Items 1 and 2 relate to the Phase I findings and potentially Phase II)*
   b. Liability determinations during the due diligence process again typically utilizing the Phase I or Phase II processes need to determine any off-site conditions caused by the facility
   c. Determinations as to previous owners / operators potential liability for off-site shipments (CERCLA defined liability for those that arrange and dispose of materials off-site)

   Commentary: *Knowing any facility used oil was sent off too is critical to understand potential liability you may be involved in so you can understand what may be required to demonstrate this activity was not done during your ownership or operational control.)*

   d. Historical and even possible ongoing liability for violations of permits, plans and any applicable compliance regulations governing operations (air, water, solid waste)

   Commentary: *This is why you must identify all violations both resolved with a confirmation and No Further Action (NFA) documentation from the agency involved and understand any agency actions currently ongoing because typically you will be responsible for all resolutions.)*

   e. Historical and possible current third party claims.

Commentary: *This area also can represent significant problems and is dependent on the conditions of the assumption of the claims. What can also be an issues is that if any claims are associated with nearby residents that could represent a potential for more problems later.)*

f. Understanding of any current owner or operator liability status with and the history of prior ownership or operators as their and if any existing liabilities still exist associated with disposal or release and those involved in arrangement for the disposal.

Commentary: *Understanding actions of current and past owners and how they may have potential liability and how the current conditions may impact your operations is critical as you may have to decide you want to have to deal with agency or contractor involvement at the site due to others past actions.*

2. It is imperative again to understand what risk avoidance, risk management, and remediate conditions will be required or who will be responsible for those activities to avoid the risks or bring facility into compliance before closing. It must be considered seriously if assuming or managing the potential risks is worth acquisition of the facility. Avoiding facilities that are contaminated or that have significant compliance issues must be part of a realistic evaluation by management and to make that happen all issues must be clearly identified for assessment.

Commentary: *This determination to avoid risk is very fact dependent as factors of exact levels of contamination where it is located at the facility and how might it be managed short and long term. Consideration for seller to remediate prior to closing, nut if they do remediate insure a formal evaluation is done by a third party as to effective remediation and obtain an agency NFA letter if agency involvement. Regulatory compliance issues are another fact dependent issues as well as your staff's ability to manage the issue or implement corrective actions.*

g. A typical and highly recommended approach in addressing known conditions and any risk of unknown conditions is to add protections in specific deal terms.
h. To be able to understand and utilize regulatory programs and defenses to address conditions and avoid liability it would be beneficial to consult counsel and or qualified environmental professionals within the process.

9. Due Diligence – Basic Concept overview.
   a. Due Diligence – Every due diligence effort is different in approach and overall complexity.

Commentary: *This is why it is so important to have qualified staff, consultants, and counsel that understand the process and issues related to the specific facility and business you are acquiring.*

   b. Consideration of requirements that can influence due diligence.

1. Management's knowledge of the property, past operations, and the associated conditions involved.

Commentary: *You and your staff's knowledge is critical related to the issues above but make sure you have the documentation to support your knowledge. Any assumption you may make about any issue that turns out wrong later can be extremely detrimental.*

2. How developed is the site and overall nature of operations conducted at the facility.
3. Time to complete due diligence and facility cost considerations.
4. Realistic understanding of any buyer's tolerance for risk as a management group and dependence on third party lenders who may not share the same risk values.
5. Elements of the deal structure and defined and understood by all parties what the seller's continued existence will consist of assessment of their viability.

Commentary: *Deal structure is obviously a business decision but the seller's* *continued existence and especially viability for any issue as part of the deal are significant considerations.*

c. Significant environmental considerations in business due diligence.

Commentary: *(The following list is significant to insure your understanding of all the Environmental issues associated with the site. Depending on the facility under consideration it may be an extremely laborious process for all involved in collection and review but it must be done and managed by qualified people to make effective determinations as to the impact on your organization in the future and to determine if any risk involved is acceptable. Each issue should be addressed even if the answer is that it does not apply or is limited in scope.)*

1. Current on-site conditions and historical releases (Phase I, and II)
2. Agency compliance, orders, citations, NOVs, not just current but any perceived or articulated by any 3rd party or agency that need to be considered as possible actions to be evaluated for consideration, Compliance agreements, schedules, consent orders, Fines or penalties, all communications with regulatory agencies that include current and historical documentation typically for at least the last five (5) years but can and often include the life of the facility.

Commentary: *(You must insure you are aware of the compliance impacts as they represent significant risk and if acceptable you must evaluate your ability to manage issues going forward. Especially significant is their relationships with agencies as a bad relationship for whatever reason can put you in an enforcement oversight position that can be significant and if a negative history exists and the best approach is to be pro-active as new owners and discuss concerns and how you plan to manage issues different. Even consider a formal Audit under the Texas Environmental, Health, and Safety Audit Privilege Act as identified in this Guide.)*

3. Assessment of any known or potential liability for off-site conditions. (Phase I and

II efforts should identify any off-site potential but you may need to seek additional environmental and legal advice on what impacts may be present or what possible impacts could occur in the future. Issues are not just related to subsurface issues but air or water issues.)

4. Presence of any environmental management system (EMS) (An effective EMS system that is utilized to its potential and in place and functioning with management support is a positive. However, it must be understood as to overall costs of operations required to continue utilization of the system, typically software based (off-the-shelf or internally developed). Smaller facilities can be effectively managed in a variety of methods but what is critical is that the process reflects the process flow required to maintain compliance. (See Flowchart 8 below for EMS review objectives.)
5. The management to assume facility control or staff to be retained must have a complete understanding of all applicable regulatory standards, audit programs, and reports as part of program elements that include sampling and monitoring, and any certifications required. (You must understand all regulations and compliance activities associated with this facility operation.)
6. Does the facility have defined practices and procedures utilized for all processes. (Do they have an up to date and complete set of Standard Operating Procedures (SOPs)? If all policies and practices are not in place are you aware of the gaps and know what is required to fill them.)
7. Environmental studies mandated or voluntary. (Insure you have all documentation as to ANY site investigation that has ever been done on site either voluntarily or required and includes remediation monitoring, release investigations, water sampling etc. ;)
8. All required permits environmental or operational (Transportation, equipment) permits or governmental authorizations, Understanding of transfer requirements of all permits and what are the qualifications for allowance of transfer by the issuing agency or certification authority. An evaluation as to the company possession of all required permits, licenses for the current operations and production levels. Facility is meeting all permit and license recordkeeping and reporting.

Commentary: *Detail, Detail, Detail is part of your due diligence in order to* *understand permit and or regulatory authorization issues applicable to this facility. Issues of permit transfer can be time consuming and problematic it is not always just a name change as indicated previously regarding UOH permits a financial responsibility document must also be established and other requirements. Understand report and fee requirements for all permits.*

Flowchart 8 – VRM Group, LLC EMS Review Objectives

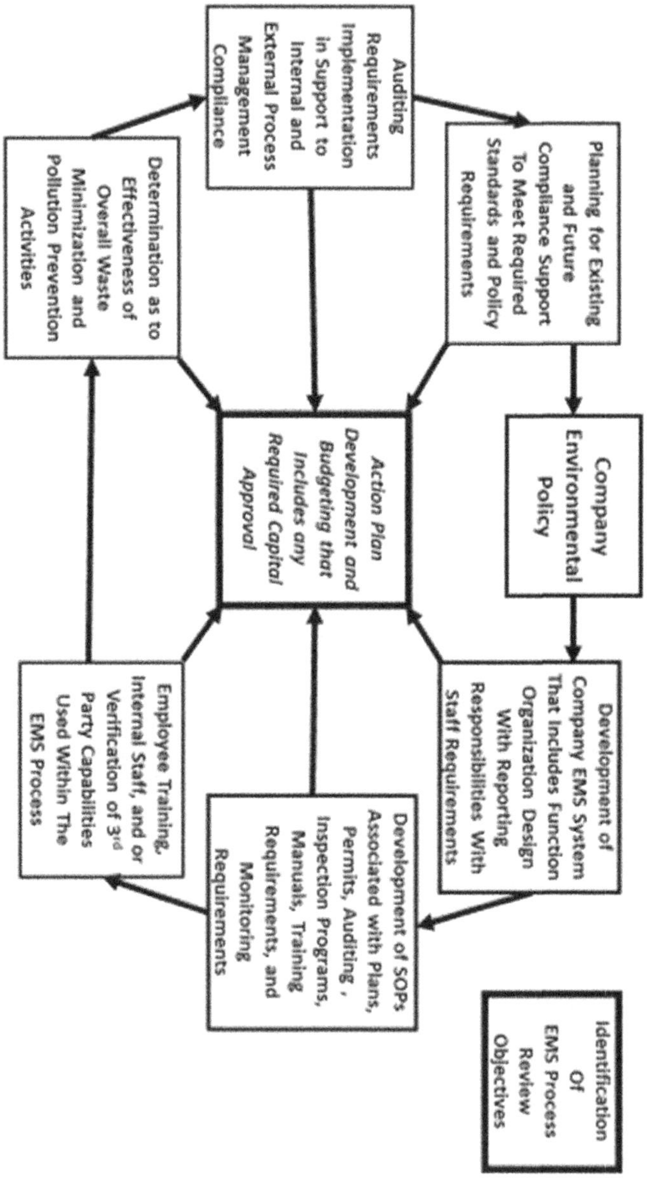

Due Diligence – Overall Scope of Work and Factor Consideration

When in the Due Diligence Process, consider the following questions. Note: Several items have been listed above, the below list is a good reference for this extensive process.

1. Is the company in compliance in all aspects associated with its operations?
2. Can amendments be obtained to define liabilities and responsibilities associated to all identified risks that are unacceptable?
3. Is the ASTM Standard identified above comprehensive enough for your project?
4. Identification of all parties relying on report
5. Are the transaction requirements to be applied to a single commercial facility or portfolio of facilities?
6. Radius of review defined to insure facility(s) impact distance are addressed.
7. All historical ownership/operations identified with identified any impacts of acquisition leasehold interest
8. Logistics associated with due diligence:
    a. Organization of records
    b. Time constraints on all parties subject to the transaction
    c. Location of sites
    d. Availability of escorts for activities to insure time constraints are addressed.
9. Confidentiality acknowledged in writing by all parties involved.
10. Evaluation of all environmental permits, certifications, and licensing.
11. Compliance status in all operational activities.
12. Current and future, operational conditions/constraints, and costs of compliance.
13. Permit, Plan, and License Transfer requirements.
14. Facility is suited for any planned modifications.
15. Conduct a detailed regulatory file review.
16. Identification and review of any previous purchase/sale agreements.
17. Lease agreements for existing equipment.
18. Local land-use and environmental permitting requirements associated with current operations and any planned modifications.
19. Identification of all current operations.
20. Likelihood of continued existence relevant to compliance requirements.

21. New entity formation impacts on existing agency and business relationships.
22. If certain assets are considered for exclusion based on due diligence findings seek concessions on price to accept associated risk or exclude the asset.
23. Accept acquisition with escrow of any estimated costs associated with potential risks identified.
24. Require indemnity for all identified or possible historical noncompliance.

The following is an example Environmental Due Diligence Request utilized for typical Used Oil Facilities acquisition. The example has been used in actual transactions, but it can vary widely based on who makes the request; it can come from legal or environmental staff associated with the acquisition. You would typically receive due diligence requests for various areas of business requiring full disclosure; the acquiring company is required to perform due diligence prior to the purchase. Therefore, it is critical that the buyer receives all requested documents and it is preferred that these are available electronically, or the buyer may request that a hard copy be provided. Understanding this is extremely time consuming for the seller but lack of complete documentation requested can delay processes. In many situations this is only a start of the process and additional documentation may be requested based on initial review.

This request also includes typical Health and Safety documentation requests as typically the process includes all Environmental Health & Safety (EH&S) requested at one time.

The facility Phase I request is outside this request. Data for the Phase I is formally requested by the assessor, but this request represents the typical data, and that is why electronic versions are so important—you will need to provide them to both requestors. Phase I preparation by the seller is also important, as it ensures any condition that may be identified as a Recognized Environmental Condition (REC) is addressed prior to the Phase I. Typical observed conditions that will impact the Phase I are surface spills, containment deficiencies, unlabeled containers, materials not stored correctly, or poor equipment condition.

Please provide copies of the following items for Texas Recycling, Inc. and (the "Company"). Understanding, that you have been sent a general listing from legal regarding due diligence documents required however, the listing below identifies specific EH&S items requested **or** to be identified as not available or not applicable (N/A). It would be appreciated if all documents are provided in electronic form where possible if not possible to create an electronic form please make a copy available during the site visit. It would greatly facilitate our document review if you reference within the electronic file identification the State, Facility and Date indicated on Document in responding to specific items of our document request. We understand that there may be no documents that are responsive to some numbered items or requested additional documents associated with a supplied document, in which case we ask that you please so indicate by writing N/A next to the request. Please send all documents electronically to Rong@VRMUS.com. In addition, the officers, counsel to the Company are requested to advise us of any areas of material importance not otherwise highlighted in this checklist. All documents now held or supplied are subject to the confidentiality agreement in place. Please call Ron Gemeinhardt at 281-788-3021 if you have any questions.

Table 14 – Example Due Diligence Request, Used With Permission From VRM Group, LLC

Document	Status/Notes/Response/Requests
A. Environmental, Health and Safety (EH&S) Documents	
Organization Flow Chart / Site Plot with all tanks, liquid storage areas, buildings and staging / parking areas.	
EH&S related description (and copies of) of judgments, settlements, decrees or orders to which the Company presently is or, during the last five years has been subject.	
Phase I, II, III performed at all facilities, all site soil or ground water investigates performed for any reason at all properties managed or owned by the Company.	
Listing of all spills / releases within the last five years and all clean-up / remediation documentation and disposal data associated with any spill or release.	
Summary of each threatened (within the last five years) or pending claim, lawsuit, arbitration, investigation or inquiry (including state and federal governmental matters such as EPA, OSHA, DOT, State Agency, County or City) involving the Company or relating to the Company's business; provide copies of any correspondence regarding the foregoing.	
Letters from the Company or from counsel to the Company or any regulatory authority in the last three years with respect to Notice of Violations, litigation, contingent liabilities, etc.	

Document	Status/Notes/Response/Requests
Details of any inspection, violations, citations or allegations regarding the failure to comply with any foreign government or any United States federal, state or local law, statute, ordinance, code, rule, regulation or governmental order including environmental regulations, OSHA, DOT or similar employee health and safety laws and regulations for the previous five years. Copies of all No Further Action (NFA) documents issued related to any required federal, state or local violation or citation.	
Copies of plans (SPCC, FRP, HazMat etc.;), permits (EPA ID, Hazardous Materials, Used Oil etc.;), Hazardous or Industrial Waste, OSHA and TIER II reports for previous 5 years, any closure plans and Material Analysis Plans associated with Used Oil permits, all other authorizations, registrations, licenses or qualifications necessary for the conduct of the Company's business, if any. Copies of all required plan / permit inspections for the last 5 years.	
All material correspondence during the past five years with the Department of Justice, Federal Trade Commission, EEOC, EPA, OSHA, Department of Labor, and any other regulatory agency (city, state or federal) to which the Company may be subject. (Includes but not limited to OSHA 200 reports, EPA Waste Reports)	
EH&S Policy, Guidance, Standard Operating Procedures, EH&S Manuals, Environmental Management System policy and identification and description of any EMS or DOT tracking systems	

Document	Status/Notes/Response/Requests
utilized with printout of data tracked.	
All air permits, Title V, Permit by Rule or any other air emission related permit associated with operations on or off site.	
All Aboveground Storage Tank (AST), Underground Storage Tank (AST), Sumps, Separators, Ponds, permits (includes tank registrations) and inspections since construction or installation. Documentation of any cleaning, maintenance and disposal of materials in support to any cleaning or maintenance. Complete tank listing with volumes	
Copies of all NPDES, POTW, SWPPP or any other water permit or one time discharge issued to the facility. The last 5 year observation and analysis documentation associated with any discharge. Plot with location of any site wells and copies of any groundwater monitoring analysis for the last 5 years.	
Copies of all EH&S Related reports related to ANY permit that includes but not limited to Used Oil, Used Oil Filters, Hazardous Waste, Industrial Waste.	
EH&S and DOT Training documentation and training program course descriptions, staff applicability determinations and source or copy of training materials utilized.	
Health and Safety Program Description, meeting documentation and incident management process description. Copies of all SDS (MSDS) utilized by facility. Insurance Rate	

Document	Status/Notes/Response/Requests
Is there a septic tank or municipal discharge? Yes – No (circle one). Documentation for any septic tank cleanout and or repair.	

Due Diligence Suggested Document You Should Review:

1. EPA, 2016, PREPARED WORKBOOK Process for Risk Evaluation, Property Analysis and Reuse Decisions, available from https://www.epa.gov/sites/production/files/2015-04/documents/munimanual.pdf

PART XXVII: ENVIRONMENTAL MANAGEMENT SYSTEMS (EMS)

When you mention Environmental Management Systems (EMS) to most people in the oil-related business they take a breath because the general perception is they are so complicated and require so much effort to maintain. However, those of us that have used various EMS know that once developed and implemented they can be extremely positive from a compliance management perspective but especially to the bottom line because they avoid so many problems and associated costs. The term EMS is actually a little misleading because many systems allow you to manage all Environmental, Health and Safety (EH&S) activities including training.

EMS come in many forms, from company built, to off the shelf software (a listing is provided at the end of this section for your reference) and most of you can do the work yourself with a few needing help in set-up but after established can be managed internally. The biggest advantage to the commercial systems are that most drive the requirements you need to input and have a notice feature to let you know when something is to be done. The biggest value too many is that they are regulatory driven with references and update you when changes in regulations occur that impact your compliance. Many systems also have the ability to store all data in the "cloud" which is secure and confidential and protects all your data in the event your facility gets damaged as well as typically being accessible from anywhere.

Costs vary widely based on the complexity of the system you need and can be only a couple thousand dollars per year to several thousand, most require licensing fees that are typically required to be renewed each year. Primary costs are associated with staff or consultant time to load all data to the system for start-up.

You can of course manage the requirements with spreadsheets and internal regulatory oversight and back-up your data to the cloud or store copies of data off-site but you should review the overall costs and compare them to a commercial system. No matter how you decide to approach your EMS remember one non-compliance fine and possible associated legal costs can more than pay for any expenditures associated with an effective EMS program. You do not have to be an environmental professional to manage an EMS but should have some help from one to assist in getting the system set up or at least review the set-up prior to going into use.

The following are guidelines and process comments from EPA, Stapleton, J., Glover, M., Davis, S., 2001, (72) Environmental Management Systems: An Implementation Guide for Small and Medium-Sized Organizations, Second Edition, TCEQ, 2011 A Guide to Developing an Environmental Management System for a Small Business, EPA, 2000, Integrated Environmental Management Systems, A Company Manual Template A Company Manual Template for Small Business, (25) EPA Website, (19) Environmental Management Systems (EMS) as identified in the section are EMS Documents For Your Review which has complete access information. Additional comments are added italicized. The process discussion below is focused on EPA guidelines with TCEQ statements as noted. If you follow either recommended development process it will get you too an effective EMS. Please note that by having an effective demonstrated EMS you can get credit on your TCEQ compliance history score, once your EMS has been in place one year, which in the event something does happen it can impact any non-compliance penalties.

EMS Overview

An Environmental Management System (EMS) is a set of processes and practices that enable an organization to reduce its environmental impacts and increase its operating efficiency.

The EPA Website provides information and resources related to EMS for small businesses and private industry, as well as local, state and federal agencies.

Building an Environmental Management System (EMS) might sound like an overwhelming task for a smaller organization, but it need not be. Taken in steps, it is a job that small and medium-sized organizations can tackle. These pages (EPA Document) will take you through basic steps as they are outlined in the 2001 Second Edition of Environmental Management Systems: An Implementation Guide for Small and Medium Sized Organizations.

EPA Perspective on EMS

An EMS helps an organization address its regulatory demands in a systematic and cost-effective manner. This proactive approach can help reduce the risk of non-compliance and improve health and safety practices for employees and the public. An EMS can also help address non-regulated issues, such as energy conservation, and can promote stronger operational control and employee stewardship. Basic Elements of an EMS include the following:

1. Reviewing the organization's environmental goals;
2. Analyzing its environmental impacts and legal requirements;
3. Setting environmental objectives and targets to reduce environmental impacts and comply with legal requirements;
4. Establishing programs to meet these objectives and targets;
5. Monitoring and measuring progress in achieving the objectives;
6. Ensuring employees' environmental awareness and competence; and,
7. Reviewing progress of the EMS and making improvements.
8. Costs and Benefits of an EMS:
9. Internal - Staff/manager time (represents the bulk of EMS resources expended by most organizations) and Other employee time.
10. External - Potential consulting assistance, Outside training of personnel
11. Potential Benefits:
    a. Improved environmental performance
    b. Enhanced compliance
    c. Pollution prevention
    d. Resource conservation
    e. New customers/markets
    f. Increased efficiency/reduced costs
    g. Enhanced employee morale
    h. Enhanced image with public, regulators, lenders, investors
    i. Employee awareness of environmental issues and responsibilities

TCEQ Perspective on EMS - Why Implement an EMS?

Developing an EMS is strictly voluntary. Carrying out an EMS has inherent benefits, and, in addition, your business can receive regulatory incentives if the TCEQ approves your EMS.

1. **Inherent Benefits:**

   An EMS for your business will reduce risk and liability, increase efficiency in using environmental resources, and facilitate compliance with environmental rules. An effective EMS improves performance by helping your business to achieve the following benefits:

   a. reduce costs,
   b. prioritize environmental issues,
   c. identify potential problems,
   d. improve environmental compliance,
   e. use materials more efficiently,
   f. streamline operations,
   g. improve internal communication, and
   h. enhance employee morale.

2. **Incentives:**

   The following incentives may be available to those who choose to have their EMS formally verified by the TCEQ:

   a. credit on TCEQ compliance history score, once your EMS has been in place one year;
   b. modification of state regulatory requirements that do not change emission or discharge limits;
   c. changes to how and how often compliance inspections are scheduled and conducted; and
   d. exemption from the reporting requirements of the Waste Reduction Policy Act.

3. **What are the basics of an EMS?**

   The following points provide an overview of an EMS. This is not a comprehensive list of what you need to do if you decide to have your EMS formally approved by the TCEQ.

   a. **Develop an environmental policy.** The environmental policy describes your organization's environmental goals and its commitment to the environment.
   b. **Assign responsibilities.** Your EMS will only succeed if it is clear to all employees that the success of the EMS is very important to the top manager; other-

wise, your EMS does not have a chance. Everyone in your organization has a role and responsibility in an EMS. Use an EMS as an opportunity to ensure that every employee understands what he or she has to do in their job when it comes to the environment.

c. **Identify and prioritize environmental aspects and impacts.** The process of identifying environmental aspects and impacts is the most technically challenging task in creating an EMS. The task requires an analysis of each of your business's activities, products, and services. An environmental aspect is an element of your activities, products, or services that can or does interact with the environment. A significant environmental aspect is one who's potential or actual impacts are significant.
d. **Set and pursue goals for continuous improvement in environmental performance and compliance.** Once you have identified your significant environmental aspects, or "hot spots," you can determine which ones will have goals. When you establish goals, keep in mind compliance, continuous improvement, and pollution prevention.
e. **Document and demonstrate results.** Results include reduced risk, enhanced compliance, and reduced pollution.
f. **Evaluate EMS performance.** Measure and monitor your activities to evaluate whether you are making progress toward achieving your environmental goals.
g. **Make your EMS results-oriented.** This is a feature the TCEQ will use to evaluate an EMS. It means that your EMS should actually reduce risk and help your business be compliant, while continually improving environmental performance.

EPA EMS Process Development

1. Plan Steps:

Time and resources are limited in any organization, so it is important to use resources wisely. The information below illustrates 10 steps in the EMS planning process. Take the time to figure out what needs to be done, how to do it, and who must be involved.

Step 1: Define Organization's Goals for EMS

The first step in EMS planning is to decide why you are pursuing the development of an

EMS. Are you trying to improve your environmental performance (e.g., compliance with regulations or prevent pollution)? Typically at least the two items are applicable to a UOH but of significance is improved productivity of those involved in the process.

Write your goals down and refer back to them frequently as you move forward. As you design and implement the EMS, ask the following questions: How is this task going to help us achieve our goals? How should we define the project scope? (i.e., What is the "organization" (Workgroup such as operations, processing, transportation etc. ;) that the EMS will cover? One location or multiple locations? Should we "pilot" the EMS at one location then implement the system at other locations later?)

> ***Commentary:*** *Suggest very strongly that you always "pilot" the system at one location first if you have multiple sites with similar requirements. In most situations if you already have any system in use you would also continue to use that system as well through this process that is typically also identified as the "Beta" test.*

Step 2: Secure Top Management Commitment

One of the most critical steps in the planning process is gaining top management's commitment to support EMS development and implementation. Management must first understand the benefits of an EMS and what it will take to put an EMS in place. To develop this understanding, explain the strengths and limitations of your current approach and how those limitations can affect the organization's financial and environmental performances. Management also has a role in ensuring that the goals for the EMS are clear and consistent with other organizational goals. Management's commitment should be communicated across the organization.

> ***Commentary:*** *This step cannot be emphasized enough because without managements support and without communication throughout the organization this can become a nightmare.*

Step 3: Select an EMS Champion

Not all small or medium-sized organizations have the luxury of choosing among multiple candidates, but your choice of project champion is critical. The champion should have the necessary authority, an understanding of the organization, and project management skills. The champion should be a "systems thinker" (ISO 9000 or ISO 14001 experience can be a plus, but is not necessary), should have the time to commit to the EMS-building process, and must have top management support.

> ***Commentary:*** *This is an often neglected step but is so critical to your success and will even provide a challenge for an employee you may want to see what project and technical skills they can utilize. When you select a "champion" you must also arrange to give them the time needed aside from their normal job responsibilities, yes it means you may need to shift work or add resources for a short time but it will be worth it.*

Step 4: Build an Implementation Team

A team with representatives from key management functions can identify and assess issues, opportunities, and existing processes.

> ***Commentary:*** *Environmental staff must involve operations staff at all levels during the process at different stages and a failure to communicate through the process can minimize your support and ultimately impact effective utilization.*

TCEQ - Management Representative

The EMS management representative is a member of the top plant management group. This job involves the following: ensuring that all tasks relating to the EMS are identified and completed in a timely manner, reporting periodically to the management group on the progress and results of the EMS, and demonstrating to all employees that top management is committed to the EMS.

TCEQ - EMS coordinator. The EMS coordinator is responsible for: identifying, assigning, scheduling, ensuring the necessary support for, and ensuring completion of all tasks relating to the EMS; maintaining the EMS manual under the leadership of the management representative; and leading the EMS team. In a small company, the management representative and the EMS coordinator might be the same person.

TCEQ - EMS team. The EMS team is composed of a supervisor and one to two employees from each major operation within Lone Star. Each team member or representative is responsible for ensuring that EMS activities in their respective areas are carried out and for reporting the results of these activities to the team. In addition, the team itself is responsible for: carrying out certain EMS activities, such as selecting significant environmental aspects; and meeting on a regular basis, usually monthly.

Step 5: Hold Kick-Off Meeting

Once the team has been selected, hold a kick-off meeting to discuss the organization's objectives in implementing an EMS, the initial steps that need to be taken and the roles of team members. If possible, get top management to describe its commitment to the EMS at this meeting.

> ***Commentary:*** *The management commitment statement from the highest level possible is very important.*

The kick-off meeting is also a good opportunity to provide some EMS training for team members. Follow-up this meeting with a communication to all employees.

Step 6: Conduct Preliminary Review

The next step is for the team to conduct a preliminary review of your current compliance and other environmental programs/systems, and to compare these against the criteria for your EMS. Evaluate your organization's structure, procedures, policies, environmental impacts, training programs and other factors. Consider utilizing an ISO 14001 self-assessment tool or incorporating Safer Choice for gap analysis tools.

> ***Commentary:*** *This Guide will not address the ISO standards for EMS because it is rarely a choice for a UOH and is a significant process to develop and manage typically not required by a UOH. Even if you wanted to take the ISO step to meet customer or governmental requirements it is strongly suggested you start with a non ISO EMS and then after securing that process within your organization move to an ISO system. The ISO 14001 Checklist reference is included in the document listing below.**(84)***

Step 7: Prepare Budget and Schedule

Based on the results of the preliminary review, prepare a project plan and budget. The plan should describe in detail what key actions are needed, who will be responsible, what resources are needed, and when the work will be completed. Keep the plan flexible, but set some stretch goals. Think about how you will maintain project focus and momentum over time. Look for potential "early successes" that can help to build momentum and reinforce the benefits of the EMS.

Step 8: Secure Resources, Assistance

The plan and budget should be reviewed and approved by top management. In some cases, there may be outside funding or other types of assistance that you can use (from a trade association, a state technical assistance office, etc.). See Appendix F of the Environmental Management Systems: An Implementation Guide for Small and Medium Sized Organizations for more ideas on possible sources of help.

Step 9: Involve Employees

Ownership of the EMS will be greatly enhanced by meaningful employee involvement in the EMS development process. Employees are a great source of knowledge on environmental, and health and safety issues related to their work areas as well as on the effectiveness of current processes and procedures. These employees can help the project team in drafting procedures.

Step 10: Monitor and Communicate Progress

As you build the EMS, be sure to regularly monitor your progress against the goals and project plan, and communicate this progress within the organization. Be sure to communicate the accomplishments that have been made and describe what happens next. Build on small successes. Be sure to keep top management informed and engaged, especially if additional resources might be required.

Before beginning the "Do" section, review the key elements of an EMS with worksheets and examples, see pages 15-77 of Environmental Management Systems: An Implementation Guide for Small and Medium Sized Organizations.

2. DO

Step 1: Identify Legal and Other Requirements

A first step in the EMS-building process is understanding the legal and other requirements that apply to your products, activities and services. This step is important for understanding compliance obligations and how these obligations affect the overall EMS design. For example, you might have an operation that is covered by an air quality permit, or clean water permit, used oil management, and might provide a service that results in the generation of regulated wastes. Your EMS should include processes to ensure that legal requirements are addressed.

Step 2: Identify Environmental Aspects and Related Products, Operations, and Activities

Once you understand legal requirements, you should assess how your organization interacts with the environment. Identify environmental aspects and impacts, and determine which are significant. Some environmental aspects may be regulated, while others may not be. For example, if you identify the generation of a particular air emission as a significant environmental aspect, it would help to know which operation(s) generate such air emissions. It might also help to know whether these air emissions are monitored or otherwise measured in some manner.

Collecting this information at an early stage will help you implement subsequent EMS elements. You can use a form to capture this information (such as Figure 15 found on page 81 in Environmental Management Systems: An Implementation Guide for Small and Medium Sized Organizations). One caveat: simply because you identify an existing control and/or monitoring activity related to a particular operation or activity, do not assume that these controls are adequate for EMS purposes. The adequacy of these controls will depend on several factors, including the EMS goals and objectives.

Step 3: Define Views of Interested Parties

Gather information on the views of your "stakeholders" or interested parties. Stakeholders might include your neighbors, interest groups, customers and others. Their views might address how your organization affects the environment, how well you are meeting environmental obligations, and whether your organization is a "good neighbor," among other topics. Gathering this information allows you to consider stakeholder input in the development of your environmental policy. Since you have already assessed your legal and environmental aspects, you should be in a good position to have meaningful dialogues with these stakeholders.

Step 4: Prepare Environmental Policy

At this point, you should have a sound basis for developing (or amending) an environmental policy. Using the information developed in the previous three steps allows your organization to prepare a policy that is relevant to the organization and the key issues that it faces. You should understand how well you are currently managing key issues. For example, information on the views of your stakeholders might be valuable in developing an environmental policy.

Step 5: Define Key Roles and Responsibilities

Once the environmental policy has been written, you can begin to define key roles and responsibilities within the EMS. At this stage of implementation, focus on "higher-level" responsibilities, such as the roles and responsibilities of senior management, key functional leaders and environmental staff (if one exists). Once the key roles and responsibilities have been defined, obtain the input of these individuals in the next step of the process - establishing objectives and targets. EMS responsibilities for other specific jobs or functions will be identified in a later step.

Step 6: Establish Objectives and Targets

These objectives should be consistent with your environmental policy and the analyses you carried out on legal requirements, environmental aspects and impacts, and the views of interested parties, etc.

You have identified the operations and activities related to environmental aspects and impacts, and key roles and responsibilities. This information will help you to determine the relevant functions within the organization for achieving objectives and targets. For example, if you set an objective to reduce hazardous waste generation by 10 percent this year, you also should know which parts of the organization must be involved in order to meet this objective.

Step 7: Develop Environmental Management Programs, Identify Operational Controls, and Identify Monitoring and Measurement Needs

This brings us to one of the most challenging (and potentially most valuable) steps in developing an EMS. You are ready to tackle several EMS elements simultaneously. These elements include the design of environmental management programs (EMPs), the initial identification of necessary operational controls, and the initial identification of monitoring and measurement needs. You should already have a head start on this step, since you identified operations and activities related to your significant environmental aspects (as well as existing control and monitoring processes) several steps ago.

One reason for combining these steps is that they can often overlap significantly. For example, you're EMPs for achieving a certain objective (such as maintaining compliance with regulations) might consist of a number of existing operational controls (procedures) and monitoring activities. Achieving an objective might require a feasibility study or the implementation of certain "new" operational controls. Determining progress on achieving objectives often requires some form of monitoring or measurement.

Compile a list of your operational control and monitoring needs. As you develop your EMPs, ask yourself the following questions:

- How do we control this operation or activity now?
- Are these controls adequate to meet our objectives and to ensure compliance?
- If additional controls are needed, what types of controls make sense?
- What type of monitoring / measurement is needed to track our progress in achieving objectives and to ensure that operational controls are implemented as designed?

This process is usually repetitive. You might need to revisit your management programs, operational controls, and monitoring processes over time to ensure they are consistent and up-to-date.

Step 8: Establish Corrective Action, Document Control, and Records Management Processes

At this stage of implementation, your EMS will begin to generate some documents (such as procedures and forms) and records that demonstrate that various processes are being carried out. You need an effective way to manage the records that your EMS generates.

Establish procedures for corrective/preventive action, document control, and records management. These three processes are essentially "system maintenance" functions. As you develop and implement procedures, work instructions, and other EMS documents, you will need a process for controlling the generation and modification of these documents to ensure that you can correct problems when they occur, as well as manage records properly (e.g., monitoring activities).

Step 9: Establish Operational Controls & Monitoring Processes

Refer to the list of operational control and monitoring needs from Step 7. Use a template for the development of work instructions or standard operating procedures. Employees that work in relevant operations or activities can provide support here.

Step 10: Define Job-Specific Roles and Responsibilities

Roles and responsibilities should address the specific operational controls and monitoring processes discussed above. You might want to document these responsibilities in a responsibility matrix or in some other form that is easily communicated to employees.

Step 11: Plan and Conduct Initial Employee Awareness

Initial employee awareness training should be conducted to promote understanding of the organization's EMS efforts and the progress made to date. As a first step, train employees on the environmental policy and other elements of the EMS. Discuss the environmental impacts of their activities, any new or modified procedures, the organization's objectives and targets, as well as their EMS responsibilities. If you have contractors or others at your site who are not employees of your organization, consider whether these other individuals should be included in these EMS awareness training sessions.

Step 12: Establish Other System-Level Procedures

Some system-level procedures (such as the procedures for identification of environmental aspects) were developed at earlier stages of the process. At this point, you can establish any other procedures required for the EMS.

These other system-level procedures might include, for example:

- employee training and awareness,

- internal and external communication,
- emergency preparedness and response,
- EMS auditing, and
- management review.

Step 13: Prepare EMS Documentation (Manual)

Once you have established roles and responsibilities and defined all of your system-level procedures, preparing the EMS manual should be a relatively simple matter. The manual should summarize the results of your efforts. It should describe the processes developed, define the roles and responsibilities, as well as other EMS elements. It is important to describe the links among system elements and provide direction to other system documents. Keep the manual simple - there is no need to provide great detail on any particular system process.

Step 14: Plan and Conduct Specific Employee Training

Once the procedures and other system documentation have been prepared, you are ready to conduct specific employee EMS training. As a first step, identify specific training needs. Employee training should be designed to ensure understanding of (1) key system processes, (2) operational controls related to their specific jobs, and (3) any monitoring or measurement for which they are responsible. At this point, you should have sufficient EMS processes in place to begin to "Check" your EMS.

3. CHECK

Checking, including monitoring and corrective action

As discussed earlier, your EMS should be built on the "Plan, Do, Check, Act" model to ensure that environmental matters are systematically identified, controlled, and monitored. Using this approach will help to ensure that performance of your EMS improves over time and that you meet your goals for implementing an EMS in the first place.

This stage continues the step-by-step action plan for developing and implementing the elements of an EMS. By this time, you should have sufficient EMS processes in place to begin to "check" your EMS. One approach is discussed below.

Conduct Internal EMS Audits:

Once internal auditors have been selected and trained, you should design and initiate the internal auditing process. At this point, you should have sufficient EMS processes in place to conduct meaningful audits. Many organizations find that it is easier to start with smaller, more frequent audits than to audit the entire EMS at once. These early audits can serve as a learning tool. Audit records should be managed in accordance with the records management process. Once the audit results are known, use the corrective and preventive action process to address any identified problems.

4. ACT

Reviewing, including progress reviews and acting to make needed changes to the EMS

This stage continues the step-by-step action plan for developing and implementing the elements of an EMS. At this point in the EMS Cycle, you may have identified problems with your EMS and should act to resolve these issues. One approach is discussed below.

Conduct Management Reviews:

Use the results of your internal audits (along with other information on the EMS) to conduct management reviews. The management of your organization should consider the need for any changes to the EMS based on your review, and make assignments for any revisions needed. Such assignments should be consistent with the roles and responsibilities previously established. After acting on the results of the management review, tasks performed in the "Plan" stage should be revisited, thus continuing the "full circle" process.

EMS SOFTWARE

The tables below identify possible EMS software providers from VRM Group, LLC with permission and from the EPA Environmental Management Systems: An Implementation Guide for Small and Medium-Sized Organizations, Second Edition (identified as one of the references).

*Table 15 – **Table 15EMS Commercial Systems***

COMMERCIAL SYSTEMS					
SYSTEM COMPANY	DAKOTA	ENVIANCE	IHS MARKET	LOGICAL DATA SOLUTIONS	INTELEX
PRODUCT NAME	DAKOTA	ENVIANCE	IHS ESSENTIAL SUITE 7.5	EPOCH	INTELEX
CAPABILITY					
RECURRING TASK MANAGEMENT	YES	YES	YES	YES	YES
WEB BASED	YES	YES	YES	WEB ACCESS TO STAND ALONE	YES
REGULATORY CHANGE NOTICE PROVIDED	YES	YES	YES	NO	YES
TASK APPLICABILITY	YES	YES	YES	NO	YES
EMAIL NOTIFICATION REGARDING TASKS AND ALERTS	YES	YES	YES	YES	YES
DATA LINK TO REAL TIME DATA	NO	YES	YES	YES	YES
EVALUATION OF DATA TO CREATE ALERT	NO	YES	YES	YES	YES
CALCULATIONS	NO	YES	YES	YES	YES
HANDHELD					

FIELD DATA ENTRY	YES	YES	YES	YES	YES
AUDIT CHECK-LISTS GENERA-TION	YES	YES	YES	NO	YES
AUDIT TRACK-ING	YES	YES	YES	YES	YES
TASK GENERA-TION REFER-ENCE	YES	YES	YES	YES	YES
TASK TRACKING	YES	YES	YES	YES	YES
INCIDENT MANAGEMENT	YES	YES	YES	YES	YES
WEB ADDRESS	https://www.dakotasoft.com/	https://www.enviance.com/	https://ihsmarkit.com/products.html	http://www.logicalds.com/	https://www.intelex.com/products/environment

Table 16 – EPA EMS Software Packages, Source - EPA Environmental Management Systems: An Implementation Guide for Small and Medium-Sized Organizations, Second Edition, Appendix F

EMS SOFTWARE PACKAGES		
Organization	Contact Info	Description
Greenware	1-800-474-0627 www.greenware.com	Provides ISO 14001 assessment, implementation and audit software
EMSoft2000	1-800-241-3618 www.rmtinc.com	Software package based on LotusNotes to support EMS implementation
ISOXpert	1-800-ISO-EASY	Built on LotusNotes platform. Customizable document formats.
ISOSoft 14001	416-679-0119 www.isogroup.simplenet.com/soft14k	Provides ISO 14001 assessment, implementation and audit software. Co-developed with BSI.

EMS Documents for Your Review

1. EPA, Stapleton, J., Glover, M., Davis, S., 2001, Environmental Management Systems: An Implementation Guide for Small and Medium-Sized Organizations, Second Edition, available from https://www.epa.gov/sites/production/files/2015-07/documents/ems_an-implementation-guide-for-small-and-medium-sized-organizations_2nded.pdf (72)
2. United States Environmental Protection Agency Position Statement on Environmental Management Systems (EMSs) (77)
3. TCEQ, 2011 A Guide to Developing an Environmental Management System for a Small Business available from https://www.tceq.texas.gov/assets/public/comm_exec/pubs/gi/gi-304.pdf (91)

4. EPA Website, Environmental Management Systems (EMS) available from https://www.epa.gov/ems (19)
5. EPA, 2000, Integrated Environmental Management Systems, A Company Manual Template A Company Manual Template for Small Business, available from https://www.epa.gov/sites/production/files/2013-12/documents/iems_manual_template_small_business.pdf (25)
6. International Organization for Standardization, 2010, ISO14001 ENVIRONMENTAL MANAGEMENT SYSTEMS An easy-to-use checklist for small business. Are you ready? available from https://www.iso.org/files/live/sites/isoorg/files/archive/pdf/en/pub100329.pdf (84)

APPENDIX A

Acronyms List

A

AC: advisory circular
AFVO: animal fat and/or vegetable oil
APA: Administrative Procedures Act
APD: Air Permits Division
API: American Petroleum Institute
AQCR: Air Quality Control Region
ASME: American Society of Mechanical Engineers
ASNT: American Society for Non-Destructive Testing
AST: aboveground storage tank
ASTM: American Society for Testing and Materials

B

BACM : best available control measures
BACT: best available control technology
BART: best available retrofit technology
BIF: boilers and industrial furnaces
BMP: Best Management Practice
BOP: blowout preventer
BTEX: benzene, toluene, ethyl benzene, and xylene
Btu: British thermal units
Btu/scf: British thermal units per standard cubic foot/feet

C

CAA: Clean Air Act
CAM: Compliance-assurance monitoring
CERCLA: Comprehensive Environmental Response, Compensation, and Liability Act
CFR: Code of Federal Regulations

CMP: (Texas) Coastal Management Plan
CN: customer identification number
Corps: U.S. Army Corps of Engineers
CRDM: Continuous Release Detection Method
CRP: Clean Rivers Program
CWA: Clean Water Act of 1972 (Federal Water Pollution Control Act)
CZARA: Coastal Zone Act Reauthorization Amendments
CZM: Coastal Zone Management

D

DOE: Department of Energy
DOI: U.S. Department of Interior
DOT: U.S. Department of Transportation
DPS: Department of Public Safety

E

EA: Environmental Assessment
EIS: Environmental Impact Statement
EO: executive order
EORRA: Edible Oil Regulatory Reform Act
E&P: Exploration and Production
EPA: U.S. Environmental Protection Agency
ERNS: Emergency Response Notification System
ESA: Endangered Species Act

F

FAA: Federal Aviation Administration
FCAA: Federal Clean Air Act
FDA: Food and Drug Administration
FEC: Formal Enforcement Conference
FEMA: Federal Emergency Management Agency
FHWA: Federal Highway Administration
FR: Federal Register
FRP Facility Response Plan
FW: Freshwater Inflow and Bay Circulation Action Plan

G

GAO: General Accounting Office
GBC: Galveston Bay Council
GBF: Galveston Bay Foundation
GBP: Galveston Bay Program of the TNRCC
GBNEP: Galveston Bay National Estuary Program Texas
GLO: General Land Office
GOP: general operating permit
GPR: General Pretreatment Regulations

H

H-GAC: Houston-Galveston Area Council
HGCSD: Harris-Galveston Coastal Subsidence District
HL&P: Houston Lighting & Power Company
HP: Habitat Protection Action Plan
HSWA: Hazardous and Solid Waste Amendments (to RCRA)
HWIR: Hazardous Waste Identification Rules
HUD: U.S. Department of Housing and Urban Development

I

IBC: intermodal bulk container
ICP: Integrated Contingency Plan
IFR: internal floating roof tank
IHW: Industrial and Hazardous Waste
IM: intermodal
ISO: International Standards Organization

J

K

L

M

MIC: microbial influenced corrosion
MMS: Minerals Management Service
MSDS: Material Safety Data Sheet
MSW: municipal solid waste
MOU: memorandum of understanding

N

NACE: National Association of Corrosion Engineers
NCP: National Contingency Plan
NDE: Non-Destructive Examination
NEPA: National Environmental Policy Act
NFPA: National Fire Protection Association
NMFS: National Marine Fisheries Service
NOAA: National Oceanic and Atmospheric Administration
NOV: Notice of Violation
NPDES: National Pollutant Discharge Elimination System
NPL: National Priorities List
NPS: Nonpoint Source
NRC: National Response Center

O

OPA: Oil Pollution Act of 1990
OSHA: U.S. Occupational Safety and Health Administration
OSPRA: Texas Oil Spill Prevention and Response Act

P

PBR: permit by rule
PCB: polychlorinated biphenyl
PE: Professional Engineer
PH: Public Health Protection Action Plan
PMAA: Petroleum Marketers Association of America
POTW: publicly owned treatment work
ppb: parts per billion
PPE: Public Participation and Education Action Plan
ppm: parts per million
PS: point source
PSM: Process Safety Management
PSTR petroleum storage tank remediation

Q

QA: quality assurance

R

RA: Regional Administrator

RBI: Risk Based Inspection
RP: Recommended Practice
RCRA: Resource Conservation and Recovery Act
RMS: Release Management Systems
RN: *(regulated entity)* reference number
RP: responsible party
RRC: Texas Railroad Commission
RAP: Research Action Plan

S

SAMP: Special Area Management Plan
SARA: Superfund Amendments and Reauthorization Act
SBEA: Small Business and Environmental Assistance
SCADA: Supervisory Control and Data Acquisition
SCS: Soil Conservation Service
SD: Spill Dumping Action Plan
SIC: source identification code *or* Standard Industrial Classification
SIP: State Implementation Plan
SM: Shoreline Management Action Plan
SMP: State Management Plan
SOP: Site Operating Permit
SOP: Standard Operating Procedure
SP: Species Population Protection Action Plan
SPCC: Spill Prevention, Control, and Countermeasure
STEERS: State of Texas Environmental Electronic Reporting System
STI: Steel Tank Institute
SWANCC: Solid Waste Agency of Northern Cook County
SWCB: Texas State Soil and Water Conservation Board
SWDA: Solid Waste Disposal Act

T

TAC: Texas Administrative Code
TAMU: Texas A&M University
TCAA: Texas Clean Air Act
TCLP: toxicity characteristic leaching procedure

TCMP: Texas Coastal Management Program
TDA: Texas Department of Agriculture
TDH: Texas Department of Health
TEA: Texas Education Agency
TED: Turtle Exeluder Device
Title III: Emergency Planning and Community Right-to-Know Act of 1986
TLAP: Texas Lab Accreditation Program *or* Texas Land Application Permit
TMDL: Total Maximum Daily Load
TNRCC: Texas Natural Resource Conservation Commission
TNRIS Texas Natural Resource Information System
TPWD: Texas Parks and Wildlife Department
TCC: Texas Chemical Council
TSC: Technical Support Committee
TSCA: Toxic Substances Control Act
TWC Texas Water Code
TWDB Texas Water Development Board
TWRI Texas Water Resources Institute
TXDOT Texas Department of Transportation
TxREG: Texas Register

U

UH: University of Houston
UIC: Underground Injection Control
UL: Underwriters Laboratory
USCG: U.S. Coast Guard
USDOT: U.S. Department of Transportation
USFWS: U.S. Fish and Wildlife Service
USGS: U.S. Geological Survey
UST: Underground Storage Tank
UT: University of Texas
UST: Underground Storage Tank
UT: ultrasonic thickness
UTS: ultrasonic thickness scans
UTT: ultrasonic thickness testing

V

VOC: volatile organic compound

VRU: vapor recovery unit or system

VTCS: Vernon's Texas Civil Statutes

W

WCID: Water Control and Improvement District

WEF: Water Environment Federation

WMP: Waste Management Plan

WQS: Water Quality Standard

WRPA; Waste Reduction Policy Act

WSQ; Water and Sediment Quality Action Plan

X

Y

Z

Acronyms References:

TCEQ Acronyms and Other Abbreviations, available from https://www.tceq.texas.gov/about/acron.html (90)

EPA, Terms of Environment, Glossary, Abbreviations and Acronyms, available from https://nepis.epa.gov/Exe/ZyPDF.cgi/200081E1.PDF?Dockey=200081E1.PDF (74)

EPA, NEPIS Environmental Acronyms, Abbreviations and Glossary of Terms, available from https://nepis.epa.gov/Exe/ZyPDF.cgi/9100VQRW.PDF?Dockey=9100VQRW.PDF (31)

APPENDIX B

TCEQ Contact List (151)

Listed below are contact numbers for various groups and support areas that can help on compliance issues or provide information associated with typical Used Oil facility operations.

The TCEQ does not have a general toll-free number. The main switchboard can be contacted by dialing 512-239-1000.

Using the main switchboard is helpful if you have a question on who you should be talking to depending on your general issue such as Used Oil permitting or RCRA issues. Experience with the system over the years is that even if they are not the person to help you, they will try to direct you as best as they can.

Small Business Assistance: 800-447-2827

Hotline for small businesses to obtain information, advice and confidential assistance from the TCEQ. Answered from 8 a.m. to noon and from 1 p.m. to 5 p.m. on business days. As identified previously this is a number you should know and not hesitate to call if you have a question.

TCEQ Used Oil Program: 512/239-6832 (select Option 2)

The number directly to staff that manage the used oil recycle program.

TCEQ Municipal Solid Waste Permits Section: 512/239-2334

TCEQ Industrial and Hazardous Waste Permits Section: 512/239-6412

TCEQ Industrial and Hazardous Waste Program: 512/239-6413

TCEQ Used Oil Collection Center Information Line: 800-CLEAN-UP

This is a prerecorded, menu-driven system that lets you to listen to selected information.

Railroad Commission of Texas (RRC): 512/463-6887

In the event you have dealings with the oil and gas industry that are under the jurisdiction of the RRC, even though not typical for most Used Oil Handlers but it can happen and you should insure anything you do is in compliance. This guide does not deal with regulations associated with RRC functions so will need to contact the Commission directly or start with the Small Business Assistance group identified above. The basic description of the RRC follows: (7) "The Railroad Commission of Texas (Commission) is the state agency with primary regulatory jurisdiction over the oil and natural gas industry, pipeline transporters, natural gas and hazardous liquid pipeline industry, natural gas utilities, the LP-gas industry, and coal and uranium surface mining operations. The Commission exists under provisions of the Texas Constitution and exercises its statutory responsibilities under state and federal laws for regulation and enforcement of the state's energy industries. The Commission also has regulatory and enforcement responsibilities under federal law including the Surface Coal Mining Control and Reclamation Act, Safe Drinking Water Act, Pipeline Safety Acts, Resource Conservation Recovery Act, and Clean Water Act."

Public Participation in the Permitting Process: 800-687-4040

One-stop calling for the general public to inquire about permitting activity within the TCEQ.

Superfund Community Relations Line: 800-633-9363

Local citizens may call with questions and concerns regarding state and federal superfund sites in their area.

Customer Service Compact with Texans: 855-685-8237

For use by the public to contact TCEQ regarding customer service concerns.

Environmental Complaints Hot Line: 888-777-3186

Texas residents can use this line to report environmental complaints. Calls will be routed automatically to the closest TCEQ regional office. Callers after business hours may leave a recorded message.

Fraud Reporting Hot Line: 877-901-0700

For use by the public to report allegations of fraud, waste, or abuse related to TCEQ operations.

Spill Reporting (24-Hour): 800-832-8224 or 512/239-2507

Used to report spills, unpermitted discharges or releases to the environment, and SARA/EPCRA report notices. Supported by several Texas agencies, this line is answered

24 hours a day and serves as the TCEQ spill reporting line during the day and the State Emergency Response Commission (SERC) line at night. This number is extremely important to keep available just in case. More detail regarding spill reporting will be provided later in this guide.

Non-Spill Emergencies Hot Line: 888-777-3186

Texas residents can use this line to report environmental emergencies that are not spills. Calls are routed automatically to the closest TCEQ regional office. After business hours, callers may leave a recorded message or be forwarded to an after-hours pager or answering service.

Tier II: (EPCRA) 800-452-2791 or 202/554-1404:

For companies in Texas that submit Tier II reports and have questions about complying with state and federal community right-to-know laws, or Texas residents and local emergency planning and response agencies who have questions or concerns about hazardous chemicals in their communities.

EPA Interaction With TCEQ:

EPA Region 6 Office: 214/665-7224

Texas is located in EPA Region 6, and even though Texas is authorized to administer its RCRA program the EPA still provides oversight. The EPA can get involved under certain circumstances and also administers federal regulations for which Texas is yet to be authorized. The Hazardous Waste Enforcement Branch, along with the state, is responsible for enforcing these regulations. The Branch conducts inspections, reviews, and data analysis to assess compliance; the Branch takes enforcement actions and assesses the nature and extent of contamination and clean up.

APPENDIX C

Selected Terms

Familiarize yourself with the following selected terms regarding the CWA from Federal and Texas regulation and guidance. Where terms are materially different, both federal and state definitions are listed. This listing is not intended to represent all terms you may need to know.

(Citations (terms) are directly from reference listing above)

1. Federal - Discharge or to discharge – To deposit, conduct, drain, emit, throw, run, allow to seep, or otherwise release or dispose of, or to allow, permit or suffer any of these acts or omissions. For purposes of this permit, an allowable discharge of wastewater can only occur adjacent to water in the state, by evaporation.
2. Federal - Effluent - Wastewater discharged from an industrial facility, sewage treatment plant, or other point source discharge (see definition below).
3. Federal - Effluent limit/limitation - A restriction on quantities, discharge rates, and concentrations of pollutants (chemical, physical, biological) discharged from a point source into surface waters under the jurisdiction of the CWA. Effluent limits are set forth as enforceable requirements in NPDES permits.
4. Texas - Evaporation Pond – A type of surface impoundment that stores and evaporates wastewater.
5. Federal - General permit - With regard to both the CWA 402 (NPDES) and 404 permit programs, a permit authorizing specified discharges of pollutants to waters of the United States from more than one source or activity. General permits apply to similar sorts of sources/activities taking place within a specified geographic area. Essentially the same requirements apply to all sources covered by a given general permit.
6. Federal - Indirect discharger - An facility that, instead of discharging pollutants or wastewater directly into a water of the United States, sends them into a municipal sewer system. Though not covered by the CWA's NPDES program, indirect discharges of toxic chemicals from industrial operations are subject to the CWA's pretreatment program (see below).

7. Texas - Liner – Any barrier in the form of a layer, membrane, or blanket, either naturally existing, constructed, or installed; to prevent a significant hydrologic connection between liquids contained in retention facilities and water in the state.
8. Federal - Nonpoint source (NPS) - A diffuse source of water pollution that is not collected and discharged through a discrete man-made conveyance. (i.e, any source of pollutants not considered a point source under the CWA.) Atmospheric deposition and hydromodification are also nonpoint sources of pollution.
9. Federal - NPDES - National Pollutant Discharge Elimination System. A CWA-established national program regulating direct discharges of pollutants from point sources to the waters falling under the jurisdiction of the CWA.
10. Texas – TPDES - Texas Pollutant Discharge Elimination System. The state program for issuing, amending, terminating, monitoring, and enforcing permits, and imposing and enforcing pretreatment requirements, under the CWA § 307, 402, 318 and 405, 40 CFR §122.2, TWC, and TAC regulations. The term includes an "approved program."
11. Federal - Outfall - A place where wastewater or stormwater, or a combination thereof, is discharged into a waterbody via a point source.
    a. Texas - Outfall. For the purpose of this permit, a point source at the point where stormwater runoff associated with industrial activity, and certain non-stormwater discharges listed in this permit, exits the facility and discharge(s) to surface water in the state or a municipal or private separate storm sewer system. An outfall from a diffuse point source includes the point or points where the diffuse point source discharges to surface water in the state or a municipal or private separate storm sewer system.
12. TEXAS - Operator. A person responsible for the management of an industrial facility subject to the provisions of this general permit. Industrial facility operators include entities with operational control over industrial activities, including the ability to modify those activities; or entities with day-to-day operational control of activities at a facility necessary to ensure compliance with the permit (e.g., the entity is authorized to direct workers at a facility to carry out activities required by the permit).
13. Federal - Point source - A discernable, discrete, confined man-made conveyance whereby pollutants are discharged directly to surface waters falling under the juris-

diction of the CWA. Such conveyances include pipes, ditches, channels, tunnels, conduits, wells, discrete fixtures, containers, rolling stock (wheeled vehicles), concentrated animal feeding operations, landfill leachate collection system, and vessels and other floating craft.

a. TEXAS - Point Source - Any discernible, confined, and discrete conveyance, including but not limited to, any pipe, ditch, channel, tunnel, conduit, well, discrete fissure, container, rolling stock, concentrated animal feeding operation, landfill leachate collection system, vessel or other floating craft from which pollutants are or may be discharged. This term does not include return flows from irrigated agriculture or agricultural stormwater runoff. For the purpose of this permit, a point source includes any identifiable conveyance from which pollutants might enter surface water in the state, including a diffuse point source as defined in this section.

14. Federal - Pollutant - Industrial, municipal, and agricultural waste, including sewage, chemical wastes, dredged spoil, sand, sewage sludge, solid waste, garbage, rock, heat, filter backwash, discarded equipment, munitions, and radioactive and biological materials discharged into water.
15. Federal - Pretreatment program - Applicable to indirect industrial dischargers (see definition above), this program sets technology-based limits on levels of toxic pollutants in wastewater from such facilities that are discharged into municipal sewer systems.
16. Federal - Process wastewater - Pollutant-carrying water resulting from contact with materials used in industrial manufacturing processes.
17. Federal - Publicly owned treatment works (POTW) - A sewer collection and treatment system owned by a municipality or a state. Includes sewers, pipes, conveyances, and devices and systems used to store, treat, recycle, and reclaim sewage or industrial wastewater.
18. Federal - Sanitary sewer system - A system of pipes and other conveyances that collects and carries waste from homes, offices, industrial facilities, and other kinds of buildings and operation to a municipal sewage treatment facility. Sanitary sewers to not carry urban stormwater runoff. (See also "MS4s" and "CSOs".)
19. Federal - Secondary treatment - Treatment of wastewater employing a combina-

tion of physical and biological processes, such as "activated sludge". Refers specifically to wastewater treatment requirements applying to discharges of BOD5, total suspended solids (TSS), and pH from municipal sewage treatment facilities.

20. Texas - Stormwater Discharge Associated with Industrial Activity - The discharge from any conveyance that is used for collecting and conveying stormwater and that is directly related to manufacturing, processing or raw materials storage areas at an industrial facility. For the purpose of this general permit, the term includes, but is not limited to, stormwater discharges from industrial plant yards; immediate access roads and rail lines used or traveled by carriers of raw materials, manufactured products, waste material, or by-products used or created by the facility; material handling areas; refuse/waste disposal areas; sites used for the application or disposal of process waste waters; sites used for the storage and maintenance of material handling equipment; sites used for residual treatment, storage, or disposal; shipping and receiving areas; manufacturing buildings; storage areas (including tank farms), intermediate products, and final products; similar areas where stormwater can contact pollutants related to industrial activity; and areas where industrial activity have taken place in the past and significant materials remain and are exposed to stormwater. For the purposes of this definition, materials handling areas include storage, loading and unloading, transportation, or conveyance of any raw material, intermediate product, final product, by-product or waste product. The term excludes areas located at industrial sites that are separate from the facility's industrial activities, such as office buildings and accompanying parking lots, as long as the drainage from the excluded areas is not mixed with stormwater drained from areas of a facility that are covered by this general permit. This term includes discharges from facilities described under this general permit that are operated by federal, state, or municipal entities. For the complete regulatory definition, including the categories of industrial activity, see 40 CFR §122.26(b)(14).
21. Texas - Stormwater and Stormwater Runoff. Rainfall runoff, snowmelt runoff, and surface runoff and drainage.
22. Federal - Surface impoundment –A natural topographic depression, man-made excavation, or diked area formed primarily of earthen materials (although it may

be lined with man-made materials), which is designed to hold an accumulation of liquid wastes or wastes containing free liquids.

23. Texas - Surface Water in the State. Lakes, bays, ponds, impounding reservoirs, springs, rivers, streams, creeks, estuaries, wetlands, marshes, inlets, canals, the Gulf of Mexico inside the territorial limits of the state (from the mean high water mark (MHWM) out 10.36 miles into the Gulf), and all other bodies of surface water, natural or artificial, inland or coastal, fresh or salt, navigable or nonnavigable, and including the beds and banks of all water-courses and bodies of surface water, that are wholly or partially inside or bordering the state or subject to the jurisdiction of the state; except that waters in treatment systems that are authorized by state or federal law, regulation, or permit, and that are created for the purpose of waste treatment are not considered to be water in the state.
24. Federal - SWPPP - A written document that identifies the industrial activities conducted at the site, including any structural control practices, which the industrial facility operator will implement to prevent pollutants from making their way into stormwater runoff. The SWPPP also must include descriptions of other relevant information, such as the physical features of the facility, and procedures for spill prevention, conducting inspections, and training of employees. The SWPPP is intended to be a "living" document, updated as necessary, such that when industrial activities or stormwater control practices are modified or replaced, the SWPPP is similarly revised to reflect these changes.
25. Texas - Waters of the United States (from 40 CFR §122.2). Waters of the United States or waters of the U.S. means: (a) all waters that are currently used, were used in the past, or may be susceptible to use in interstate or foreign commerce, including all waters that are subject to the ebb and flow of the tide; (b) all interstate waters, including interstate wetlands; (c) all other waters such as intrastate lakes, rivers, streams (including intermittent streams), mudflats, sandflats, wetlands, sloughs, prairie potholes, wet meadows, playa lakes, or natural ponds that the use, degradation, or destruction of which would affect or could affect interstate or foreign commerce including any such waters: (1) that are or could be used by interstate or foreign travelers for recreational or other purposes; (2) from which fish or shellfish are or could be taken and sold in interstate or foreign commerce; or (3) that are used or could be used for industrial purposes by indus-

tries in interstate commerce; (d) all impoundments of waters otherwise defined as waters of the U.S. under this definition; (e) tributaries of waters identified in paragraphs (a) through (d) of this definition; (f) the territorial sea; and (g) wetlands adjacent to waters (other than waters that are themselves wetlands) identified in paragraphs (a) through (f) of this definition.

26. Texas - Waste treatment systems, including treatment ponds or lagoons designed to meet the requirements of the CWA (other than cooling ponds as defined in 40 CFR §423.11(m) that also meet the criteria of this definition) are not waters of the U.S. This exclusion applies only to manmade bodies of water that neither were originally created in waters of the U.S. (such as disposal area in wetlands) nor resulted from the impoundment of waters of the U.S. [See Note 1 of this section.] Waters of the U.S. do not include prior converted cropland. Notwithstanding the determination of an area's status as prior converted cropland by any other federal agency, for the purposes of the CWA, the final authority regarding CWA jurisdiction remains with EPA.
27. Federal - Wetland - For regulatory purposes under the Clean Water Act, the term wetlands means those areas that are inundated or saturated by surface or groundwater at a frequency and duration sufficient to support, and that under normal circumstances do support, a prevalence of vegetation typically adapted for life in saturated soil conditions. Wetlands generally include swamps, marshes, bogs and similar areas.

APPENDIX D

CERCLA References for Your Review

1. EPA, 2001, RCRA, Superfund & EPCRA Call Center Training Module, Used Oil (40 CFR Part 266, Subpart E and Part 279) available from https://www.epa.gov/sites/production/files/2015-09/documents/uoil.pdf (46)
2. ENVIRONMENTAL PROTECTION AGENCY, 40 CFR Part 312 [EPA–HQ–SFUND–2014–0474; FRL–9917–28–OSWER], Amendment to Standards and Practices for All Appropriate Inquiries, Federal Register /Vol. 79, No. 193 /Monday, October 6, 2014 /Rules and Regulations, available from https://www.gpo.gov/fdsys/pkg/FR-2014-10-06/pdf/2014-23399.pdf (6)
3. EPA, FactSheet All Appropriate Inquiries Rule: Reporting Requirements and Suggestions on Report Content, available from https://www.epa.gov/sites/production/files/2015-05/documents/aai_reporting_factsheet.pdf (5)
4. EPA, Memorandum 2003, Interim Guidance Regarding Criteria Landowners Must Meet in Order to Quality for Bona Fide Prospective Purchaser, Contiguous Property Owner, or Innocent Landowner Limitations on CERCLA Liability ("Common Elements"), available from https://www.epa.gov/sites/production/files/documents/common-elem-guide.pdf (30)
5. EPA - Assessing Contractor Capabilities for Streamlined Site Investigations -- Additional Information Regarding All Appropriate Inquiries and Hiring an Environmental Professional, available from https://www.epa.gov/sites/production/files/2015-09/documents/hiringep_addendum_factsheet.pdf (7)
6. Cornell Law School, Comprehensive Environmental Response, Compensation and Liability Act (CERCLA), available from https://www.law.cornell.edu/wex/comprehensive_environmental_response_compensation_and_liability_act_cercla (3)

7. EPA, Criminal Provisions of the Comprehensive Environmental Response, Compensation - and Liability Act (CERCLA), available from https://www.epa.gov/enforcement/criminal-provisions-comprehensive-environmental-response-compensation-and-liability-act (12)
8. EPA, Memorandum, 1987, Scope of the CERCLA Petroleum Exclusion Under Sections 1O1(14) and 104(a)(2), available from https://www.epa.gov/sites/production/files/2013-09/documents/petro-exclu-mem.pdf (29)
9. EPA Fact Sheet, All Appropriate Inquiries Rule: Definition Of Environmental Professional, available from https://www.epa.gov/sites/production/files/2015-09/documents/ep_deffactsheet.pdf (5)

APPENDIX E

CERCLA Related Definitions

De Minimis Condition: Means a condition that does not threaten human health or environment and generally would not be subject to enforcement action (Per the ASTM Standard)

Joint and Several Liabilities: Liability that may be apportioned either among two or more parties or to only one or a few select members of the group, at the adversary's discretion. Thus, each liable party is individually responsible for the entire obligation, but a paying party may have a right of contribution or indemnity from nonpaying parties. (Black's Law Dictionary, Fifth Edition, 2016)

Recognized Environmental Condition (REC): As defined in ASTM 1527-05, this means "the presence or likely presence of any hazardous substances or petroleum products on a property under conditions that indicate an existing release, a past release, or a material threat of a release of any hazardous substances or petroleum products into structures on the property or into the ground, ground water, or surface water of the property." The term includes hazardous substances or petroleum products under conditions in compliance with laws (e.g., permitted discharges). Under this definition, a recognized environmental condition could not only relate to spills, releases, or other unauthorized disposal of hazardous substances or petroleum products, but also permitted or otherwise authorized discharges or disposal activities.

Service Station Dealers (SSD) CERCLA Section 101(37)(A) defines "service station dealer" as any person (i) who owns or operates a motor vehicle service station, filling station, or garage, or similar retail establishment engaged in the business of selling, repairing, or servicing motor vehicles, where a significant percentage of the gross revenue of the establishment is derived from the fueling, repairing, or servicing motor vehicles, and (ii) who accepts for collection, accumulation, and delivery to an oil recycling facility, recycled oil that (I) has been removed from the engine of a light duty motor vehicle or appliance, and (II) is presented, by such owner, to such person for collection, accumulation, and delivery to an oil recycling facility.

Strict Liability: Liability that does not depend on proof of negligence or intent to do harm but that is based instead on duty to compensate the harms proximately caused by the activity or behavior subject to the liability rule. (Black's Law Dictionary, Fifth Edition, 2016)

REFERENCES

1. Arner, R. (2006) Used Oil Recycling in America, Presented to The Twenty-Second International Conference on Solid Waste Technology and Management March 20, 2007 Philadelphia, PA U.S.A.
2. Boulding, K. E. (1978) Ecodynamics: A New Theory of Societal Evolution, Management of Used Oil
3. Comprehensive Environmental Response, Compensation and Liability Act (CERCLA), Cornell Law School, retrieved from https://www.law.cornell.edu/wex/comprehensive_environmental_response_compensation_and_liability_act_cercla (2017)
4. De Blasi, M. (2014), LIABILITY OF PARENT CORPORATIONS, OFFICERS, DIRECTORS, AND SUCCESSORS: When Can CERCLA Liability Extend Beyond the Company? retrieved from http://arizonastatelawjournal.org/wp-content/uploads/2014/07/DeBlasi_Final.pdf
5. EPA, All Appropriate Inquiries Rule: Reporting Requirements and Suggestions on Report Content, FactSheet, retrieved from https://www.epa.gov/sites/production/files/2015-05/documents/aai_reporting_factsheet.pdf (2015)
6. EPA, Amendment to Standards and Practices for All Appropriate Inquiries, 40 CFR Part 312 [EPA–HQ–SFUND–2014–0474; FRL–9917–28–OSWER], Federal Register /Vol. 79, No. 193 /Monday, October 6, 2014 /Rules and Regulations, retrieved from https://www.gpo.gov/fdsys/pkg/FR-2014-10-06/pdf/2014-23399.pdf (2014)
7. EPA, Assessing Contractor Capabilities for Streamlined Site Investigations -- Additional Information Regarding All Appropriate Inquiries and Hiring an Environmental Professional, retrieved from https://www.epa.gov/sites/production/files/2015-09/documents/hiringep_addendum_factsheet.pdf (2015)
8. EPA, CAA, Website, retrieved from https://www.epa.gov/clean-air-act-overview/clean-air-act-text#what (2017)
9. EPA, CAA, The Plain English Guide to the Clean Air Act, retrieved from

https://www.epa.gov/sites/production/files/2015-08/documents/peg.pdf
10. EPA, Clean Air Act Title V – Permits, retrieved from https://www.epa.gov/clean-air-act-overview/clean-air-act-title-v-permits (2017)
11. EPA, Copeland C. (2016) EPA - Clean Water Act: A Summary of the Law, retrieved from https://fas.org/sgp/crs/misc/RL30030.pdf
12. EPA, Criminal Provisions of the Comprehensive Environmental Response, Compensation - and Liability Act (CERCLA), retrieved from https://www.epa.gov/enforcement/criminal-provisions-comprehensive-environmental-response-compensation-and-liability-act (2017)
13. EPA, CWA, Website retrieved from https://www.epa.gov/laws-regulations/summary-clean-water-act (2017)
14. EPA, Designation of Hazardous Substances, CFR 40, Subchapter J, §302.4 (2017)
15. EPA, Developing Your Stormwater Pollution Prevention Plan, A Guide for Industrial Operators (2015)
16. EPA, Developing Your Stormwater Pollution Prevention Plan: A Guide for Industrial Operators (2009) retrieved from https://www3.epa.gov/npdes/pubs/industrial_swppp_guide.pdf
17. EPA, Emergency Planning and Community Right-to-Know Act (EPCRA), retrieved from https://www.epa.gov/epcra (2017)
18. EPA, Environmental Enforcement, A Citizen's Guide (1990)
19. EPA, Environmental Management Systems (EMS), Website, retrieved from https://www.epa.gov/ems (2017)
20. EPA, Environmental Regulations and Technology, Managing Used Motor Oil (1994)
21. EPA, Facility Response Plans, 40 C.F.R. Part 112, Subpart D, §112.20 (2017)
22. EPA, Facility Response Planning Compliance Guide, retrieved from https://www.epa.gov/sites/production/files/2014-04/documents/frpguide.pdf (2002)
23. EPA, Guidance And Summary Of Information Regarding The RCRA Used Oil Rebuttable Presumption, Publication Number 905-R03-005 (2005)
24. EPA, (2017) History: Resource Conservation and Recovery Act, retrieved from US EPA, retrieved from https://www.epa.gov/history/epa-history-resource-

conservation-and-recovery-act

25. EPA (2000) Integrated Environmental Management Systems, A Company Manual Template A Company Manual Template for Small Business, retrieved from https://www.epa.gov/sites/production/files/2013-12/documents/iems_manual_template_small_business.pdf
26. EPA , Introduction to the Clean Water Act, retrieved from https://cfpub.epa.gov/watertrain/pdf/modules/introtocwa.pdf (2017)
27. EPA, Key Elements to Include in a Facility Response Plan (FRP), 40 C.F.R. Part 112, Subpart D, Appendix F (2017)
28. EPA, Managing Used Oil, Advice For Small Businesses (1966)
29. EPA, Memorandum (1987), Scope of the CERCLA Petroleum Exclusion Under Sections 101(14) and 104(a)(2), retrieved from https://www.epa.gov/sites/production/files/2013-09/documents/petro-exclu-mem.pdf
30. EPA, Memorandum 2003, Interim Guidance Regarding Criteria Landowners Must Meet in Order to Quality for Bona Fide Prospective Purchaser, Contiguous Property Owner, or Innocent Landowner Limitations on CERCLA Liability ("Common Elements"), retrieved from https://www.epa.gov/sites/production/files/documents/common-elem-guide.pdf
31. EPA, NEPIS Environmental Acronyms, Abbreviations and Glossary Of Terms, retrieved from https://nepis.epa.gov/Exe/ZyPDF.cgi/9100VQRW.PDF?Dockey=9100VQRW.PDF (2017)
32. EPA, Notification of RCRA Subtitle C Activity - Instructions and Form 8700-12, (2012) retrieved from https://www.epa.gov/hwgenerators/instructions-and-form-hazardous-waste-generators-transporters-and-treatment-storage-and
33. EPA, Oil Pollution Prevention, Spill Prevention, Control, and Countermeasure Plan requirements for onshore facilities (excluding production facilities), 40 C.F.R. Subchapter D, Subpart B §112.8
34. EPA, Oil Pollution Prevention, Memorandum of Understanding Between the Secretary of Transportation and the Administrator of the Environmental Protection Agency, 40 C.F.R. Part 112 Subpart D Appendix A (2017)
35. EPA, Oil Pollution Prevention, Response Requirements, Substantial Harm

Criteria, 40 C.F.R. Subpart D Appendix C (2017)

36. PA, Oil Pollution Prevention, Response Requirements, Determination and Evaluation of Required Response Resources for Facility Response Plans, 40 C.F.R. Subchapter D, Part 112, Subpart D, Appendix E (2017)

37. EPA, Oil Pollution Prevention, Response Requirements, Facility response training and drills/exercises, 40 C.F.R. Subchapter D, Part 112, Subpart D, §112.21 (2017)

38. EPA, Oil Pollution Prevention, Response Requirements, Facility-Specific Response Plan, 40 C.F.R. Subchapter D, Part 112, Appendix F (2017)

39. EPA, Oil Pollution Prevention, Response Requirements, Tier I Qualified Facility SPCC Plan, 40 C.F.R., Part 112, Subpart D, Appendix G (2017)

40. EPA, Oil Pollution Prevention, Requirements for Petroleum Oils and Non-Petroleum Oils, Except Animal Fats and Oils and Greases, and Fish and Marine Mammal Oils; and Vegetable Oils (Including Oils from Seeds, Nuts, Fruits, and Kernels) Contents, 40 C.F.R. Subchapter D, Part 112, Subpart B, §112.8 Spill Prevention, Control, and Countermeasure Plan requirements for onshore facilities (excluding production facilities), §112.9 Spill Prevention, Control, and Countermeasure Plan Requirements for onshore oil production facilities (excluding drilling and workover facilities), §112.10 Spill Prevention, Control, and Countermeasure Plan requirements for onshore oil drilling and workover facilities, §112.11 Spill Prevention, Control, and Countermeasure Plan requirements for offshore oil drilling, production, or workover facilities. (2017)

41. EPA, Oil Pollution Prevention, Response Requirements Contents, 40 C.F.R. Part 112, Subpart D, §112.20 Facility response plans, §112.21 Facility response training and drills/exercises, Appendix A to Part 112—Memorandum of Understanding Between the Secretary of Transportation and the Administrator of the Environmental Protection Agency, Appendix B to Part 112—Memorandum of Understanding Among the Secretary of the Interior, Secretary of Transportation, and Administrator of the Environmental Protection Agency, Appendix C to Part 112—Substantial Harm Criteria, Appendix D to Part 112—Determination of a Worst Case Discharge Planning Volume, Appendix E to Part 112—Determination and Evaluation of Required Response Resources for Facility Response Plans, Appendix F to Part 112—Facility-Specific Response Plan, Appendix G to Part 112—Tier I Qualified Facility SPCC Plan (2017)

42. EPA, Oil Pollution Prevention, Requirements for Animal Fats and Oils and Greases, and Fish and Marine Mammal Oils; and for Vegetable Oils, including Oils from Seeds, Nuts, Fruits, and Kernels Contents, 40 C.F.R. Chapter I, Subchapter D, Part 112, Subpart C, §112.12 Spill Prevention, Control, and Countermeasure Plan requirements (2017)
43. EPA, Oil Pollution Prevention, Applicability, Definitions, and General Requirements for All Facilities and All Types of Oils, 40 C.F.R. Subchapter D, Part 112, Subpart A, §112.1 General applicability, §112.2 Definitions, §112.3 Requirement to prepare and implement a Spill Prevention, Control, and Countermeasure Plan, §112.4 Amendment of Spill Prevention, Control, and Countermeasurc Plan by Regional Administrator, §112.5 Amendment of Spill Prevention, Control, and Countermeasure Plan by owners or operators, §112.6 Qualified Facilities Plan Requirements, §112.7 General requirements for Spill Prevention, Control, and Countermeasure Plans.(2017)
44. EPA (2016) Prepared Workbook, Process for Risk Evaluation, Property Analysis and Reuse Decisions, retrieved from https://www.epa.gov/sites/production/files/2015-04/documents/munimanual.pdf
45. EPA, RCRA Orientation Manual (2014) retrieved from
46. https://www.epa.gov/sites/production/files/2015-07/documents/rom.pdf EPA (2001), RCRA, Superfund & EPCRA , Call Center Training Module, Introduction to: Used Oil
47. EPA (2001), RCRA, Superfund & EPCRA Call Center Training Module, Used Oil (40 CFR Part 266, Subpart E and Part 279) retrieved from https://www.epa.gov/sites/production/files/2015-09/documents/uoil.pdf
48. EPA, RCRA Training Module, Introduction to United States Environmental Protection Agency Generators (40 CFR Part 262) retrieved from https://www.epa.gov/sites/production/files/2014-12/documents/gen05.pdf
49. EPA, Report Spills and Environmental Violations, retrieved from https://www.epa.gov/pesticide-incidents/how-report-spills-and-environmental-violations
50. EPA, (2005) SPCC Guidance for Regional Inspectors Version 1.0
51. EPA, (2013) SPCC Guidance for Regional Inspectors
52. EPA, (2006) Spill Prevention, Control and Countermeasure (SPCC) Rule

Amendments, Option for Qualified Facilities that Store or Handle 10,000 Gallons or Less of Oil

53. EPA, (2006) Spill Prevention, Control and Countermeasure (SPCC) Rule Amendments, Amended SPCC Requirements Finalized in December 2006
54. EPA, (2011), Spill Prevention, Control, and Countermeasure (SPCC) Rule Streamlined Requirements for Tier I and II Qualified Facilities, FactSheet, retrieved from https://www.epa.gov/sites/production/files/2013-08/documents/qualfac_fs.pdf
55. EPA (2013) Spill Prevention, Control and Countermeasure Plan (SPCC) Program, Bulk Storage Container Inspection Fact Sheet, retrieved from https://www.epa.gov/oil-spills-prevention-and-preparedness-regulations/spill-prevention-control-and-countermeasure-11
56. EPA Standards For The Management Of Used Oil, Subpart A, Definitions, 40 C.F.R. §279.1 (2017)
57. EPA Standards For The Management Of Used Oil, Subpart B, Applicability, 40 C.F.R. §279.10 (2017)
58. EPA Standards For The Management Of Used Oil, Subpart B, Used Oil Specifications, 40 C.F.R. §279.11 (2017)
59. EPA Standards For The Management Of Used Oil, Subpart B, Prohibitions, 40 C.F.R. §279.12 (2017)
60. EPA Standards For The Management Of Used Oil, Subpart C, Hazardous Waste Mixing, 40 C.F.R. §279.21 (2017)
61. EPA Standards For The Management Of Used Oil, Subpart C, Used Oil Storage, 40 C.F.R. §279.22 (2017)
62. EPA Standards For The Management Of Used Oil, Subpart C, On-site burning in space heaters. 279.23 (2017)
63. EPA Standards For The Management Of Used Oil, Subpart C, Off-site shipping, 279.24 (2017)
64. EPA Standards For The Management Of Used Oil, Applicability, 40 C.F.R. Subchapter C §279.20 (2017)
65. EPA Standards For The Management Of Used Oil, Definitions, 40 C.F.R. Subpart C §279.1 (2017)
66. EPA Standards for Used Oil Generators, 40 C.F.R. Subpart C §279.20

Applicability, §279.21 Hazardous waste mixing, §279.22 Used oil storage, §279.23 On-site burning in space heaters, §279.24 Off-site shipments (2017)

67. EPA Standards For The Management Of Used Oil, Applicability, 40 C.F.R. Subchapter C, §279.50-59 (2017)
68. EPA Standards For The Management Of Used Oil, Standards for Used Oil Generators, 40 C.F.R. Subpart C, §279.20 Applicability, §279.21 Hazardous waste mixing. §279.22 Used oil storage, §279.23 On-site burning in space heaters, §279.24 Off-site shipments (2017)
69. EPA Standards for Used Oil Collection Centers and Aggregation Points Contents, 40 C.F.R. Part 279 Subpart D §279.30 Do-it-yourselfer used oil collection centers, §279.31 Used oil collection centers, §279.32 Used oil aggregation points owned by the generator.
70. EPA Standards for Used Oil Transporter and Transfer Facilities, CFR 40, Subpart E, §279.40 Applicability, §279.41 Restrictions on transporters who are not also processors or re-refiners., §279.42 Notification, §279.43 Used oil transportation, §279.44 Rebuttable presumption for used oil, §279.45 Used oil storage at transfer facilities, §279.46 Tracking, §279.47 Management of residues (2017)
71. EPA Standards for Used Oil Fuel Marketers Contents, 40 C.F.R. Part 279 Subpart H §279.70 Applicability, §279.71 Prohibitions, §279.72 On-specification used oil fuel, §279.73 Notification, §279.74 Tracking, §279.75 Notices (2017)
72. EPA, Stapleton, J., Glover, M., Davis, S., (2001) Environmental Management Systems: An Implementation Guide for Small and Medium-Sized Organizations, Second Edition, retrieved from https://www.epa.gov/sites/production/files/2015-07/documents/ems_an-implementation-guide-for-small-and-medium-sized-organizations_2nded.pdf
73. EPA, Superfund: CERCLA Overview, retrieved from https://www.epa.gov/superfund/superfund-cercla-overview (2017)
74. EPA, Terms of Environment, Glossary, Abbreviations and Acronyms, retrieved from https://nepis.epa.gov/Exe/ZyPDF.cgi/200081E1.PDF?Dockey=200081E1.PDF (2017)
75. EPA, (2015) Texas: Final Authorization of State Hazardous Waste Management Program Revision, Federal Register / Vol. 80, No. 203

76. EPA (2012) Toxic Chemical Release Inventory Reporting Forms and Instructions
77. EPA, United States Environmental Protection Agency Position Statement on Environmental Management Systems (EMSs) (2017)
78. EPA (1994) Used Oil Questions and Answers, A Collection of Questions Compiled by the Hotline
79. EPA, When Are You Required to Report an Oil Spill and Hazardous Substance Release? What Information is Needed When Reporting an Oil Spill or Hazardous Substance Release? retrieved from https://www.epa.gov/emergency-response/when-are-you-required-report-oil-spill-and-hazardous-substance-release (2017)
80. General Land Office, Facility Planning Guide - Required Items List, All Facility Plans 31 TAC, §19.12, §19.13(c), Intermediate Facilities 31 TAC §19.13 (d), Large Facilities 31 TAC §19.13(e), §19.14 (2002)
81. General Land Office, Requirements for Discharge Prevention and Response Plans, 31 TAC, Chapter 19, Subchapter B §19.13 (2017)
82. General Permit To Dispose Of Wastewater, TAC, Water Code 30, Chapter 26, Chapter 205, retrieved from https://www.tceq.texas.gov/assets/public/permitting/wastewater/wqg100000_draft_permit.pdf (2017)
83. Hazardous Substances Releases, Liability, Compensation Act, Title I, retrieved from https://www.niehs.nih.gov/research/supported/assets/docs/r_s/srp_about_2010_508.pdf
84. ISO 14001 Environmental Management Systems, An easy-to-use checklist for small business. Are you ready?, International Organization for Standardization, (2010), retrieved from https://www.iso.org/files/live/sites/isoorg/files/archive/pdf/en/pub100329.pdf
85. Jones, M. T. (2001) The disposal of spilled oils and sorbent materials
86. Khawaja, M., Aban, A., Mahmoud, M, (1996), Characteristics of used lubricating oils, their environmental impact and survey of disposal methods, (Pg 23-32)
87. Schiller, A. (2010), Why People Don't Recycle, retrieved from Earth911.com
88. Szczepanski, M. (2016) The Challenges Low Oil Prices Bring to the Waste and Recycling Industry

89. TCEQ, Aboveground Storage Tanks, TAC 30 Chapter 334, Subchapter F, §334.121. Purpose and Applicability for Aboveground Storage Tanks (ASTs), §334.122. Definitions for Aboveground Storage Tanks (ASTs), §334.123. Exemptions for Aboveground Storage Tanks (ASTs), §334.124. Exclusions for Aboveground Storage Tanks (ASTs), §334.125. General Prohibitions and Requirements for Aboveground Storage Tanks (ASTs), §334.126. Installation Notification for Aboveground Storage Tanks (ASTs), §334.127. Registration for Aboveground Storage Tanks (ASTs), §334.128. Annual Facility Fees for Aboveground Storage Tanks (ASTs), §334.129. Release Reporting and Corrective Action for Aboveground Storage Tanks (ASTs), §334.130. Reporting and Recordkeeping for Aboveground Storage Tanks (ASTs), §334.132. Other General Provisions for Aboveground Storage Tanks (ASTs) (2008)
90. TCEQ Acronyms and Other Abbreviations, retrieved from https://www.tceq.texas.gov/about/acron.html (2017)
91. TCEQ, A Guide to Developing an Environmental Management System for a Small Business (2011), retrieved from https://www.tceq.texas.gov/assets/public/comm_exec/pubs/gi/gi-304.pdf
92. TCEQ, Air PBR 106.183: Boilers, Heaters, and Other Combustion Devices, retrieved from https://www.tceq.texas.gov/permitting/air/permitbyrule/subchapter-g/big_boilers.html (2017)Integra
93. TCEQ, Air Quality Permitting, Fact Sheet, retrieved from http://tceq.net/permitting/air/guidance/newsourcereview/permit_fact5_09.html/at_download/file
94. TCEQ, Annual Enforcement Report Fiscal Year 2016, retrieved from http://www.tceq.texas.gov/enforcement/reports/AER/annenfreport.html (2016)
95. TCEQ, Boilers, Heaters, and Other Combustion Devices Air Permits by Rule (PBR) Checklist Title 30 TAC § 106.183, retrieved from https://www.tceq.texas.gov/permitting/air/permitbyrule/subchapter-g/big_boilers.html (2017)
96. TCEQ, Comparison of Tier II, TRI, and 112(r) Requirements, retrieved from https://www.tceq.texas.gov/assistance/resources/tierIIchart.html (2017)
97. TCEQ, Evaporation Pond General Permit, TPDES General Permit No.

WQG100000, effective March 28, 2014 – March 28, 2019; retrieved from https://www.tceq.texas.gov/assets/public/permitting/wastewater/general/wqg100000_draft.pdf (2017)

98. TCEQ, Fact Sheet and Executive Director's Preliminary Decision General Permit WQG100000 (2014) retrieved from https://www.tceq.texas.gov/assets/public/permitting/wastewater/general/wqg100000_factsheet.pdf

99. TCEQ, General Permit To Discharge Under The Texas Pollutant Discharge Elimination System (TPDES) retrieved from https://www.tceq.texas.gov/assets/public/permitting/stormwater/txr050000.pdf (2017)

100. TCEQ, Guidelines For The Classification And Coding Of Industrial And Hazardous Wastes Texas Commission On Environmental Quality Waste Permits Division RG-022 Revised 11/14 retrieved from https://www.tceq.texas.gov/assets/public/comm_exec/pubs/rg/rg-022.pdf (2014)

101. TCEQ, Occupational, Health, Hazardous Chemical Right-To-Know, TAC 25, Part 1, Chapter 295, Subchapter H, §295.181 General Provisions and Definitions, §295.182 Responsibilities and Requirements, §295.183 Compliance and Fees (2017)

102. TCEQ, Permit by Rule Applicability Checklist (Title 30 TAC § 106.4) retrieved from https://www.tceq.texas.gov/assets/public/permitting/air/Forms/PermitsByRule/Checklists/10149.pdf (2017)

103. TCEQ, Soil Prevention And Control, TAC 30, Part 1, Chapter 327, §327.1 Applicability, §327.3 Notification Requirements, §327.4 Reportable Quantities (2017)

104. TCEQ, Solid Waste Act, General Provisions, Subchapter A, TAC 30 Subchapter A, General Provisions §361.001 (2017)

105. TCEQ, Spills: Reportable Quantities, retrieved from https://www.tceq.texas.gov/response/spills/spill_rq.html (2017)

106. TCEQ, Strategic Plan Fiscal Years 2015 – 2019, Submitted to the Governor's Office of Budget, Planning and Policy and the Legislative Budget Board July 2014

107. TCEQ, Storage Tank and Change of Service Air Permits by Rule (PBR)

Checklist Title 30 TAC § 106.478 retrieved from https://www.tceq.texas.gov/assets/public/permitting/air/Forms/PermitsByRule/Checklists/10144.pdf (2017)

108. TCEQ, The State of Texas Environmental Electronic Reporting System (STEERS), retrieved from https://www3.tceq.texas.gov/steers/. (2017)
109. TCEQ, The Used Oil Recycling Handbook (2005)
110. TCEQ, Tier II Chemical Reporting Program, retrieved from https://www.tceq.texas.gov/permitting/tier2/
111. TCEQ, Types of New Source Review Authorizations, retrieved from https://www.tceq.texas.gov/permitting/air/guidance/authorize.html (2017)
112. TCEQ, Underground And Aboveground Storage Tanks, General Provisions, TAC 30, Part 1, Chapter 334, Subchapter A, §334.1 Purpose and Applicability, §334.2 Definitions, §334.3 Exemptions for Underground Storage Tanks (USTs) and UST Systems, §334.4 Exclusions for Underground Storage Tanks (USTs) and UST Systems, §334.5 General Prohibitions for Underground Storage Tanks (USTs) and UST Systems, §334.6 Construction Notification for TCEQ, Underground Storage Tanks (USTs) and UST Systems, §334.7 Registration for Underground Storage Tanks (USTs) and UST Systems, §334.8 Certification for Underground Storage Tanks (USTs) and UST Systems, §334.9 Seller's Disclosure, §334.10 Reporting and Recordkeeping, §334.12 Other General Provisions, §334.14 Memorandum of Understanding between the Attorney General of Texas and the Texas Natural Resource Conservation Commission, §334.15 Limits on Liability of Lender, §334.16 Limits on Liability of Corporate Fiduciary, §334.17 Privatization of Storage Tank Program, §334.18 Limits on Liability of Taxing Unit §334.19 Fee on Delivery of Petroleum Product (2017)
113. TCEQ, Underground And Aboveground Storage Tanks, Release Reporting And Corrective Action, TAC 30, Part 1, Chapter 334, Subchapter D, §334.71 Applicability and Deadlines §334.72 Reporting of Suspected Releases, §334.73 Investigation Due to Off-Site Impacts, §334.74 Release Investigation and Confirmation Steps, §334.75 Reporting and Cleanup of Surface Spills and Overfills, §334.76 Initial Response to Releases, §334.77 Initial Abatement Measures and Site Check, §334.78 Site Assessment, §334.79 Removal of Non-Aqueous Phase Liquids (NAPLs), §334.80 Investigation for Soil and

Groundwater Cleanup, §334.81 Corrective Action Plan, §334.82 Public Participation, §334.83 Emergency Orders, §334.84 Corrective Action by the Agency, §334.85 Management of Wastes (2017)

114. TCEQ, Underground and Aboveground Storage Tanks, Technical Standards, TAC 30, Chapter 334, 334.41. Applicability, §334.42. General Standards. 334.43. Variances and Alternative Procedures, 334.44. Implementation Schedules. §334.45. Technical Standards for New Underground Storage Tank Systems. 334.46. Installation Standards for New Underground Storage Tank Systems. §334.47. Technical Standards for Existing Underground Storage Tank Systems. 334.48. General Operating and Management Requirements. §334.49. Corrosion Protection. §334.50. Release Detection. 334.51. Spill and Overfill Prevention and Control. 334.53. Reuse of Used Tanks. §334.54. Temporary Removal From Service. 334.55. Permanent Removal from Service. 334.56. Change to Exempt or Excluded Status (2004)

115. TCEQ Used Oil Recycling, TAC Title 30, Chapter 324, Subchapter A, §324.1 Federal Rule Adoption by Reference, §324.2 Definitions, §324.3 Applicability, §324.4 Prohibitions, §324.6 Generators, §324.7 Collection Centers, §324.11 Transporters and Transfer Facilities, §324.12 Processors and Re-refiners, §324.13 Burners of Off-specification Used Oil for Energy Recovery, §324.14 Marketers of Used Oil Fuel, §324.15 Spills, §324.16 Polychlorinated Biphenyls (PCBs), §324.21 Suspension or Revocation of Registration, §324.22 Soil Remediation Requirements for Used Oil Handlers (2013)

116. TCEQ Used Oil Filter Management And Recycling, Definitions, TAC 30, Chapter 328, Subchapter D, Definitions §328.21, Applicability §328.22, Registration and Reporting §328.24, Limitations on Storage §328.26

117. TCEQ Used Oil Standards, Used Oil Recycling, Polychlorinated Biphenyls (PCBs) TAC 30 Chapter 324 Subchapter A §324.16 (2017)

118. TCEQ, Used Oil Standards, Subchapter A, Used Oil Recycling, TAC 30 Definitions §324.2 (2017)

119. TCEQ Used Oil Standards, Used Oil Recycling, Spills TAC 30 Chapter 324 Subchapter A §324.15 (2017)

120. TCEQ, Used Oil Standards, Used Oil Recycling, TAC 30 Subpart A §§324.1 - 324.4, 324.6 324.7, 324.11 - 324.16, 324.21, 324.22 (2013)

121. Texas, A Guide to the Texas Environmental, Health, and Safety Audit Privilege Act (2013)
122. Texas Community Right-to-Know Acts (TCRAs), Manufacturing Facility Community Right-to-Know Act, Public Employer Community Right-to-Know Act, Non-manufacturing Facilities Community Right-to-Know Act, TAC Health and Safety Code, Chapters 505, 506, 507 (2017)
123. Texas, General Provisions, Health and Safety Code Title 5, Sanitation and Environmental Quality, Subtitle C Air Quality, Chapter 382, Clean Air Act, Subchapter A (2017)
124. Texas, Title 30 Texas Administrative Code (TAC) Chapter 305 (2017)
125. Texas Used Oil Collcction, Management, and Recycling Act, TAC 30 Subchapter A, General Provisions §371.001 (2017)
126. Texas, Vernon's Civil Statutes, Environmental Health, And Safety Audit Privilege Act, Title 71, Chapter 1, Art. 4447cc (2017)
127. Texas Water Code (TWC) §26.001 (2017)
128. The Public Health and Welfare. United States Code 42, 2010 Edition Title 42, Chapter 85, Air Pollution Prevention and Control, retrieved from https://www.gpo.gov/fdsys/pkg/USCODE-2010-title42/html/USCODE-2010-title42-chap85.htm
129. Thompson & Knight, (2013) Conducting Environmental Due Diligence and Audits, retrieved from http://www.tklaw.com/files/Publication/814c352d-2d78-40df-8a38-2c852bcabbd1/Presentation/PublicationAttachment/82993da8-c144-4cb8-88f0-6e7193e3c9e7/2013%20Conducting%20Environmental%20Due%20Diligence%20and%20Audits%20Presentation.pdf
130. United States Department of Agriculture , National Resources Conservation Services, Clean Water Act retrieved from https://www.nrcs.usda.gov/Internet/FSE_DOCUMENTS/stelprdb1187162.pdf (2017)
131. Water Pollution Prevention And Control, 33 U.S.C., Chapter 26, Sections 1251-1387 (2017)
132. Witkin, J. (2002), Environmental Aspects of Real Estate and Commercial Transactions: From Brownfields to Green Buildings, American Bar Association,

133. Wolfe, P. R., (1992), Economics of used oil recycling: still slippery
134. EPA, Texas: Final Authorization of State Hazardous Waste Management Program Revision, Federal Register / Vol. 80, No. 203, October 21, 2015
135. TCEQ, Flowchart of TCEQ and its Predecessor Agencies, retrieves from https://www.tceq.texas.gov/about/tceqhistory.html (2017)
136. TEXAS, General Provisions, HSC, Subtitle B, Solid Waste, Toxic Chemicals, Sewage, Litter, And Water, Chapter 361, Solid Waste Disposal Act Subchapter A
137. EPA, Used Oil Management Standards, FR 41566, Vol. 57. No. 176 , September 10, 1992
138. EPA, Used Oil Management Standards Training (1994)
139. Vertex Energy Incorporated, CFO Letter Authorizing Content Use (2017)
140. Take Care of Texas, retrieved from http://takecareoftexas.org/conservation-tips/reduce-your-waste
141. Texas TCEQ Small Business Support, retrieved from https://www.tceq.texas.gov/assistance
142. EPA, Hazardous Waste Management System; Burning of Waste Fuel and Used Oil Fuel In Boilers and Industrial Furnaces, FR 49164, Vol. 50, No. 230, November 29, 1985
143. EPA, Used Oil Management Standards, 51 FR 41900; November 19, 1986.
144. EPA, Used Oil Management Standards, FR 48000, Vol. 56. No. 184 , September 23, 1991
145. Hazardous Waste Management System; Identification and Listing of Hazardous Waste; Used Oil, FR 21524, Vol. 57. No. 98 , May 20, 1992
146. EPA, Identification And Listing Of Hazardous Waste, Exclusions, §261.4
147. EPA, RCRA Section 3014, retrieved from http://legcounsel.house.gov/Comps/RCRA.PDF
148. EPA, Toxic Substance Control Act, 40 C.F.R, Chapter I, Subchapter R, Part 761, §761.1 Applicability, §761.2 PCB Concentration Assumption For Use, §761.3 Definitions, §76.19 References (2017)
149. TCEQ, Special Waste Regulations in Texas (2006)
150. TCEQ, History of the TCEQ and Its Predecessor Agencies, retrieved from https://www.tceq.texas.gov/about/tceqhistory.html (2017)
151. TCEQ, Mission Statement, Organization, retrieved from

https://www.tceq.texas.gov/about (2017)
152. TCEQ, Texas Enviro Help - Small Business and Local Government Assistance, retrieved from https://www.tceq.texas.gov/assistance (2017)
153. EPA, Standards for Use as a Dust Suppressant and Disposal of Used Oil, 40 C.F.R. Subpart I, Part 279, § 279.80 Applicability. § 279.81 Disposal. § 279.82 Use as a dust suppressant (2017)
154. Texas, General Provisions, HSC, TITLE 5, SUBTITLE D, SUBCHAPTER A, §401.415
155. EPA, Used Oil Management Standards, FR 41578, Vol. 57, No. 176, September 10, 1992
156. TCEQ, Annual Reporting Form For Used Oil Collection Centers. Retrieved from https://www.tceq.texas.gov/assets/public/permitting/rrr/forms/0567.pdf Unknown
157. EPA, The Benefits of Reusing and Recycling used Oil, retrieved from https://www.epa.gov/recycle/managing-reusing-and-recycling-used-oil#02 (2004)
158. EPA, Managing, Reusing, and Recycling Used Oil, retrieved from https://www.epa.gov/recycle/managing-reusing-and-recycling-used-oil
159. API, Used Oil Motor Oil Collection and Recycling, retrieved from http://www.recycleoil.org/ (2017)
160. MADEP, Nonpoint Source Pollution Education: Motor retrieved from http://www.mass.gov/eea/agencies/massdep/water/watersheds/nonpoint-source-pollution-education-motor-oil.html
159. EPA, Industrial Furnaces, 40 CFR Part 260 , Definitions §260.10 (2017)
160. TCEQ, Transporters and Transfer Facilities, TAC 30, Part 1, Chapter 324, Subchapter A, §324.11 (2017)
161. TCEQ, Used Oil Filter Management And Recycling, Chapter 328, §328.21. Definitions, §328.22. Applicability, §328.23. General Requirements, §328.24. Registration and Reporting, §328.25. Shipment Records, §328.26. Limitations on Storage, §328.27. Variances, §328.28. Spill Prevention and Control (2014)
162. EPA, Standards Applicable To Generators Of Hazardous Waste, 40 C.F.R., Subchapter I, Part 262, §262.10 Purpose, scope, and applicability, §262.11 Hazardous waste determination and recordkeeping, §262.13 Generator Category Determinations (2017)
163. TCEQ, CORE Data Form and Instructions, retrieved from

https://www.tceq.texas.gov/permitting/central_registry/guidance.html (2017)
164. TCEQ, Instructions for Registration for Used Oil Handler and/or Used Oil Filter Handlers, retrieved from https://www.tceq.texas.gov/assets/public/permitting/rrr/forms/10062inst.pdf (2017)
165. TCEQ, Collection Centers Form retrieved from https://www.tceq.texas.gov/permitting/registration/used_oil/collection.html (2017)
166. EPA, Instructions and Form for Hazardous Waste Generators, Transporters, and Treatment, Storage and Disposal Facilities to Obtain an EPA Identification Number (EPA Form 8700-12), retrieved from https://www.epa.gov/hwgenerators/instructions-and-form-hazardous-waste-generators-transporters-and-treatment-storage-and (2017)
167. TCEQ, Biennial Reporting Form for Used Oil Processors, Re-Refiners and Filter Handlers, TCEQ-0648 Form, retrieved from https://www.tceq.texas.gov/assets/public/permitting/rrr/forms/0648.pdf, (2017)
168. EPA, Operating Record and Reporting, 40 C.F.R. Part 279, §279.57 (2012)
169. TCEQ, Used Oil: Collection Basic requirements for Used Oil and Used Oil Filter Collection Centers, retrieved from https://www.tceq.texas.gov/permitting/registration/used_oil/collection.html (2017)
170. TCEQ, Financial Assurance, retrieved from http://tceq.state.tx.us/adminservices/financial-assurance/revenue/used_oil_recycling.html/#overview
171. TCEQ, Financial Assurance For Used Oil Recycling, Chapter 37, Subchapter L, §37.2001. Applicability, §37.2003. Definitions, §37.2011. Financial Assurance Requirements for Used Oil Handlers, §37.2013. Financial Assurance Mechanisms for Used Oil Handlers, §37.2015. Submission of Documents, §37.2021. Financial Assurance Requirements for Transporters of Used Oil, retrieved from, https://www.tceq.texas.gov/assets/public/legal/rules/rules/pdflib/37l.pdf (2003)
172. TCEQ, Financial Assurance Mechanisms for Closure, Post Closure, And Corrective Action, Chapter 37, Subchapter C, §37.200. Applicability, §37.201. Trust Fund, §37.211. Surety Bond Guaranteeing Payment, §37.221. Surety Bond Guaranteeing Performance, §37.231. Irrevocable Standby Letter of Credit, §'37.241. Insurance, §37.251. Financial Test, §37.261. Corporate Guarantee,

§37.271. Local Government Financial Test , §37.281. Local Government Guarantee, retrieved from https://www.tceq.texas.gov/assets/public/legal/rules/rules/pdflib/37c.pdf (2006)

173. TCEQ, MAP Requirements, Chapter 324, Subchapter A, retrieved from https://www.tceq.texas.gov/assets/public/legal/rules/rules/pdflib/324a.pdf (2013)
174. EPA, Used Oil Management, Analysis Plan, 40 C.F.R. Chapter I, Subchapter I, Part 279, Subpart F, §279.54, §279.55
175. EPA, Closure and Post Closure, 40 C.F.R. Part 265 Subpart G, §265.110 Applicability, §265.111 Closure performance standard, §265.112 Closure plan; amendment of plan, §265.113 Closure; time allowed for closure, §265.117 Post-closure care and use of property §265.118 Post-closure plan; amendment of plan, §265.119 Post-closure notices, §265.120 Certification of completion of post-closure care, §265.145 Financial assurance for post-closure. (2017)
176. EPA, Closure and Post Closure Care, 40 C.F.R. Part 265 Subpart N, §265.310 (2017).
177. EPA – Spill Prevention, Control, And Countermeasure (SPCC) Regulation 40 CFR part 112 A Facility Owner/Operator's Guide to Oil Pollution Prevention (2010)
178. EPA, The Federal Pollution Control Act 1972, retrieved from https://www.epa.gov/laws-regulations/history-clean-water-act
179. EPA, Substantial Harm Criteria, 40 C.F.R. Chapter I, Subchapter D, Part 112, Subpart D, Appendix C
180. EPA, Facility Response Plan (FRP) Overview, retrieved from https://www.epa.gov/oil-spills-prevention-and-preparedness-regulations/facility-response-plan-frp-overview
181. EPA, Facility Response Plan Form A Model Plan, retrieved from https://www.ecfr.gov/cgi-bin/textidx?SID=934dd060745fb05152cd5fea2224d24b&mc=true&node=ap40.22.112_121.f&rgn=div9
182. TCEQ, AST Registration Form, retrieved from https://www.tceq.texas.gov/assets/public/permitting/rrr/forms/0659.pdf
183. Federal, Hazardous Waste Operations and Emergency Response, Part 1910, OSHA, Subpart H, Hazardous Materials, §1910.120

184. EPA, The Practical Guide to Environmental Management for Small Business, retrieved from http://dnr.wi.gov/topic/SmallBusiness/documents/air/PracticalGuidetoEMforSB.pdf

INDEX

ALL TCEQ USED OIL HANDLERS FORMS FOUND IN THIS BOOK ARE PROVIDED AT NO COST IN PDF FORM FROM WWW.VRMUS.COM USE CODE TX1218 AS PASSWORD TO OBTAIN FORMS AFTER ENTERING EMAIL FOR INSTANT ACCESS AND A COPY EMAILED TO EMAIL ENTERED. (FORMS WILL BE AVAILABLE AS LONG AS BOOK IS IN PUBLICATION)

Made in United States
North Haven, CT
10 November 2026